Food Processing and Preservation

ABOUT THE AUTHOR

Dr. (Mrs.) Neelam Khetarpaul presently working as Professor and Head, Department of Foods and Nutrition at CCS Haryana Agricultural University, Hisar is the recipient of many awards viz., Distinguished teacher Award-2000, Young Investigator Award, Ms. Manju Utreja Gold Medal and cash award for doing the best research work in the University and Best research Paper Award by AFST (I), Hisar Chapter. She is the recipient of various Visiting Fellowships abroad funded by different national and international agencies and visited USA, Australia, UK, Hungary and Netherlands for academic pursuits. She has published more than 150 research papers in various journals of national and international repute and 12 books in the discipline of Food Science and Human Nutrition.

Food Processing and Preservation

Neelam Khetarpaul
Professor (Foods & Nutrition)
Department of Foods & Nutrition
CCS Haryana Agricultural University
Hisar – 125 004 (Haryana), India

2015
Daya Publishing House®
A Division of
Astral International (P) Ltd
New Delhi 110 002

First Impression, 2005
Reprinted, 2015

ISBN: 978-93-5124-059-4 (International Edition)

Published by : **Daya Publishing House®**
A Division of
Astral International Pvt. Ltd.
– ISO 9001:2008 Certified Company –
4760-61/23, Ansari Road, Darya Ganj
New Delhi-110 002
Ph. 011-43549197, 23278134
E-mail: info@astralint.com
Website: www.astralint.com

Laser Typesetting : **Classic Computer Services**
Delhi - 110 035

Printed at : **Replika Press Pvt. Ltd.**

PRINTED IN INDIA

Preface

India is world's second largest producer of food and has the potential to become number one in due course of time with sustained efforts. The growth potential of this sector is enormous and it is expected that the food production will double in the next 10 years and the consumption of value added food products will grow at a fast pace. In contrast to past situation, today there is a problem of food abundance. Our silos are full and there is a need to process and preserve the various food items using various food preservation techniques so as to avoid food wastage and utilize and enjoy all kinds of seasonal foods throughout the year. Hence, in the times to come, Food Processing Industry will be of enormous significance for India's development because of the vital linkages and synergies that it promotes between the two pillars of the economy, namely Industry and Agriculture. This growth of the Food Processing Industry will bring immense benefits to the economy, raising agricultural yields, meeting productivity, creating employment and raising the standard of very large number of people through out the country, specially, in the rural areas.

A large amount of food items, especially the perishable ones, is wasted every year due to improper storage conditions in India which cause microbial spoilage and make the food unfit for human consumption. However, application of modern advances in most of the food processing and preservation techniques *viz.* drying, freezing, canning, controlled and modified atmosphere, hurdle technology, food irradiation etc. have led to elimination of microbial and toxicological hazards to a large extent in the preserved foods while maintaining their nutritional and sensory attributes. Lot of advancement leading to refinement in these processing technologies is taking place. Through this book, an effort has been made to transfer this latest knowledge to the readers.

This book entitled 'Food Processing and Preservation' deals with fairly advanced knowledge on importance and scope of food preservation, food spoilage and its causes, prevention of microorganisms, IMF concept in food preservation, hurdle technology, freeze drying of foods, modern techniques applied in food industries, physiological study of perishable foods, atmospheric modifications, CA storage of fruits, preservation of tree nuts and dried nuts, storage and disinfestation of grains, food additives, food irradiation, food freezing, application of HACCP in food industry, National Food Processing Policy of India etc.

I sincerely hope that this book would serve as a technical guide and handy reference not only for the undergraduate and postgraduate students, research scientists and teachers but also for the personnel engaged in food processing industries.

I am grateful to my family members and friends who have directly or indirectly helped me in completing this book.

Finally, I would like to express my special thanks to Mr. Anil Mittal of Daya Publishing House for his continued interest in bringing out this edition.

Neelam Khetarpaul

Contents

Part I
Status of Food Processing Industry

Chapter 1

Status of Food Processing Industry in India

Today, the single most important problem facing the country perhaps is providing remunerative price to the farmers for their produce without incurring additional burden of subsidy through minimum support prices or some such measures. This problem could be solved largely if the surplus production of cereals, fruits, vegetables, milk, fish, meat and poultry etc. are processed and marketed aggressively both inside and outside the country. Food processing adds value, enhances shelf life, and encourages crop diversification. It has the largest employment generation potential per unit of investment. Food processing coupled with marketing has thus the potential of solving the basic problems of agricultural surpluses/wastages and providing rural jobs/ensuring better prices to the growers etc.

A strong and dynamic food processing sector thus plays a vital role in diversification and commercialization of agriculture, ensures value addition to the agricultural produce, generates employment, enhances income of farmers and creates surplus for export of agro foods.

Economic liberalization and rising consumer prosperity is opening up new opportunities for diversification in Food Processing Sector. Liberalization of world trade will open up new vistas for growth. In a recent study on the potentials of food processing industry it has been found that there is enormous potential in India to build profitable business in food processing. The food processing industry ranks fifth in size in the country and employs 16 lakh workers which consists 19 per cent of the country's industrial labour force. It accounts for 14 per cent of the total industry output with 5.5 per cent of the GDP. The turnover of this industry is estimated to be Rs.144,000 crores (approx.) of which Rs.111.200 cores (approx.) persists in the unorganized sector. The industry has started producing many new items like ready to eat food, beverages, processed and frozen fruit and vegetable products, marine and meat products, IQF products etc. Indian consumers are being fast introduced to more and more high quality food products produced by induction of latest technologies in various food processing sectors. State of art technology has been introduced in food processing industries for improving the competitive edge.

The Food Processing Industry has been identified as a thrust area for development. This industry is included in the priority lending sector. Most of the Food Processing Industries have been exempted from the provisions of industrial licensing under Industries (Development and Regulation) Act, 1951 with the exception of beer and alcoholic drinks and items reserved for Small Scale Sector, like vinegar, bread, bakery etc. As far as foreign investment is concerned automatic approval for even 100 per cent equity is available for majority of the processed food items.

Food processing involves any type of value addition to the agricultural produce starting at the postharvest level. It includes even primary processing like grading, sorting, cutting, seeding, shelling packaging etc. Processing of foods especially fruits and vegetables is another important way to avoid glut and to make their products available to the consumers in all seasons.

The sector comprises of the following major areas:

Fruit and Vegetable

Major Products

Beverages, Juices, Concentrates, Pulps, Slices, Frozen and Dehydrated products, Wine, Potato Wafers/Chips etc.

Fisheries

Major Products

Frozen and Canned products mainly in fresh form.

Meat and Poultry

Major Products

Frozen and packed mainly in fresh form, Egg Powder (only a couple of units)

Milk and Dairy

Major Products

Whole Milk Powder, Skimmed milk powder, Condensed milk, Ice cream, Butter and Ghee

Grain and Cereals

Major Products

Flour, Bakeries, Biscuits, Starch Glucose, Cornflakes, Malted Foods, Vermicelli, Pasta Foods, Beer and Malt extracts, Grain based Alcohol.

Consumer Industry

Major Products

Chocolates, Confectionery, Soft/Aerated Beverages/Drinks.

Plantation

Major Products

Tea, coffee, cashew, cocoa, coconut etc.

Broad Categorisation of Food Processing Industries and Their Present Status

Food processing industries have a crucial role to play in reduction of postharvest losses. The most important point in the food industry is that a substantial portion being rural based it has a very high employment potential with significantly lower investment. From the energy conservation aspect also, promotion of food processing industry could be rewarding as it can reduce the wastage of energy, which occurs in home cooking.

It has been felt at the national level that priority should be given to processing and enhancing shelf life of perishable items so as to reduce wastage and encourage value addition in that sector. Considering that rice/pulses/flour are consumed in the processed form only and primary processing in these sectors adds little to shelf life, wastage control and value addition, it has been decided to discontinue assistance to this sector from the financial year 2004-05. Priority will be given to fruit and vegetables, dairy, meat and poultry, fish, bakery, confectionary and consumer goods.

Fruit and Vegetable Sector

India is the second largest producer of fruits and vegetables in the world, next to China, contributing 10 and 13.4 per cent of the total world production of fruits and vegetables, respectively. India ranks number one in respect of mango and banana in the world and as far as vegetables are concerned, India occupies first position in the production of cauliflower, second in onion and third in cabbage in the world. The country's tropical, subtropical and temperate climate and varieties of soils are amply suited for cultivation of over 40 vegetables and 30 fruit trees. Though horticultural crops cover only 6.1 per cent of the gross cropped area, it contributes 18.8 per cent gross value of agricultural output and 14.5 per cent of foreign exchange earned through agriculture. Thus, diversification of agriculture to horticulture is very essential from the angle of minimising risk, enhancing profit, generating employments and optimising as well as conserving natural resources, providing nutritional and economic security and healthy environment. Development of horticulture is a pre-requisite as 6.7 per cent annual growth rate of horticulture is a must, if agricultural growth rate of 4.0 per cent is to be achieved.

Realizing the need and urgency, the 10th Five Year Plan has projected the demand of fruits and vegetables in the country at 2,496 lakh tonnes in the year 2007-08 indicating a rise by 89.71 per cent over the production of 1315.7 lakh tonnes in the year 1998-99.

The installed capacity, production and growth over the previous year for the last ten years in this sector as on 1st January are indicated as under.

Table 1.1: Area, Production and Productivity of Fruits and Vegetables

As on Jan., 1	*Year*											
	1993	*1994*	*1995*	*1996*	*1996*	*1997*	*1998*	*1999*	*2000*	*2001*	*2002*	*2003*
Capacity (Lakh tones)	11.08	12.60	14.02	17.60	19.1	20.4	21.00	20.08	21.00	21.10	21.98	23.28
Production (Lakh tones)	4.69	5.59	6.76	8.50	9.6	9.1	9.8	9.4	9.8	9.9	10.3	10.72
Growth in production over the previous year (per cent)	30.23	20.00	20.93	25.74	12.94	5.2	4.26	3.3	4.26	1.0	4.04	4.08

Source: Annual Report of Ministry of Food Processing and Industries, 2003-04

Due to the short shelf-life of these crops, as much as 30-35 per cent of fruits and 20-25 per cent of the vegetables perish during harvest, storage, grading, transport, packaging and distribution. It is estimated that out of the total production of fruits and vegetables in the country, only 0.5 to 1 per cent of the raw material is processed, which compares very unfavourably with countries such as Brazil and the USA with 70 per cent utilisation, the Philippines with 78 per cent, Malaysia with 83 per cent and Thailand with 30 per cent. The fruit production in India has recorded a growth rate of 4.08 per cent, whereas the fruit processing sector has grown at about 20 per cent per annum.

The growth rates have been extensively higher for frozen fruits and vegetables and dehydrated fruits and vegetables (24 per cent). There exist over 4000 fruit processing units in India with an aggregate capacity of more than 1.2 million MT (less than 4 per cent of total fruits produced).

The number of fruit and vegetable processing units, the installed processing capacity and production of processed items are going up steadily in the last six years. As a result of withdrawal of excise duty on Fruits and Vegetable Products in the Union Budget of 2001, the installed capacity and growth of this sector has been increasing steadily. The growth in terms of production during 2003 is approximately 4 per cent compared to 2002.

The installed capacity of fruit and vegetable processing industry has been increased from 11.08 lakh tons on 01.01.1993 to 23.28 lakh tons as on 01.01.04. The estimated installed capacity of fruit and vegetable processing industries has increased from 20.8 lakh tonnes in 1998 to 23.28 lakh tonnes in 2003-04. The production of processed fruits and vegetables in the country has increased from 9.4 lakh tonnes in 1998 to 10.72 lakh tonnes in 2003-04. During the same period the number of license issued under Fruit Product Order (FPO), 1955 has increased to 5875 from 5112. There is 4.08 per cent growth in production during 2003-04 when compared to that of the previous years (Annual Report 2003-04, Ministry of Food Processing Industries).

The utilization of fruits and vegetables for processing in the organized and unorganized sectors is estimated to be around 2 per cent of the total production. Of the various products, fruit juices and fruit pulp account for 27 per cent, followed by ready-to-serve beverages and pickles being between 12 to 13 per cent each, jams and jellies 10 per cent and synthetics 8 per cent. The other products are squash, tomato products, canned vegetables, etc. Recent additions to these products are frozen pulps and vegetables, freeze-dried fruits and vegetables, fruit concentrates, aromas, packed vegetable curries, canned mushrooms and mushroom products, tomato paste, potato chips, etc. Over the last few years, there has been a positive growth in ready to serve beverages, fruit juices and pulps, dehydrated and frozen fruits and vegetable products, tomato products, pickles, convenience vegetable-spice pastes, processed mushrooms and curried vegetables.

There have been fluctuations in production of canned and bottled fruits and vegetable products, fruit juice concentrates. The domestic consumption of value added fruits and vegetable products is also low compared to the primary processed food in general and fresh fruits and vegetables in particular, which is attributed to higher incidence of tax and duties including that on packaging material, lower capacity utilisation, non-adoption of cost effective technology, high cost of finance, infrastructural constraints, inadequate farmers-processors linkage leading to dependence upon intermediaries. The smallness of units and their inability for market promotion is also another main reason for inadequate expansion of the domestic market.

India's share of world trade in this sector is only around one per cent. It is estimated that around 20 per cent of the production of processed fruits is meant for exports, the rest caters to the defence, institutional sectors and household consumption. Mango and mango based products constitute 50 per cent of exports. In 1997-98, processed fruits and vegetables exported valued Rs. 5240 million. India's major exports are fruit pulps, pickles, chutneys, canned fruits and vegetables, concentrated pulps and juices, dehydrated vegetables, and frozen fruits and vegetables. This sector has attracted proposals worth total investment of Rs. 75000 crores in the last six years *i.e.* since the initiation of the liberalization process including foreign investment of Rs. 8800 crores.

Overall, there is a further need for maximum commercial utilisation of fruits and vegetables and to adapt production and marketing activities to the requirements of the world market and to cater to domestic demand which over the past few years has been increasing because of various socio-economic factors. If the nutritive value of the processed food products could be maintained, this sector will emerge as a major value-added horticulture industry. To boost up fruit and vegetable processing sector, MFPI is extending financial assistance in the form of grant-in-aid for setting up new units, modernisation and upgradation of existing units. During 2003-04 (as on 31.03.2004), an amount of Rs. 3.65 crore has been approved/ sanctioned to 21 such industrial units in 10 different States other than North East.

Meat and Poultry Industry

Meat

India has the world's largest number of livestock and ranks first in the cattle population and comprises about 50 per cent of the buffalo population and one-sixth of the total goat population of the world. Such a large livestock population clearly presents a challenge to retain existing traits of productivity by application of modern science and technology. Rigorous efforts are, therefore, being made to improve vast population of livestock by providing basic infrastructure and adopting latest levels of technology.

The present production of various types of meat and poultry products is estimated to be about 4.42 million tonnes as per the FAO's estimate. Only 11.1 percentage of the total buffalo population, and 6.4 per cent of the total cattle population, 33.1 per cent of sheep population, and 38.1 per cent of goat population is culled for meat. Currently, a small percentage of the meat produced is converted into value added products and most meat is purchased by the consumers in the country in the fresh/frozen form and converted into various meat products at homes, restaurants etc. The maximum conversion takes place only in case of pork products.

With growing urbanization and increasing quality consciousness, the market for scientifically produced meat products is expected to grow rapidly. There is also a growing demand for ready to eat and semi-processed meat products on account of changing life styles as also for export to neighbouring countries, especially in the Middle-east. The current level of exports of meat products from India is Rs.8000 million, the major destinations being the countries in the Middle East and South East Asia.

Meat processing in India has recently become a thrust sector for Indian industry and a number of processing centres with the latest technology are being set up. The areas where India needs technical co-operation are in building up organized facilities for rearing meat producing animals and proper storage and refrigerated transport system. This sector has attracted a total investment of Rs. 9000 million in the last six years *i.e.* since the initiation of the liberalization process including foreign investment of Rs. 5000 million.

Poultry

India ranks fifth in world's egg production and produces about 1.61 million tonnes every year. Yet the per capita availability is very low. However, with the changing food habits and increasing availability of eggs, there has been an increase in demand which is growing at about 10 per cent a year.

Over the past 30 years, egg production has shown an average annual growth rate of 16 per cent while that for broilers is even higher at 27 per cent per annum. During this period, Indian poultry industry has made spectacular progress transforming itself from backyard farming into a dynamic and sophisticated agri-based industry.

The increasing awareness of the need for balanced nutrition has led to changes in eating habits with vegetarians accepting eggs as part of their diet. Simultaneously, there has been an increase in purchasing power and more money is available for spending on food. With the changing food habits and increasing availability of eggs, there has been an increase in demand which is growing at about 10 per cent a year. Despite this, the egg industry experiences periodic slumps. There are five modern integrated poultry processing plants functioning in the country. Besides, there are good numbers of small plants although not very modern working in the country. These plants are producing dressed frozen chicken and cut parts. While the poultry industry is gradually taking shapes, poultry dressing and processing is still in its infancy in this country.

There are about 15 pure-line and grand parent franchise projects in India. There are 115 layer and 280 broiler hatcheries both in the private and the Government sector producing 1.3 million-layer parents and 2.6 million broiler parents which, in turn, supply about 95 million hybrid layer and 275 million broiler, day-old chick.

In India, five egg powder plants have also been established to produce whole egg, yolk and albumen powders. The demand for egg powder is increasing every year. Each project will process about a million eggs daily. Only egg powder plants have been in operation in Bombay since the 1970's. These egg powder plants will certainly help to boost egg production.

Milk and Milk Products

India is having the largest livestock population and the most important constituent of its being milch cows and milch buffaloes. India has emerged today as the largest milk producer in the world (72 million tonnes). Dairy development in India has been acknowledged the world over as one of the modern India's most successful developmental program. India, with its status as the largest milk producer in the world, is on the verge of assuming an important position in the global dairy industry. Many international dairy companies are viewing India with an eye to tapping its vast growing market for dairy products. While world milk production declined by 2 per cent in the last three years, according to FAO estimates, Indian production increased by 4 per cent.

Currently, consumption of liquid milk accounts for about 46 per cent of the total production of milk. The remaining 54 per cent is utilized for conversion to milk products. Organized dairy industry accounts for less than 15 per cent of the milk produced in India. The rest of the milk is either consumed at farm level, or is sold as fresh, non-pasteurised milk through unorganized channel. The share of organized industry is expected to rise rapidly-especially in the urban regions. The rise in the market for dairy products is likely to witness the fastest growth at over 20-30 per cent per annum. Nine units have been provided financial assistance under the Plan Scheme of the Ministry.

Technologies for primary processing *viz.* refrigerated chilling centres, compact milk chilling emits etc. are available in the country though not much in use. The processing of liquid milk is largely done by pasteurization whereas technologies like Ultra High Temperature treatment and Extended Shelf Life are likely to emerge in the future.

Among the products manufactured by the organized sector are ghee, butter, cheese, ice creams, milk powders, malted milk food, condensed milk, infants' foods etc. Of these, ghee (clarified butter) alone accounts for 85 per cent. Today industry has also introduced a number of new products such as casein, lactose, dairy whiteners, and certain milk products. The production of various processed milk products is showing a rising trend in India and there lies enormous scope for products like cheese, weaning foods, yogurt, lactose, caseinate etc. For by-products processing technology, research efforts are needed in obtaining value-added products from sour whey.

The value of Indian dairy produce is expected to rise from Rs. 288000 million in 1990 to Rs. 1,000,000 million by 2005. In the last six years total investment in this sector was in the order of Rs. 16,000 million of which foreign investment was Rs. 3600 million.

Consumer Industries

Consumer food industry includes pasta, breads, cakes, pastries, rusks, buns, rolls, noodles, corn flakes, rice flakes, ready to eat and ready to cook products, cocoa products, biscuits, soft drinks, beer, alcoholic beverages (non-molasses based), mineral and packaged water. Bread and biscuits constitute the largest segment of consumer foods. Their production is about 3.70 million tons per year.

Manufacturing of bread is reserved for SSI sector. Out of the total production of bread, 40 per cent is produced in the organized sector and the remaining 60 per cent in the unorganised sector. Similarly, production of biscuits in the organized sector is about 11.50 lakh tons and quantity of biscuits produced in the unorganised sector is about 3.80 lakh tons. According to available information, production of flakes is around 14,000 tons. Production of pasta products has registered a marginal growth in the organized sector but its growth in the unorganised sector is comparatively higher. Confectionery output grew at the compound rate of six to seven per cent in recent years. Chocolate production is growing at the rate of 10 to 15 per cent per cent per annum.

This sector has been successful in attracting substantial investment of about Rs. 1,28,000 million since liberalization of which foreign investment is about Rs. 50,000 million.

MFPI has framed and implemented several Plan Schemes for establishment and development of food processing industries.

During the 9th Five Year Plan, MFPI sanctioned financial assistance to the extent of Rs. 6.97 crore for establishment, expansion and modernization of consumer food industry. During the Tenth plan period 2002-03 and 2003-04, 47 food processing units availed loan/ financial assistance from MFPI and an amount of Rs. 4.24 crore was approved/ sanctioned to 23 such industrial units.

India produces a large range of cocoa and non-cocoa based confectionery items. Various other cocoa-based products are also manufactured and marketed as consumer products.

The production of confectionery items other than chocolates is reserved for the small scale sector. However, there are several large companies with established market presence and brands in both the cocoa and non-cocoa based confectionery product markets.

Aerated Soft Drinks and Packaged Drinking Water

The aerated soft drinks industry in India comprises over 100 plants across all States. It provides direct and indirect industry related employment to over 125,000 employees. It has attracted one of the highest foreign direct investments in the country amounting to around Rs.4700 crore. It contributes over Rs.1200 crore annually by way of excise duty, sales tax and related taxes. It has strong forward and backward linkages with over Rs.1000 crores relating to glass, plastic, refrigeration, sugar and transportation industry. Total export earnings are over Rs.700 crore per annum. Presently there are more than 7,000,000 outlets in the country.

The soft drinks constitute the 3rd largest packaged foods regularly consumed after packed tea and packed biscuits. The estimated production of soft drinks has registered a gradual increase as follows:

Year	*Bottles (in Million)*
1998–1999	5670
1999–2000	6230
2000–2001	6450
2001–2002	6600
2002–2003	10000
2003–2004	11040

Currently, it is estimated that 65 per cent prefer the non-carbonated drinks. In India lemon drinks continue to be very popular.

There are 215 companies, of packaged drinking water which have been granted licence for manufacturing packaged drinking water and 3 for manufacturing packaged natural mineral water by the MFPI in 2003-04. The Bureau of Industrial Standards have issued IS 13428: 1998 specifications for packaged drinking water. There has been a spurt in growth for the last 3-4 years, which can largely be attributed to a range of various packaged sizes to suit the consumers. 80 per cent of the packaged water sale comes from the bulk containers (5 litres and above).

Alchoholic Beverages

Liquor made in India is generally categorized as beer, country liquor and Indian Made Foreign Liquor (IMFL). Country liquor is made from a variety of raw materials and is known by different names in different parts of the country.

IMFL production primarily comprises wine, Vodka, Whiskey, Gin, Rum, Brandy etc. Pre-mixed drinks are being introduced in India now, *e.g.* Gin and Lime and Rum and Cola. Draught beer is another recent introduction and has done well where introduced. Canned beer has been introduced recently.

India is the third largest market for alcoholic beverages in the world. The demand for spirits and beer is estimated to be around 373 million cases. There are 12 joint venture companies having a licensed capacity of 33919 Kilo-litres per annum for production of grain based alcoholic beverages. 56 units are manufacturing beer under licence from the Government of India with production capacity of 1216 lakhs dozen per annum. Total investment in this sector in the last six years has been of the order of Rs. 11,000 million of which foreign investment is Rs. 7,000 million.

The Indian beer market estimated currently at Rs. 7,000 million a year, has been growing by 15 per cent per annum and today all world famous brands of liquor are available in the Indian market. India has 212 distilleries with an installed capacity of 1,933 million litres per annum. However, there are only 24 units, which produce IMFL, and 31 units which produce country liquor. Alcohol produced by the balance units is either sold as industrial alcohol or sold in bulk as potable alcohol to other distilleries, for bottling or for use in making bottled alcohol which is estimated at 927.82 million litres. Raw materials like molasses, barely, maize, potatoes, grapes, yeast and hops for beer and alcoholic products industry are abundantly available in India.

Wine industry India has made a beginning to establish itself as a quality conscious industry. The demand per annum for wine in the domestic market is estimated to be around 6 million bottles (750 ml). Against this, the estimated annual production of wine by the domestic industry is over 2.4 million bottles.

The market is estimated to grow at around 25 per cent per annum in the next five years. The wine industry in India provides considerable opportunities for value addition and employment generation in the agro-processing sector. It offers:

1. Possibilities for value addition to grapes/fruits, which could be cultivated on a large scale as a measure of agri diversification offering higher incomes to farmers
2. Opportunities for employment generation
3. Possibilities of weaning people away from hard liquor and
4. Opportunities for increased domestic production to counter large imports of wine

In order to promote wine industry in the country, MFPI has taken initiatives to support the wine industry in different ways.

Fisheries

With its long coast line of over 8000 kms, 50600 sq, kms. of continental shelf area and 2.2 million sq. km. of Exclusive Economic Zone, India is endowed with rich fishery resources. Total marine and inland production increased from 4.78 million tones to 5.65 million tonnes in 2001-02.

Considerable infrastructure facilities for processing of marine products have been developed over a period of 50 years. At present, there are over 402 freezing units with a daily processing capacity of 8500 tons and 485 frozen storage with a capacity of 109090 tons. Apart from the above there are 5 surimi units, 13 canneries, 559 pre processing unit and dry fish storages. 72 numbers of the freezing units mentioned above have cooking facilities and 84 have IQF machinery.

Export of marine products from India at present is about $ 1.3 billion. Indian marine products are exported to over 64 countries. Japan is by far the most important market for our marine products in terms of value followed by USA, South East Asia and European Union. India is ranked 17th among the marine products exporting countries in the world, though it has a vast coast line of 8000 kms., and huge fresh water fishery resources and a huge potential for export led growth in fresh water and marine fish and related industry. The export of fisheries products during the next 5 years could be increased US$ 3 to 4 billion with intensive efforts.

With a view to achieving this objective, Government of India proposes to have a comprehensive survey-cum study done and integrated and comprehensive action plan prepared for all the coastal States in fish and fish processing with a special focus on Andaman and Nicobar and Lakshadweep Islands. Such a plan of action is envisaged to identify (*i*) major goals, (*ii*) policy, programme, legal and other constraints, (*iii*) location specific infrastructure and other facilities for fishery, fish processing and diversification of products for high value addition, marketing etc., (*iv*) specific measures to increase India's share in the global marine trade covering cost and quality issues, compliance with global standards, trade barriers, country specific export opportunities and challenges.

Grain Processing

The current production of food grains (including rice, jawar, bajra, maize, ragi, wheat, barely, gram and pulses) in the country is estimated to be about 225 million tonnes per annum. The most salient feature in food production scenario is that India has achieved 'food security'–that signifies a true independence in terms of ability of feeding the entire population in India. Cereals account for the largest share, about one fifth of the consumption expenditure in India.

Grain processing is the biggest component in the food sector sharing over 40 per cent of the total value. The basic feature of the sector is the predominance of primary processing sector sharing 96 per cent of the total value with secondary and tertiary sector contributing about 4 per cent of the total value addition. This area needs to be looked into as it has a high potential for growth.

Grains are mainly consumed in the form of products obtained from primary processing, such as rice from paddy and 'atta' (flour) from wheat. Products from secondary processing are yet to gain mass acceptability in the country, possibly due to non-availability, taste and price factors. However, with increasing purchasing power, rising middle class population, increasing number of women in the work force, proliferation of media etc. cereal processing, both primary and secondary, are expected to gain momentum to cater to changing lifestyles.

Indian basmati rice commands a premium in the international market. The export of basmati and non-basmati rice has been increasing continuously and as against export of Rs.3538.3 million in 1988-89, the export has increased to Rs.62000 million in 1998-99.

Presently, total milling capacity of 186.00 million tonnes (in terms of paddy) is already available in the country of which 85 million tonnes is with traditional mills and 101 million is with modern mills.

Plantation

Tea coffee, cashew, cocoa etc. are the major plantation crops in the country. They are grown in different parts of the country as their plantations require specific agro-climatic conditions.

India's principal plantation crops accounted for around 5 to 6 percentage of India's aggregate export earnings. Production of major plantation crops has gone up considerably over the three decades; so also has their domestic consumption.

India continues to be world's largest tea producer, consumer and exporters of black tea. Cashew is an important crops of the country and earns valuable foreign exchange (about Rs.16000 million) India is the world's leading producer and exporter of cashew kernels (75000 tonnes annually) and accounts for about 31 per cent of the world's production of raw cashew and nearly 48 per cent of the world's export of cashew kernels.

Coffee is the oldest among the plantation crops in India.

Number of Licensed Units under FPO

The prominent processed items are fruit pulps and juices, fruit based ready-to-serve beverages, canned fruits and vegetables, jams, squashes, pickles, chutneys and dehydrated vegetables.

More recently, products like frozen pulps and vegetables, frozen dried fruits and vegetables, fruit juice concentrates and vegetable curries in retortable pouches, canned mushroom and mushroom products have been taken up for manufacture by the industry.

The fruit and vegetable processing industry in India is highly decentralized. A large number of units are in the cottage/home scale and small scale sector, having small capacities upto 250 tonnes/annum. But big Indian and multinational companies in the sector have large capacities in the range of 30 tonnes per hour or so.

Since liberalisation and withdrawal of excise duty on fruit and vegetable products there has been significant rise in the growth rate of the industry. Number of F and V P units licensed under FPO exceeds 5198, inclusive of over 650 sweetened aerated water units.

Challenges, Constraints and Concerns

India is already a major producer of food (first in cereals, livestock population, milk and second in fruits and vegetables), producing over 600 million tons of food products, and in case the immense untapped potential of growth is achieved the country can emerge as the largest producer of major food items.

1. Processing level presently being extremely low, the wastage levels are very high resulting in colossal wastage of national wealth running in thousands of crores.
2. Value addition to the raw produce in the country is only seven per cent, compared to as much as 23 per cent in China, 45 per cent in the Philippines and 188 per cent in the U.K.
3. Despite a large production base, the availability of raw materials *i.e.* fruits and vegetables to the industry continues to be a major constraint. This has a strong relation with the per capita production of these commodities and India's share in the world exports of processed fruits and vegetables. For countries whose share in the world trade of processed fruits exceeded 3 per cent, the per capita production of fruits was three to 12 times that of India. Similarly, for countries whose share in the world export of processed vegetables exceeded 2 per cent, the per capita vegetable production was two to five times that of India.

In countries like Israel, New Zealand, Brazil, Argentina, Chile, Turkey, Australia, etc, utilisation of fruits and vegetables for processing is much higher than that in India and their share in world exports is also equally high as compared to India.

4. The quality of the material available in the country for fruits and vegetables processing is another limiting factor. The Indian production is made up of produce of different varieties and, therefore, lacks uniformity in physio-chemical characteristics. Added to this, all varieties produced in large quantities are not suitable for processing, which in turn contributes to the poor output from a unit weight of the raw material used. It is reported that 16 tonnes of Indian pineapple are needed to produce 1 tonnes of its concentrate, as against only 8 tonnes of the Philippine pineapple. Similarly, 7 tonnes of Indian tomato will produce about 1 tonne of paste as against only 4 tonnes of Italian tomato.
5. The price of the raw materials is another key factor contributing to its low availability to the industry. According to the Annual Survey of Industries (ASI), raw material input constitutes as much as 83 per cent of the value of the output. Indian prices are about two to four times higher than those prevailing in the international markets with respect to tomatoes, pineapples and oranges.
6. The small scale and unorganized sectors today account for 75 per cent of the total industry having only local presence without much access to knowledge, technology and marketing network.
7. The differential between the farmer's realization and the final consumer price is very high in our country even in the fresh produce.
8. In processed food products, the high price on account of cumulative effect of low productivity, high cost of raw material, spoilage due to poor infrastructure, inefficient and costly transportation, high cost of finance and high incidence of taxes and duties, leads to the vicious cycle of low demand, low capacity utilization, high per unit cost and low demand.

Despite the existence of a strong and wide network of R&D institutions (CSIR labs, ICAR institutions, ICMR Establishments, Universities and Private institutions), their linkage with the users like farmers and industry, is not well established resulting in lack of technology flow, pure and academic research rather than applied and commercial, lack of involvement of industry in research work, and resource crunch.

The unattractive nature and the high risk profile of food processing industry has impeded required flow of credit from financial institutions who are yet to acquire the proper understanding of this sector to attain the requisite levels of appraising skills.

Low margins, seasonality and high perishability being the distinct features of this industry, the access to seed capital and working capital is not easy. Despite having been declared a priority-lending sector, there is hardly any growth in capital flow to this industry.

Despite vast domestic market size, the present level of processed food marketability is very low but by doing massive awareness and educational campaigns this market could grow higher enough to consume substantial part of any quantum of our processed foods.

Indian brands are yet to establish in the international markets calling for a concerted effort to capture world market share in tune with our standing in the production front.

With the coming in of WTO regime the country has to prepare for meeting the requisite quality standards in order to compete with imported goods in the domestic market itself. This calls for adoption of high tech machine and technologies as also development of entire chain of the infrastructure.

1. Weak database and lack of market intelligence are the prevailing features of this sector.
2. There is poor infrastructure of not only processing but even transportation, ports, airports, storage and handling etc. India occupies first position in the world in fruit production, but our export figures for fresh fruits and their products are negligible. This is mainly because of quality standards and postharvest handling of fruits where little attention has been paid so far. Fruits being a living entity,

their quality deteriorates even after their harvest. In developing countries like India, extent of these losses ranges from 20-45 per cent, which is every high. As area is limiting factor more attention is needed towards decreasing these losses besides boosting production per unit area. Postharvest losses are primarily due to faulty handling, packaging, storage and transportation. Due emphasis has to be given to reduce transportation and storage losses by adopting cold chain technology and use of modified atmosphere storage, devising proper packaging and use of postharvest chemicals which are helpful in enhancing shelf life.

3. The backward linkage between the farmer and the processor is yet to take proper shape to tide over the impediments which exist on account of fragmented and small land holdings, erratic production due to natural factors, non-uniformity and inconsistent supply of raw material and longer chain of intermediaries.
4. Multiplicity of laws and regulatory authorities throttle the industry in its further growth calling for harmonization of laws, development and administration of standards in consonance with international standards like Codex through a single authority.
5. Prevailing packaging system lacks requisite quality and presentability parameters creating handicap as compared to the imported products.
6. Cooperative institutions and other parastatal organizations are weak and peoples' participation, either through Panchayat Raj Institutions or NGOs or farmers' organizations, industries' association in food sector remains far from adequate.

Chapter 2

Importance and Scope of Food Preservation

Importance of Food Preservation

Food preservation can be defined as the science which deals with the process of prevention of decay or spoilage of food and helps it to be stored in a fit condition for future use. Throughout the man's history, certainly since the beginning of civilization, the preservation of food material has played a vital role in avoiding wide-spread starvation and it continues to do so till date, even in industrially developed world. Food supply has to keep pace with the needs of the population. There is always shortage of food in developing countries like India because of demand of increasing population. Therefore, it is important to improve and expand facilities for storage and preservation of food.

1. Food preservation helps in increasing shelf life of food, thus, increasing its supply. Delay in the use of fresh foods alters their freshness, palatability and nutritive value but by preserving these foods, spoilage can be avoided. Thus, perishable foods can be made available throughout the

year by preservation. In this way, preservation helps to add variety in our diet and better balanced.

2. By preservation and proper storage of food, it can be saved for future use at the time of scarcity, natural drought, etc. Besides this, preparation of meals becomes easier, ultimately, it helps in saving time by reducing preparation time and energy of housewives.
3. Availability of food can be increased throughout the year and wastage can also be minimized. In this way, preservation of food also helps in stabilizing the prices of the food by making the availability of seasonal foods throughout the year so it helps to improve the nutrition of the people.

Scope of Food Preservation in India

Fruits and vegetables are an important supplement to the human diet as they provide the essential minerals and vitamins and fibre required for maintaining health. Fortunately, in the production of fruits and vegetables, our country stands on the second rank in the world. The combined production of fruits and vegetables is about 50 million tonnes annually. However, for various reasons, this abundance of production is not utilized and about 20-30 per cent of it is wasted because of spoilage due to insufficient transport facilities, lack of good roads and poor availability of packaging materials, therefore, surplus cannot be taken quickly to the markets in urban areas and also cannot be stored for sale in the off-season because of inadequate local cold storage facilities.

Therefore, the spoilage can be prevented during the glut season at the producing centres by converting them into new categories of processed products. In view of globalization of Indian economy, there is ample scope for exports of agricultural commodities in general and processed foods in particular. With increasing urbanization, rise in middle class purchasing power, change in food habits, there is increasing demand for factory-made jams, jellies, fruit beverages, dehydrated foods, pickles etc. in the domestic market also. Moreover, there is considerable demand for some of these products in foreign markets, *e.g.* mangoes (fresh and canned), fruit juices, salted cashew are good foreign exchange earners.

In spite of all this, the fruit and vegetable preservation industry at present is able to utilize less than 1 per cent of the total production for conversion into products as against 40 per cent in developed countries. Thus, there is considerable scope for expansion of the industry by developing and exporting value-added products and earn more foreign exchange. A wide variety of products can be made from fruits and vegetables as shown in Table 2.1. The choice of the products to be manufactured will depend on the readily availability of raw materials in the area where the processing unit is proposed to be set up and the consumer preference in the neighbouring markets.

Table 2.1: Products Prepared from Fruits and Vegetables

Sl.No.	*Name of Fruits/ Vegetables*	*Product*
Fruits		
1.	Mango	Juice, RTS, squash, jam, preserve, toffee, *amchur*, pickle, chutney, canned mango, mango powder, mango concentrate
2.	Guava	Jelly, cheese, toffee, nectar, canned guava, squash, vinegar
3.	Aonla	Preserve, jam, candy, syrup, pickle, chutney, dried shreds, triphla, chyawanprash
4.	Pomegranate	Juice, squash, syrup, *anardana* (dried product)
5.	Pineapple	Canned pineapple, juice, squash, syrup, jam
6.	Litchi	Juice, syrup, canned litchi
7.	Papaya	Jam, candy, nectar, pickle, sauce, canned papaya, papain
8.	Grape	Wine, juice, raisin, munakka
9.	Karonda	Pickle, jelly, candy, preserve
10.	Banana	Canned banana, dried banana, toffee
11.	Fig	Dried fig
12.	Loquat	Jam, jelly, canned loquat
13.	Ber	Candy, preserve, canned ber, jam
14.	Bael	Preserve, nectar, squash, canned bael, cider
15.	Phalsa	Juice, squash, syrup
16.	Jackfruit	Pickle
17.	Citrus fruits	Juice, pickle, marmalade, squash, cordial, barley water, candy
18.	Jamun	Jelly, syrup, vinegar
19.	Strawberry	Jam, juice

Contd...

Table 2.1–Contd...

Sl.No.	*Name of Fruits/ Vegetables*	*Product*
20.	Mulberry	Juice, squash
21.	Apple	Jam, preserve, juice, chutney, cider
22.	Cherry	Jam, candy, canned cherry, dried cherry
23.	Peach	Jam, chutney, canned and dried peach
24.	Pear	Jam, chutney, pickle, preserve, canned pear
25.	Plum	Jam, chutney, sauce, dried plum
26.	Apricot	Jam, chutney, canned and dried apricot
27.	Date	Dried date
Vegetables		
1.	Tomato	Sauce, chutney, pickle, puree, paste, canned tomato, juice, soup, jam
2.	Cauliflower	Pickle, dried and canned cauliflower
3.	Carrot	Jam, preserve, pickle, candy, canned carrot
4.	Cabbage	Sauerkraut, dried cabbage
5.	Peas	Canned peas, dried peas, pickle
6.	Brinjal	Pickle
7.	Green chilli	Pickle
8.	Beetroot	Pickle, canned beetroot
9.	Radish	Pickle
10.	Turnip	Pickle, canned and dried turnip
11.	Pointed gourd (Parwal)	Canned and dried parwal
12.	Ash gourd (Petha)	Candy
13.	Bitter gourd (Karela)	Pickle, dried bitter gourd
14.	Onion	Pickle, dried onion
15.	Garlic	Pickle, powder
16.	Beans	Canned and dried beans
17.	Spinach	Canned spinach
18.	Ginger	Pickle, preserve, candy, dried ginger, RTS, syrup
19.	Cucumber	Pickle
20.	Water melon	Juice, squash
21.	Musk melon	Juice, squash
22.	Mushroom	Pickle, sauce, canned mushroom. dried mushoom
23.	Potato	Chips, papad, starch, canned potato

Source: Srivastava and Kumar (1998).

Preservation of the foods has to be established in developing countries for one or another reasons also such as to generate both rural and urban employment, improve farmers' nutrition by allowing them to consume their own processed fruit and vegetables during the off-season, to stimulate agricultural production by obtaining marketable products, reduction of imports and meeting of export demands and to develop new value added products. In this way, we have to strengthen such areas which are of high value addition. Therefore, various factors that have been taken into consideration in setting up a fruit and vegetable processing industry are as follows:

Product Mix

The product profile of the fruit and vegetable processing industry has remained static and is dependent only on a few fruits and vegetables like mango, pineapple, citrus, tomato and pea. The production of new products are necessary for the survival and growth of the processing industry would also meet new taste and demand in home as well as export market.

Availability of Raw Materials

A wide variety of fruits and vegetables are cultivated in India and are available in abundance in the season. Some of these fruits like mango, guava, sapota, banana, cashewnut, pineapple, aonla etc. do not grow in many countries and their products are likely to find a ready market in foreign countries.

Manpower

India is in advantageous position, as compared to developed countries in having a large resrvoir of manpower, but there is short supply of skilled manpower which affect the productivity. Proper training, in-factory or institutional, good working conditions and reasonable wages would go a long way to increase productivity.

Lack of Awareness

Most commercial fruit and vegetable growers are not aware of the market for preserved products and do not have the necessary technical knowledge to undertake processing themselves. The Central and State Governments have started a number of projects to impart different levels of training in canning and preservation.

Marketing Facilities

However, there is demand for preserved products which is likely to grow in future, these are not readily available in small towns due to reluctance of shopkeepers in stocking such items. The establishment of grower's cooperatives would help in the marketing of such products and Government should encourage the establishment of such cooperatives.

Transport Facilities

Earlier, there was a serious problem due to lack of good roads and shortage of trucks and rail wagons for rapid transportation of fruits and vegetables in good condition from one part of the country to another part. But now, there is considerable improvement in both road and transportation of fruits and vegetables from producing areas and that will be connected to processing factories in distant parts of the country.

Availability of Containers

Bottles and cans are the two major types of containers required by the food processing industry. At present there is a great difficulty in the availability of cans since there are very few factories for their manufacture. Therefore, there is need for setting up more factories to meet the demands for cans for food processing industries.

Publicity

Proper publicity is the only way to attract the consumer and give him information about the new products in the market. Most commercial fruit and vegetable growers are not aware of the market for preserved products and do not have necessary technical knowledge to understand processing themselves. Their products can also be popularized by displaying in exhibitions and fairs by practical demonstrations and distributing samples to the public.

Role of Government

The central and state governments have started a number of projects to impart different kinds of training in canning and preservation at small scale basis. Several types of training courses have also been started. The same type of training is also given at

Central Food Technological Research Institute, Mysore and also some agricultural universities.

At various research institutions all over the country, attempts are being made to evolve better and more economical methods of preservation of fruits and vegetables under Indian conditions.

Chapter 3

National Food Processing Policy of India

Agricultural development is crucial to India's overall progress and this can be achieved by intensifying the agricultural and allied activities as well as by diversification and commercialization of agriculture. A strong and dynamic food processing sector plays a significant role in diversification of agricultural activities, improving value addition opportunities and creating surplus for export of agro-food products.

This would, however, require policies and plans for improvement of food processing infrastructure including upgradation of technology and enforcement of quality standards, promoting investment in food processing, activating domestic market with focus on exports. The Ministry of Food Processing Industries is concerned with formulation and implementation of the policies and plans for the food processing industries within the overall national priorities and objectives.

The Ministry of Food Processing Industries (MFPI) was set up in July, 1988 to give an impetus to development of food processing industries in the country. Subsequently vide Notification No. Doc. CD- 442/99 dated 15.10.99 this Ministry was made a Department and brought under the Ministry of Agriculture. It has once again been notified as Ministry of Food Processing Industries vide Cabinet Secretariat's Note No. 1/22/1/2001-Cab.(1) dated 06.09.2001.

Allocation of Business

The subjects allocated to the Ministry are as under:

1. Fruit and vegetable processing industry (including freezing and dehydration).
2. Food-grain milling industry.
3. Processing and refrigeration of certain agricultural products like milk powder, infant milk food, malted milk food, condensed milk, ghee and other dairy products, poultry and eggs, meat and meat products.
4. Processing of fish (including canning and freezing)
5. Planning, development and control of, and assistance to industries relating to bread, oilseeds, breakfast foods, biscuits, confectionery (including cocoa processing and chocolate making), malt extract, protein isolate, high protein food, weaning food and extruded food products (including other ready-to- eat foods).
6. Beer, including non-alcoholic beer.
7. Alcoholic drinks from non-molasses base.
8. Aerated waters and soft drinks.
9. Specialized packaging for food processing industries.
10. Establishment and servicing of the Development Councils for fish processing industries.
11. Technical assistance and advice to fish processing industry.

Goals and Objectives

In the era of economic liberalization where the private, public and co-operative sectors are to play their rightful role in the development of food processing sector, the Ministry acts as a catalyst for bringing in greater investment into this sector, guiding and helping the industry in the proper direction, encouraging exports and creating a conducive environment for healthy growth of the food processing industry. Within this overall objectives, the Ministry aims at:

1. Better utilization and value addition of agricultural produce for enhancement of income of farmers.
2. Minimizing wastage at all stages in the food processing chain by the development of infrastructure for storage, transportation and processing of agro-food produce.
3. Induction of modern technology into the food processing industries from both domestic and external sources.
4. Maximum utilization of agricultural residues and by-products of the primary agricultural produce as also of the processed food industry.
5. Encouraging R&D in food processing for product and process development and improved packaging.
6. Providing policy support, promotional initiatives and physical facilities to promote value added exports.

Developmental

1. Assistance under various plan schemes.
2. Widening the R&D base in food processing by involvement of various R&D institutes and support to various R&D activities.
3. Human Resource Development to meet the growing requirement of managers, entrepreneurs and skilled workers in the food processing industry.
4. Assistance for setting up analytical and testing laboratories, active participation in laying down of food standards and their harmonization with the international standards.

Promotional

In order to create awareness about the potential and prospect of food processing industries in the country, this Ministry provides,

1. Assistance for organizing workshops, seminars, exhibitions and fairs.
2. Assistance for studies/surveys etc.
3. Publications and films.

Regulatory

Vision

The vision of National food processing policy of India as formulated by Ministry of Food Processing Industries, Ministry of Agriculture, Government of India is to motivate farmers and food processors, and to provide an interactive coupling between technology, economy, environment and society for speedy development of food processing industries to build up a substantial base for production of value added agro food products for domestic and export markets with a strong emphasis on food safety and quality enabling the farmers especially to realize direct benefits of new technology and marketing network and to ensure adequate availability of quality food products for the consumer at economic prices.

Mission

To fulfill the VISION so as to contribute to all round economic and social development of India through generation of employment opportunities especially in rural areas.

The Policy

Creating Enabling Environment

The Policy will seek to create an appropriate environment for entrepreneurs to set up Food Processing Industries through:

1. Fiscal initiatives/interventions like rationalization of tax structure on fresh foods as well as processed foods and machinery used for the production of processed foods.
2. Harmonization and Simplification of food laws by an appropriate enactment to cover all provisions relating to food products so that the existing system of multiple laws is replaced and also covering issues concerning standards, Nutrition, Merit goods, futures marketing, equalisation fund etc.

3. A concerted promotion campaign to create market for processed foods by providing financial assistance to Industry Associations, NGOs/Cooperatives, Private Sector Units, State Government Organization for undertaking generic market promotion.
4. Efforts to expand the availability of the right kind and quality of raw material round the year by increasing production, improving productivity.
5. Strengthening of database and market intelligence system through studies and surveys to be conducted in various States to enable planned investment in the appropriate sector matching with the availability of raw material and marketability of processed products.
6. Strengthening extension services and to the farmers and co-operatives in the areas of postharvest management of agro-produce to encourage creation of pre-processing facilities near the farms like washing, fumigation, packaging etc.
7. Efforts to encourage setting up of agro-processing facilities as close to the area of production as possible to avoid wastage and reduce transportation cost.
8. Promotion of investments, both foreign and domestic.
9. Simplification of documentation and procedures under taxation laws to avoid unnecessary harassment arising out of mere technicalities.

Infrastructural Development

The policy will facilitate:

1. Establishment of cold chain, low cost pre-cooling facilities near farms, cold stores and grading, sorting, packing facilities to reduce wastage, improve quality and shelf life of products.
2. Application of biotechnology, remote sensing technology, energy saving technologies and technologies for environmental protection.
3. Building up a strong infrastructural base for production of value added products with special emphasis on food safety and quality matching international standards.

4. Development of Packaging Technologies for individual products, especially cut-fruits and vegetables, so as to increase their shelf life and improve consumer acceptance both in the domestic and international markets.
5. Development of new technologies in Food Processing and Packaging and also to provide for the mechanism to facilitate quick transfer of technologies to field through a net work of R&D Institutions having a Central Institute at the national level with satellite institutions located strategically in various regions to cover up the whole country and to make available the required testing facilities. This could be done by establishing a new institution or strengthening an existing one.
6. Development of area-specific Agro Food Parks dedicated to processing of the predominant produce of the area *e.g.*, apple in J&K, pineapple in North East, Litchi in Bihar, Mango in Maharashtra and Andhra Pradesh etc.
7. Development of Anchor Industrial Centre and/or linkage with Anchor Industrial Units having net work of small processing units.
8. Development of Agro-industrial multi-products units capable of processing a cluster of trans-seasonal produces.

Backward Linkage

The policy will promote:

1. Establishment of a sustained and lasting linkage between the farmers and the processors based on mutual trust and benefits by utilizing the existing infrastructure of cooperative, village panchayats and such other institutions.
2. Development of Future Market in the best interest of both the farmers and the processors ensuring a minimum price stability to the farmer and a sustained supply of raw material to the processor.
3. Mechanism to reduce the gap between the farm gate price of agro-produce and the final price paid by the consumer.
4. Setting up of an Equalisation Fund to ensure sustained supply of raw material at a particular price level and at

the same time to plough back the savings occurring in the eventuality of lower price to make the Fund self-regenerative.

Forward Linkage

The policy will promote

1. Establishment of a strong linkage between the processor and the market to effect cost economies by elimination of avoidable intermediaries.
2. Establishment of marketing network with an apex body to ensure proper marketing of processed products.
3. Development of marketing capabilities both with regard to infrastructure and quality in order to promote competitive capabilities to face not only the WTO challenge but to undertake exports in a big way.

Special Provision

The following shall receive higher priority and special consideration in policy and plans:

1. The North Eastern Region, the Hilly Areas, Islands and the ITDP areas in the country to be given not only special attention but also special consideration.
2. The fiscal incentives like excise duty/sales tax concession and the tax holidays to be provided not only to those units which are set up in these areas but also to those units which though set up outside these areas near the market center, are engaged in processing the produce coming from these areas.
3. Tax holiday for food processing units, with the exception of liquor, cigarettes and aerated drinks and similar luxury items, for a period of 10 years.

General Policy Initiatives

A number of policy initiatives have been taken from time to time to promote growth of processed food sector in the country. Some of these are:

1. Most of the processed food items have been exempted from the purview of licensing under the Industries (Development and Regulation) Act, 1951, except items reserved for small-scale sector and alcoholic beverages.
2. Food processing industries have been included in the list of priority sector for bank lending.
3. Automatic approval for foreign equity upto 100 per cent is available for most of the processed food items excepting alcohol and beer and those reserved for small scale sector subject to certain conditions.
4. Excise duty on processed fruits and vegetables has been brought down from 16 per cent to zero level in the Budget, 2001-02.

Tenth Plan Initiatives

MFPI has been operating several plan schemes for development of processed food industry since the inception of the 8th Plan period. Taking into account the recommendations of the Working Group constituted by the Planning Commission, the constraints faced by the industry, the need to accelerate growth of the sector and the priority status accorded to food processing industry by Government, MFPI operated many Plan schemes during the 9th Plan.

The Planning Commission constituted a Working Group on Food Processing Industries for the 10th Five Year Plan. Simultaneously, an exercise of zero based budgeting was undertaken by the Planning Commission to weed out such schemes as may have become irrelevant and consider other components which may be more relevant. As a result of these two exercises, a macro management approach has been adopted resulting in reduction in the number of Plan Schemes from 19 to 6. The outlay approved for the Plan period has gone up from Rs. 235 crores in the 9th Plan to Rs. 650 crores for the 10th Plan.

This approach is expected to contribute positively to growth in the sector as a whole and assist such components of each sub-sector as are most critical.

Details of approved outlays for the 10th Plan are given below:

Scheme for Infrastructure Development

1. Food Park
2. Packaging Centre
3. Modernised Abattoirs
4. Integrated Cold Chain facilities
5. Irradiation facilities
6. Value added centres

Scheme for Technology Upgradation, Establishment/ Modernization of Food Processing Industries

1. Processed food industries in general
2. Pulse milling

Scheme for Backward and Forward Integration and Other Promotional Activities

1. Backward linkage
2. Forward integration
3. Generic advertisement
4. Promotional activities such as participation in exhibition/ fairs/supporting seminars/workshops/studies and surveys.
5. Preparation of short films and material for different meetings.
6. Strengthening of Directorate of F&VP
7. Strengthening of Industry Associations
8. Fortification of wheat flour

Scheme for Quality Assurance, Codex Standards and R&D

1. Food safety and quality assurance mechanisms
2. Bar coding system
3. Strengthening the Codex cell
4. Continuous R&D
5. Setting up/upgradation of quality control laboratories

Scheme for Human Resource Development

1. Setting up of FPTC
2. Imparting training to update skills
3. EDP
4. Facilitating Universities/Institutions for running degree/diploma courses and extension services

Scheme for Strengthening of Institutions.

Part II
Food Spoilage, Prevention and Food Safety

Chapter 4

Food Spoilage and its Causes

Food spoilage can be defined as any decay or undesirable decomposition of constituents by excessive growth of microorganisms or by other physical and chemical causes. Food is unfit for human consumption if maximum limit of the microorganisms is crossed than the prescribed one. The major causes of food deterioration include the following:

1. Growth and activities of microorganisms, principally bacteria, yeasts and moulds
2. Activities of food enzymes and other chemical reactions within the food itself
3. Infestation by insects, parasites and rodents
4. Inappropriate temperatures for a given food
5. Either the gain or loss of moisture
6. Reaction with oxygen
7. Light
8. Physical stresses or abuse
9. Time

These factors can be divided into biological, chemical and physical factors. Often these factors do not operate in isolation. Bacteria, insects and light, for example, can all be operating simultaneously to spoil food in the field or in warehouse. Similarly, heat, moisture and air simultaneously affect the multiplication and activities of bacteria, as well as the chemical activities of food enzymes. At one time, many forms of deterioration may take place, depending on the food and environmental conditions.

Foods are classified into three groups on the basis of ease of spoilage.

1. The stable or non-perishable foods like sugar, flour, grains and dry products do not ordinarily spoil unless handled carelessly.
2. The semi-perishable foods like potatoes, onions, some varieties of apples, fruits and vegetables having thick skin can be protected from spoilage if properly stored.
3. The perishable foods like meat, fish, poultry, eggs, milk and fruits and vegetables spoil readily.

Major causes of food deterioration

Biological

1. Growth of bacteria, yeast, mould
2. Activity of enzymes
3. Insects, rodents and parasites

Chemical

1. Reaction with oxygen
2. Chemical reactions within food constituents
3. Light

Physical

1. Temperature
2. Physical stress or abuse

Note: All of these factors can act together.

Common Methods Employed to Preserve Foods

1. Heat (blanching, pasteurisation, sterilisation)
2. Cold (chilling, freezing)
3. Drying (partial or fully)
4. Acid (fermentation, addition)
5. Sugar and salt (addition- osmotic effect)
6. Smoke (heat+smoke- formaldehyde and others)
7. Atmospheric conditions (O_2 or CO_2)
8. Radiation (X-ray, UV, microwaves, ionising)
9. Chemicals (preservatives)

Biological Changes

Growth and Activity of Microorganisms

Most significant changes in food are due to biological factors especially microorganisms. Microorganisms are present in air, water, on most of the foods and on our hands etc. They are so small that they can only be seen through microscope.

There are three types of microorganisms which cause spoilage.

1. Bacteria
2. Yeasts
3. Moulds

Bacteria

Bacteria are found everywhere. They are minute unicellular microorganisms. The growth of bacteria depends upon a number of external factors. They require proper kind of food material, heat, pH, moisture and oxygen. Bacteria are much more difficult to be killed than moulds and yeast and are the most common causes of food spoilage. They are present in active form (vegetative stage) or resting form (spore stage). In vegetative stage, bacteria are destroyed at boiling temperature but spores are difficult to destroy and require application of heat for a long time. At boiling temperature (100°C) six hours heating may be required to destroy the spores or they can be killed in 30 min at 121°C under 10 lbs pressure.

In foods with high acid content (all fruits, tomatoes, pickles etc.), all forms of bacteria are sensitive to acids and can be killed easily in acidic pH at a temperature of boiling water. So foods with high acid content are processed at 100°C whereas low acid foods such as meat, corn, peas, beans and all vegetables except tomatoes have to be processed at higher temperature (116°C) in a steam pressure to kill bacteria at faster rate. The temperature maintained and the length of time the food is held vary with each kind of food. Some of the moist heat resistance bacteria are present in the soil, hence, preparation and processing of root vegetables require special care. *Colostridium botulinum* causes spoilage in canned foods.

Factors influencing the bacterial growth

Environmental factors that influence the growth of bacteria are:

Food

Each kind of bacteria has a definite range of food requirements. For some species, range is wide and growth takes place in a variety of substances, *e.g.* coliform bacteria, whereas for other bacteria range is narrow and they can grow in only limited number of food substances, *e.g.* many of the pathogenic bacteria.

Moisture

Concentration of moisture in food is important factor in preventing or allowing the bacterial growth in the food. Bacteria require more moisture than yeasts or moulds.

Temperature

Each bacteria has an optimal temperature at which it grows best. Temperature below and above the optimum adversely affects the growth of bacteria. There are three types of bacteria according to temperature tolerance:

1. Psychrophilic–which grow well at refrigerator temperature (below 10°C)
2. Mesophilic–grow best at median temperature between 16°C and 38°C
3. Thermophilic–the term thermophillic means 'heat loving'.

These are found in soils, in manure and compost etc. These are spore forming bacteria. They can grow best at temperature as higher as 82°C and they can grow slowly up to 100°C.

Hydrogen Ion Concentration or pH

pH determines the kind of bacteria to grow in a food and the changes produced. Most bacteria grow best at pH near neutrality, some bacteria are favoured by an acid reaction, few kinds of bacteria can grow in fairly acid or alkaline media.

Oxidation Reduction Potential

On the basis of their process of respiration, bacteria are classified as:

Aerobic Bacteria

Some bacteria and moulds require oxygen for growth called aerobic, require free oxygen

Anaerobic Bacteria

They do not require free oxygen and grow better in the absence, called anaerobic bacteria.

Facultative Bacteria

They grow either with or without free oxygen and hence called facultative bacteria.

Microaerophilic Bacteria

They require definite, but small amount of free oxygen.

Inhibitory Substances

During the process of growth, bacteria produce certain substances which will in time slow down or stop the growth of bacteria and may inhibit the multiplication of other organisms. Inhibitory substances added during the processing of foods may check the growth of most microorganisms or at least the undesirable ones, *e.g.* propionates added to bread inhibit the growth of bacteria.

Yeasts

These are microscopic unicellular organisms which are non-motile round or oval. Fungi are usually known as yeast. Yeast reproduce or multiply by a characteristic process of "budding". The bud when fully mature, breaks away from the mother cell and becomes independent and repeats the process of multiplication. Yeasts require less moisture and acidic pH to grow and do not grow in alkaline medium. They grow well in light sugar solution and acidic medium.

Yeasts are very useful in making bread, beer, wine, vinegar and many other fermented products. The yeast causes fermentation of sugar and convert them into alcohol and CO_2. This property is very important in the preparation of alcohol beverages and in baking.

$$\text{Sugar} + \text{yeast} \rightarrow \text{Alcohol} + CO_2$$

Yeasts are responsible for much of fermentation of fruits and fruit products, *e.g.*, sometimes the cork of juice bottle, in which fermentation may be going on is thrown out or the juice bottle is shattered to pieces with great force. It is due to CO_2 generated in the process of fermentation. So yeasts are undesirable when they grow on fruits, juices, squashes, sharbat, honey etc. They spoil the appearance, taste, texture and wholesomeness of fruits and fruit products. During active fermentation, yeast can be recognized by formation of bubbles or foam on the surface of the product. Some of yeasts which grow on fruits are *Saccharomyces, Candida*, etc. Yeasts grow very well under moderate temperature (25-30°C) in solution containing sugar and plenty of water. Most of the yeasts usually do not grow in media containing more than 65% of sugar or 0.5% acetic acid. Heating at 60°C for a few minutes is sufficient to destroy most species of yeasts. Boiling destroys yeast cells and spores effectively.

Moulds

Moulding of food is quite common. A slice of moist bread or a piece of orange left for a time becomes covered with a whitish or greyish cottony matter. This is mould. They are larger and more complex in structure than bacteria or yeast. Moulds are made up of mycelium and spores. Mould growth starts as fine thread known as hypha. As development proceeds interlaced hyphae form mycelium, which have appearance of cotton wool and at later stage spores may be pigmented and give appearance and colour.

They thrive best in close, damp and dark situation and require adequate supply of warmth, moisture and air for growth. They are aerobic in nature (require free oxygen) and require less available moisture and can grow well at 25-30°C. Moulds prefer sugar containing substances like jam, jelly, preserves and other sweet based products. They can grow at wide range of pH (2 to 8.5) but majority grow well at acidic pH. Therefore, they grow nicely on pickles, juices etc. They can grow on many kind of foods especially when temperature, air and humidity are favourable. Their growth can be seen only on the surface of food; they not only consume nutrients present in food thereby lowering the food value but also produce odd-by-products which spoil the flavour, taste and texture of food hence change the quality contents of the entire products.

Majority of moulds are sensitive to heat and are destroyed at 60°C when heated for 30 min. Boiling quickly destroys both moulds and their spores. Some of the common moulds are *Aspergillus, penicillium, Rhizopus* and *Helminthosporium*.

Food-borne Disease

A special kind of food deterioration that may or may not alter a food's organoleptic properties has to do with food-borne diseases. Food-borne diseases are commonly classified as food infections or food intoxications. Food infections involve microorganisms present in the food at time of consumption when they grow in the host and cause illness and disease. Food intoxication involves toxic substances produced in foods as by-products of microorganisms prior to consumption and cause disease upon ingestion.

Staphylococcus aureus and *Clostridium botulinum* produce bacterial food poisoning. *Clostridium botulinum* produces toxins. Certain moulds like *Asperigillus flavus* produce mycotoxins known as aflatoxins.

Control of Microorganisms

The most important means of controlling the growth of bacteria, yeast and mould are heat, cold, drying, acid, sugar, salt, smoke, air, chemicals and radiation.

Action of Enzymes

Enzymes produced by living things are compounds which catalyze chemical reactions. Enzymes are found in all fresh foods, plants and animals. Their action is important for ripening of fruits and vegetables. Upto certain stages, enzymes actions may be desirable but its continuous action ultimately renders the food inedible by bringing undesirable changes in food tissues. These changes include the darkening of cut surfaces, fermentation of soft spots and development of off flavours. Many foods contain natural enzymes, which under certain conditions produce significant changes. Enzymes convert starch into sugar, proteins into amino acids, and pectin into pectic acids and these change the constituents of food. Some fruits and vegetables turn brown when damaged or when their cut surfaces are exposed to air due to the presence of the enzymes such as phenolase, peroxidase and polyphenol oxidase.

Natural tendering or aging of meat is due to action of enzymes. Natural food enzymes can be inactivated by heat, cold, drying, the addition of certain inhibiting chemicals and radiation.

Enzymes may catalyze different chemical reactions in plants and animal tissues. They are responsible for the chemical changes occurring in the ripening of fruits and the maturing of vegetables. Enzymes are proteins and may be responsible for certain undesirable chemical changes in preserved foods. So boiling temperature is adequate to destroy enzymes as in case of blanching of vegetables.

Enzyme actions can be easily controlled by regulating the temperature and excluding moisture and air. Enzymes can act between (15.6°-65.6°C). The optimum temperature of reaction is usually 37°C. Enzymes are protein in nature and coagulated by high temperature. Enzymes are usually inactivated by heating at temperature between 70° and 93.3°C.

Enzymes cause chemical reactions to occur at their fastest rate when the temperature is at optimum level (15.6°-65.6°C). Enzymes are protein in nature and coagulated by high temperature.

Insects, Parasites and Rodents

Insects, worms, bugs and fruit flies may damage food stuffs such as cereal grains and fruits and vegetables and render them unfit for human consumption. The injuries caused by these insects serve as pathways by which microorganisms reach the inner tissues and cause spoilage in the food.

Purely Chemical Action

These are those reactions which are not catalyzed by enzymes of tissues or microorganisms, *e.g.* rancidity of fat. Atmospheric oxygen and even sunlight cause certain foods to undergo various undesirable changes. Oxidation results in production of off flavour as well as loss of ascorbic acid. Such reactions are called lipid oxidation or non-enzymatic browning. Non-enzymatic browning or Maillard reaction is also one of the major causes of deterioration which occurs during storage of dried foods.

The food constituents influence the type of microorganisms that can grow in it and thus determine the changes in its appearance, flavour and other qualities. Proteins are degraded by proteolytic organisms. Fats are digested by relatively few microorgnisms, mainly moulds. Fats become rancid due to hydrolytic decomposition to mal-odourous fatty acids. Carbohydrates are affected by carbhydrates' fermenting microorganisms, particularly yeasts and moulds.

Protein + proteolytic microorganisms → Amino acids + amines + ammonia + hydrogen sulphide

Fat + lypolytic microorganisms → Fatty acids + glycerol

Carbohydrate + fermenting microorganisms → Acids + alcohols + gases

Moisture is required for chemical and reactions and microbial growth foods with a high percentage of water deteriorate fast. Moisture can cause lumping, caking, crystallization and stickiness in foods.

Food having high sugar or salt concentration do not support the growth of bacteria but most likely to be spoiled by moulds.

The pH of nearly all foods is below 7.0. Foods are classified as acid or non-acid depending on whether the pH is below or above 4.5. Most foods are acid foods, while all vegetables, fish, meats and milk products are non- acid. The low pH of acid foods prevents the growth of most bacterial species but such foods are spoiled mainly by yeast and moulds. Non- acid foods are most likely spoiled by bacteria.

The deterioration includes off colour development and surface biting. Banana, tomato, lemon and squash should be stored at about 10°C for retaining their quality.

Physical/Mechanical Change

All physical changes which are caused by freezing, burning, drying, poor selection of packaging material etc. during transportation and storage of foods also cause spoilage of foods.

Sometimes fruits are bruised or scratched during harvesting and handling. If precautions are not taken, the injured spots become the points of entrance of microorganisms which cause spoilage, *e.g.* stem end root in mango, green mould in citrus fruits and blue mould in apple.

Removal of diseased, damaged and scratched fruits during grading and postharvest treatment of the fruits with fungicides like thiabendazole etc. are highly useful in reducing mechanical damage to fruits.

Storage Conditions

Temperature, oxygen, light, duration of storage are the important factors that influence the type of microbial growth and spoilage of food during storage.

Temperature

Heat and cold though playing a role in food preservation, contribute to deterioration of food if not controlled. Excessive heat brings about protein denaturation and destruction of vitamins etc. Several fruits and vegetables deteriorate even at temperature of refrigerator (4°C). They suffer discolouration, changes in texture etc. Freezing may also cause deterioration of liquid foods such as a container of milk is allowed to freeze, the emulsions will be broken and fat will separate. Freezing also denatures the milk protein. Sometimes some fruits and vegetables at refrigerator temperature (4°C) deteriorate called "chill injury".

Oxygen is necessary for the growth of moulds and therefore, it must be excluded from food during processing, vacuum packaging or in some cases by use of oxygen absorbing chemicals.

Moisture

Excessive moisture pick up or loss causes substantial deterioration in foods. Moisture required for chemical reactions not for microorganism growth; excessive moisture can accelerate these types of deterioration. Excessive loss of moisture can also have detrimental effects particularly on appearance and texture.

Oxygen

Atmospheric oxygen brings about undesirable changes in food such as destruction of food colour, flavour and vitamin A and C.

Light

Light destroys vitamins B_2, A and C and also many food colours. Foods may be protected from light during packing or keeping them in containers.

Duration

All the other food deteriorating factors are time-dependent. The longer the storage time, greater the deterioration.

Chapter 5

Prevention of Microorganisms

Sterilisation

First of all, we will take a look at microorganisms and at the techniques we can employ to inhibit their growth or destroy them altogether. This is followed by some observations on the mathematical methods needed to compare processes with one another. Next we will discuss the term 'cooking', in connection with the sterilisation of food products, and the containers used for such products.

The two factors, product and container, determine the choice of sterilising system, to which we will devote a few explanatory notes before examining the Stork Hydromatic steriliser in some detail.

Sterilisation by means of heat is one of the methods available to preserve food products and give them a longer shelf life. Other techniques that can be employed to preserve foods are irradiation, drying, deep freezing, concentration, salting and smoking.

All these processes are designed to prevent microorganisms from utilising the medium presented by the food. They will either keep the organisms from growing or destroy them completely. It is important, however, that the preservation process should have the least possible impairing effect on the product.

The keeping quality of food products depends not merely on the presence or absence of microorganisms, but also on the enzymes within the food product. Enzymes play the role of biochemical catalysts that cause food products to deteriorate and start decomposing. Thus preservation has the added object of stopping this degradation process by inactivating the enzymes present in the food.

The inactivation of both microorganisms and enzymes can be brought about by heat, a fact that was discovered by Louis Pasteur around 1880. The pasteurisation process he invented and lent his name to enable product's shelf life to be prolonged by a few days; nowadays, it still finds application in some areas, such as the dairy industry.

To prolong a product's shelf life beyond a couple of days, all thermotolerant or thermoduric organisms in the product must be killed, and the enzymes inactivated.

Thermal food processing necessitates some knowledge of how microorganisms enter and multiply in food and of the consequences. The organisms that concern us most are those which cause food spoilage and those which are harmful to man. The microorganisms found in food are moulds, yeasts and bacteria. Of these, bacteria present the greatest hazard.

Moulds occur everywhere in nature. They are found in the soil, on vegetation and even in the air around us. They multiply through spore formation, which entails a kind of chain reaction, *i.e.*, they develop spores, which grow into moulds, which then produce spores again, and so on.

Fully grown moulds cannot withstand heat treatment, but some types develop spores which are highly resistant to heat and which are known as thermoduric. They can survive exposure to 92°C for more than 1 min.

Like the moulds, yeasts occur widely in nature. They are highly useful in certain processes, such as fermentation, and the production of beer, alcohol and glycerol. Yeasts can reproduce through sporulation, but they are also capable of multiplying by division or fission. They grow at temperatures ranging from 20 to 35°C and, as a rule, cannot withstand heat treatment at the temperature needed for pasteurisation (approx. 90°C).

Bacteria are the third group of microorganisms we have to deal with. They consist of single cell and are among the smallest life forms known to man. They occur in a wide variety of shapes, *e.g.* round (cocci), elongate (rods), twisted (spirilla), etc. and reproduce only by direct cell division or fission. The growth rate of bacteria is subject to a range of factors, including the species of bacterium involved. Some bacteria also produce spores, but in contrast with those of yeasts and moulds, such spores constitute a phase of dormancy in the normal life cycle. When conditions become unfavourable for their continued growth, these bacteria develop spores which are extremely resistant to heat, cold and chemical agents and are therefore a means of ensuring the organisms' survival.

Since bacteria are living organisms, they need food for their subsistence and reproduction. This food consists of nitrogen combined with organic matter.

Another crucial element in the life of bacteria is oxygen. Aerobic bacteria can only grow in the presence of oxygen while anaerobic bacteria will only multiply in the absence of oxygen.

A further determinant of bacterial growth is the ambient temperature.

For the purpose of thermal sterilisation, bacteria can be classified as follows:

- Psychrophiles
- Mesophiles
- Facultative thermophiles
- Thermophile

At high acidity levels (pH values below 4.6) bacterial growth is hardly possible and although pasteurisation will not destroy any bacterial spores, yet this process can be applied successfully in the preservation of such acid foods as fruit, sauerkraut, pickles, some seafood products, etc. All these products have a comparatively low pH that prevents spore growth and, consequently, product spoilage.

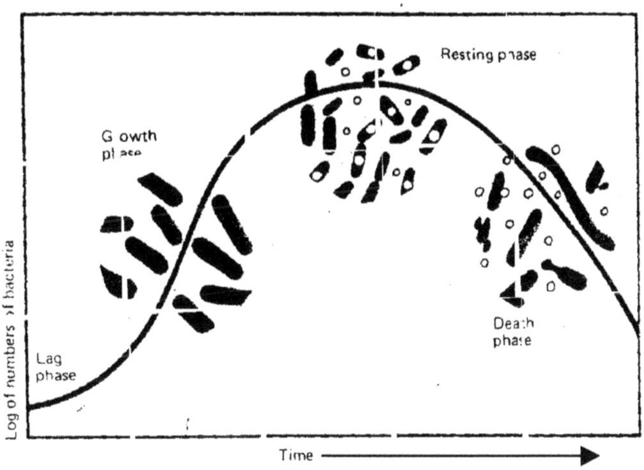

Figure 5.1: Growth of Bacteria

Sterilisation has the purpose of eliminating all microbiological activity in a given product, whilst minimising impairment of the product's quality. The process is carried out at temperatures ranging from 110°C to as high as 140°C. It can be used in case of most meat products, vegetables, babyfood, petfood and liquid dairy products, and will give them a commercial shelf life up to one year.

Thermal Death Time (TDT)

The destruction of microbiological spores can be illustrated by TDT (thermal death time) graphs. These are semilogarithmic graphs, in which the time is shown on a verticallog-scale and the temperature on a horizontal linear scale. By means of experiments, the decimal reduction time, *D.* which is the time required to eliminate 90 per cent of a given spore type, can be established for several temperature constants and then plotted on a TOT graph.

Since the death rate of spores is represented logarithmically, the curves obtained will have the form of straight lines in the semilog TOT graph. Thus for any given microorganism, the graph gives the relation between the times and temperatures at which 90 per cent of the organisms spores will be killed.

The lines result from heating tests conducted at various temperatures to determine the decimal reduction time *D*. A TDT curve provides another useful index, known as the Z value. The Z value is the number of degrees centigrade required for the TDT curve to traverse one logarithmic cycle. It is a measure of the change in death rate with a change in temperature. In Figure 5.2 the Z values for organisms A and B are indicated, respectively, by Z_A and Z_B. A third quantity frequently used in thermobacteriology is the *F* value. It represents the amount of time needed to destroy a given number of microorganisms at a given temperature.

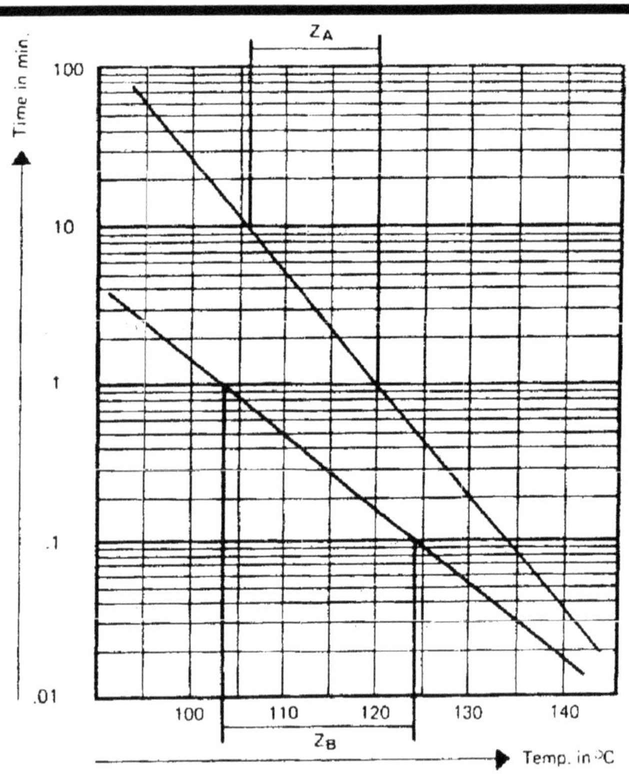

Figure 5.2: TDT Graph for Two Different Types of Organisms

As the thermal resistance of microbiological spores depends heavily on the type of product they are contained in, it has been necessary to introduce a standard unit of lethality, called the F_0 value. It is based on a decimal reduction time (*D* value) of 1 min at 121.1°C and a *Z* value of 10°C.

The F_0 value of a process is the value resulting from a constant temperature of 121.1°C applied for a certain number of minutes, based on instant heating and cooling of the product involved. At any other constant temperature (*T*), the F_0 value is found by the following formula:

$$F_0 = t \times 10\exp\left\{\frac{T - 121.1}{10}\right\}$$

in which *t* is the time in minutes and *T* is the temperature of the product during time *t*.

With the help of the F_0 value as a unit of lethality, we can compare processes conducted at different temperatures for different periods of time. Also, it enables processes involving a gradual temperature increase and decrease to be evaluated. In practice it is used as a standard by which to rate the lethality of a process, and the minimum

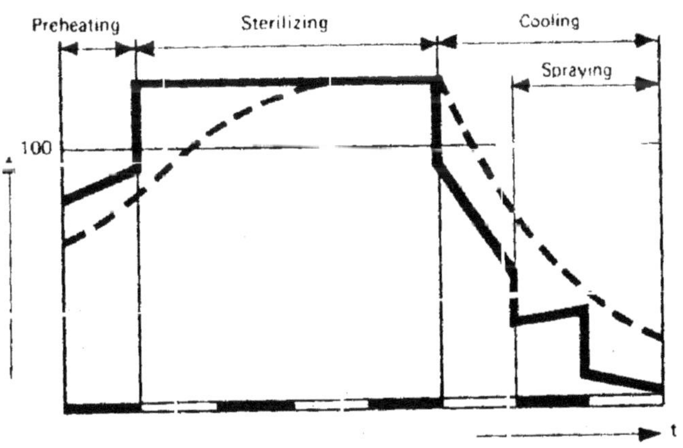

Figure 5.3: Time/Temperature Graph

value adopted for most products is seven. The F_0 value can be established through graphs or by chemical analysis, but the general point of departure is always the time temperature graph of the sterilising process.

The Importance of Sterilisation

Over the years the social function of sterilised food has undergone a marked change. Initially, sterilisation was adopted for the purpose of rendering food surpluses fit for consumption during the winter months. One example of such sterilisation is the traditional process of preserving food in jars. It can be regarded as the forerunner of commercial food sterilisation. It consists of heating products to 95–98°C in glass jars provided with rubber sealing ring. When the product and the water vapour cool down, the resulting vacuum creates a hermetic seal. Food preservation in jars was (and is) a form of domestic industry designed to cover private needs.

The start of the 20th century saw the introduction of commercial food preservation. The first canneries were built and processed such foods as peas, beans, carrots, and the like. This young industry was largely season-oriented and highly labour intensive. Sterilised food acquired a different role in that it had to supply the needs of an industrialising society with its rapidly expanding cities. The initial period of commercial food preservation ended by 1950. Since then the food processing industry has made great forward strides that resulted in today's modern food plants with outputs, for example, of up to 1800 containers per minute and running 6000 h per year on a round-the-clock basis. The product range has become much wider and now includes complete instant meals, specialities, diverse meat and fish products, soups, liquid and solid infant foods, and pet foods.

The variety of container types and materials has been broadened and today preserved food is available in cans, glass jars, aluminium pouches, plastic containers and other packages.

Given the great number of products and container types, it is obvious that each product requires its own specific heat treatment.

As has been explained earlier, the adoption of elevated temperatures involves far more rapid destruction of microbiological spores and a much slower rate of chemical change in the product. This explains the prevailing tendency to increase processing temperatures and thereby reduce the processing time, so that product quality will be improved. However, a limiting factor in applying higher temperatures is the transfer of heat from the periphery to the slowest heating point in the container. Such transfer becomes even more complicated in products containing solid particles, because here the overall transfer rate depends on the partial transfer rates from the container wall to the viscous phase α1 from the viscous phase to the solid particle α2, and inside the solid particle α5. Thus, the total heat transfer coefiicient is determined by several factors, namely:

- Viscosity of the product
- Agitation of the container
- Percentage of headspace in the container
- Difference in density between liquid and solid particles
- Ability of solid particles to move freely in the liquid phase.

For a long time the quality of heat sterilised food was judged by purely subjective standards. Since the early 1970s, however, a greater interest has been taken a finding qualifications more accurate than 'commercially sterile' and including such aspects as colour, texture and general appearance in the assessment. Most likely this interest was brought about by new types of sterilisable packages and increasing competition among food processors.

Food sterilisation not only involves the elimination of microorganisms (as reflected in the F_0 value) but also adequate inactivation of enzymes (notably peroxidase).

Some data on the behaviour of thiamine (vitamin B_1) and certain enzymes have been published by Reichert, although the figures he provides represent relative rather than absolute values. Just as for the thermal destruction of microorganisms, enzymatic inactivation and thiamine retention can be shown by straight lines in semilog graphs and resolved through the Arrhenius equation:

$$F = t.10 \exp \left\{ \frac{T - T_{ref}}{Z} \right\}$$

The influence of thermal processing conditions relative to these factors is illustrated for a few cases all based on an F_0 value of 10 min for the slowest healing point.

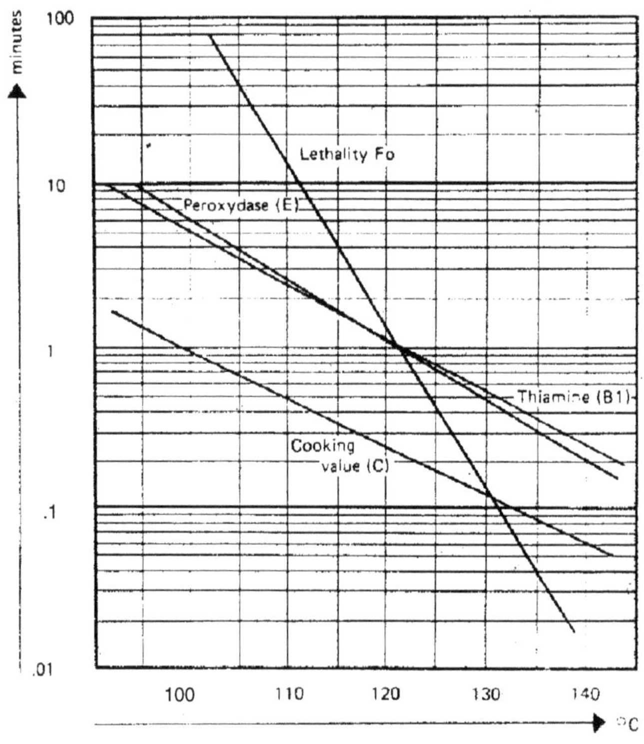

Figure 5.4: Inactivation Lines Using the Arrhenius Equation

The Use of Containers in Food Sterilisation

The main function of the container is to prevent food from being infected by the reentry of microorganisms, after sterilisation has been completed.

Next to this, the container has a number of secondary functions that may vary with container type. The oldest type of container used in industrial food processing is the can, invented by Peter Durand in 1810. The first cans were heavily tin-plated and sealed by soldering prior to sterilisation. They have no resemblance to today's cans, which consist of two pieces that are deep-drawn or have a single side-seam and come with varnished lids.

The container must be sealed hermetically to preclude any infection. Since sterilisation is performed at temperatures varying from 110 to 135°C, a temperature gradient will occur between the filling stage and the sterilising stage, which entails a pressure build-up in the container. Needless to say the amount of pressure a container can withstand depends on its design.

The pressure developing in the container is the result of several factors, chief among them being water vapour pressure. Most food products to be sterilised contain a fairly large amount of water and this water will cause a certain pressure in the container as the processing temperature increases. In fact there is a direct relationship between temperature and water vapour pressure. The pressure increase resulting from the water vapour pressure corresponds to the difference between the water vapour pressure at the filling temperature ($Pvap_{T1}$) and the water vapour pressure at the sterilising temperature ($Pvap_{T2}$). This pressure increase is depicted in Figure 5.5.

Another factor responsible for pressure increase in the container is the air entrapped in the containers headspace, which will expand as the product is heated from filling temperature to sterilising temperature. Mathematically, the resulting pressure increase follows the equation $(PV)/T = C$, according to the law of Boyle and Gay-Lussac, but in practice it is hard to establish with accuracy, because the container can itself absorb a certain increase in volume.

The third factor underlying the pressure increase arising in the container is the expansion the product undergoes as a result of the increase in temperature.

Aside from these three factors, which occur invariably regardless of the product, there may be an additional pressure increase caused by the release of gases through chemical reactions in the product. The solubility of gases in the product may change with variations in temperature and this will lead to a higher pressure in the container.

Thus, the overall pressure that will develop in the container during the sterilising process can only be calculated on an approximate basis. This pressure is reflected in the formula:

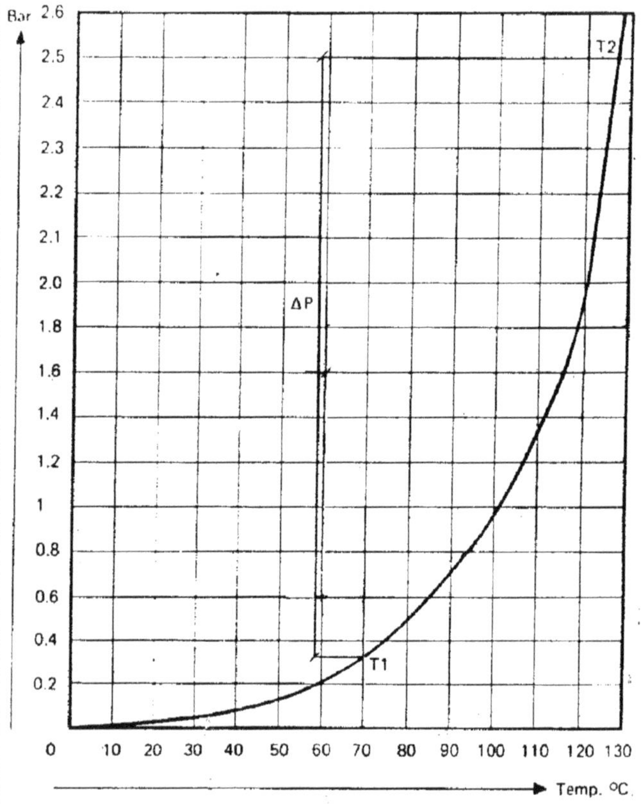

Figure 5.5: Temperature Water Vapour Pressure Relationship

$$Ptot = Pvap + Pair$$

The factor *P*vap is found by the subtraction $Pvap_{T2} - Pvap_{T1}$ where $Pvap_{T2}$ = water vapour pressure at the sterilising temperature, and $Pvap_{T1}$ = water vapour pressure at the filling temperature. The factor *P*air can be determined by the formula:

$$Pair_{T1} = Pair_{T1} \times \frac{V_{T1}}{V_{T2}} \times \frac{T_2}{T_1}$$

where,

$Pair_{T1}$ = partial air pressure during the sealing operation, which is usually 1 bar, unless container sealing takes place under a vacuum.

V_{T1} = headspace at filling temperature $T1$

V_{T2} = headspace at sterilising temperature $T2$

$T2$ = sterilising temperature in degrees Kelvin

$T1$ = filling temperature in degrees Kelvin.

V_{T1}, $T2$ and $T1$ are measurable and can be found empirically. V_{T2} depends on the following variables:

- The container's elastic deformation,
- The expansion of the product,
- The expansion of the container.

When the container is a glass bottle, as employed in the sterilisation of, say, milk or infusion fluids, the elastic deformation is negligible and the factor V_{T2} can be calculated. Owing to this absence of deformation, glass bottles can withstand very high pressures.

When the container is a can, V_{T2} practically defies calculation, because the deformation at the bottom and at the lid depends upon the external and internal pressures which produce a state of equilibrium.

It is this equilibrium which ultimately determines V_{T2}. Hence, the load on the can's seam is limited to the factor Pw, which equals P external – P internal.

In a flexible container, such as a pouch or a plastic bottle, a similar state of equilibrium Pw will occur. As the resistance inside such containers is virtually 0, we can say that P external = Ptot internal.

Here again, it is possible to calculate V_{T2} and Ptot. The three variants referred to are shown in Figure 5.6. When we plot the pressures on a graph, the differential pressure Pw becomes apparent for the various container types during the various stages of the process.

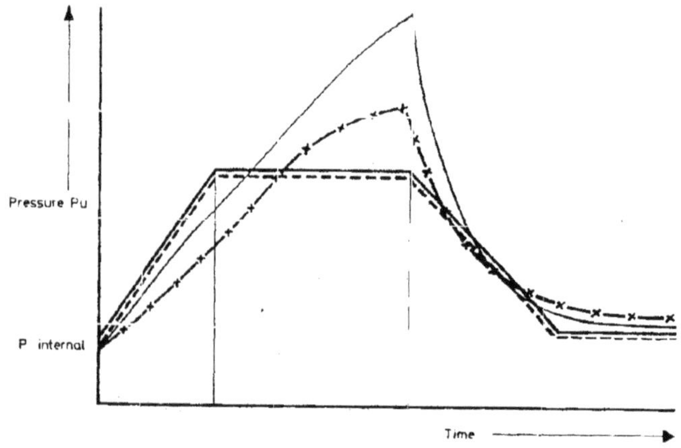

Figure 5.6: Temperature Pressure Graph

So we can come to the conclusion that the container, employed for packaging plays a vital role in the sterilisation of food products.

Now let us study about the process of parameters.

The principal parameters are:

- The process temperature
- The related processing time
- The pressure that can be applied without causing permanents container deformation.

The prime parameters are supplemented by such secondary variables as:

- The product filling temperature
- The requisite product outlet temperature
- The plant conditions (utilities).

On the basis of the process parameters and the product to be handfed, we can look at the sterilising equipment we should employ. Such equipment can be classified under two broad categories, namely equipment for batchwise sterilisation and equipment for automatic sterilisation.

Automatic Sterilising System

Looking at an automatic sterilising system from the process parameter viewpoint, we can distinguish three key functions which the system must perform, namely:

- Container infeed and discharge
- In-steriliser container transport
- Container processing (pressure time and temperature)

In a batch sterilising system based on autoclaves, these functions are:

- Loading the autoclave crates
- Taking the loaded crates to the autoclaves
- Autoclaving the loaded crates by means of steam/air/water.

In automatic sterilisation, also known as continuous sterilisation, constant flow of containers moves from filling unit to steriliser. Here the containers are transferred to the insteriliser conveyor system and taken through the sterilising zone.

The continuous system most widely adopted is the hydrostatic system. It has a sterilising zone that is in communication with the atmosphere through columns or legs containing heads of water (Figure 5.7). The steam pressure at the steam/water level is in balance with the hydrostatic pressure exerted by the heads of water.

The containers are taken through the hydrostatic legs and the sterilising zone in carriers attached to a conveyor chain.

A hydrostatic continuous steriliser's general configuration comprises:

- A preheating section
- A sterilising zone
- A cooling section.

The unit's design characteristics depend upon the product (*i.e.*, the required processing time and temperature), the containers, the output needed, and the plant conditions (utilities).

The key advantage offered by a continuous steriliser is that it makes each container undergo the same uniform, reproducible

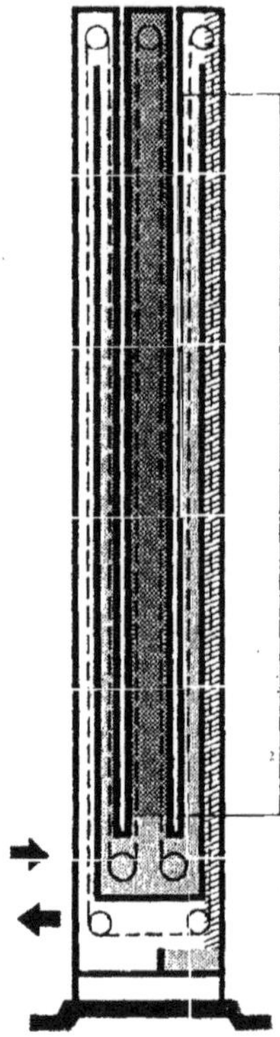

Figure 5.7: Hydrostatic Principle

process, and thus, greatly facilitates product data registration. This latter aspect is becoming more and more significant, because food processors are required to keep accurate records of the products sterilised in their plants.

The design of a continuous steriliser is established on the basis of variable data known as design criteria, which include:

1. The product
2. The container (plus closure)
3. The capacity

4. The number of units to be processed per annum
5. The plant utilities or consumables.

The product, or products, determine the shape of the carriers that are to ensure in-steriliser transport.

It is part of the steriliser manufacturer's specific capabilities to design a carrier that will take the packaged product(s) safely through the various processing sections.

To process the containers in question, an external excess pressure can be created by using a steam/air mixture or superheated water as a sterilising medium rather than saturated steam. This excess pressure will permit adequate sterilisation despite the container's low pressure resistance.

The pressure inside the container is built up by the factors mentioned earlier, *i.e.*, $P\text{vap}_{T2}$, related to the sterilising temperature and *P* volume (air), related to the increase in volume of product and headspace.

The external pressure results' from the air pressure and the water vapour pressure corresponding to the sterilising pressure. This can be illustrated most readily on the basis of a glass jar.

For flexible containers, such as PE bottles and aluminium pouches application of the calculating procedure is not warranted, because these containers have their own permissible deformation, so that in many cases it will be necessary to establish the overpressure by trial.

Frequently, overpressure is not only needed during sterilisation but also during the initial cooling phase. The external pressure induced by the saturated steam will fall off fairly rapidly, whilst the temperature inside the container is slow to drop. This causes the differential pressure to increase after sterilisation, so that cooling must be conducted at a given overpressure. The situation concerned can again be depicted by a graph (based on a hydrostatic system), as in Figure 5.8. The graph clearly shows that $P2 > P1 > P5$.

As a third design criterion we have mentioned the capacity. In conjunction with the production schedule, indicating the amount of product to be processed per unit of time for a given period of time, an optimum must be established for the continuous steriliser. The first continuous sterilisers had a capacity of 125–150 1/1-litre cans per

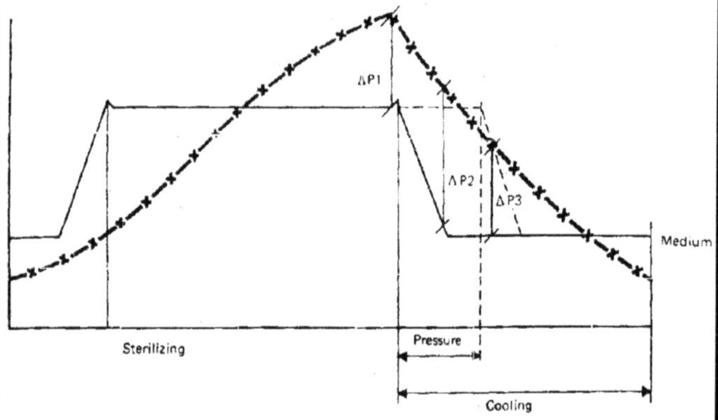

Figure 5.8: Course of Pressure in Container and Medium with and without Pressure

minute. Today, Stork Amsterdam builds sterilisers handling 1800 cans per minute, which corresponds to a capacity of 100000 units per hour, or an annual production of 726×10^6 units.

Finding the steriliser's optimum capacity is a complicated matter in which economic aspects play a dominant role. First of all the various design criteria must be reviewed and their interaction considered, whereupon the outcome of the study must be expressed in terms of system capacities.

The utilities, or plant consumables, constitute a further design criterion. Those of interest in sterilisation are water steam and the effluent resulting from the process. The latter especially is receiving more and more attention nowadays.

In a sterilising system the product is heated from the filling temperature.

In a Stork continuous steriliser (named Hydromatic), the consumption of steam and water is drastically reduced by regeneration. From the outgoing hydrostatic leg, which picks up heat from the sterilised containers, hot water is pumped to the preheating leg. The water in the latter leg imparts its heat to the incoming containers, whose temperature thus becomes higher than

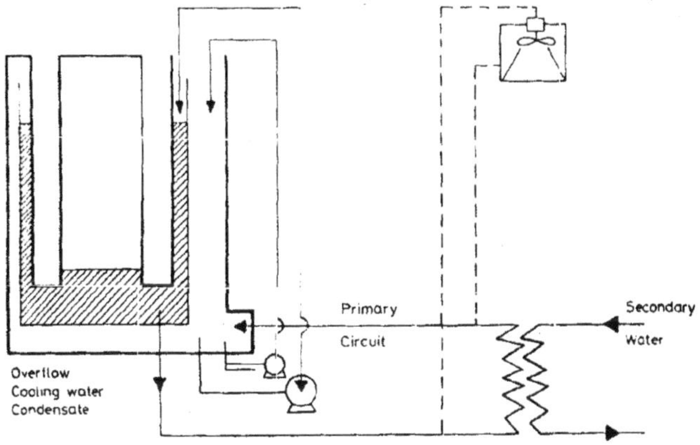

Figure 5.9: Continuous Steriliser with Closed-circuit Cooling System

the filling temperature before they enter the sterilising zone. This regenerating system makes it possible to save up to 60 or 70 per cent on steam.

The consumption of cooling water can be reduced considerably through the use of recirculation. It involves the creation by means of heat exchangers of a primary cooling circuit whose water circulates through the sterilising system and cools down the containers.

Next to this, the system has a secondary circuit which is only subject to thermal load and which either contains surface water or water intended for use elsewhere as boiler feed water, process water, product cleaning water, etc. after it has served its cooling function.

Some times, open cooling towers are employed to discharge the heat absorbed, in which case a minimal amount of makeup water will be needed.

Each sterilising system involves a certain risk potential in that the product may be understerilised or reinfected after sterilisation in the final container. Such understerilisation or reinfection can ensure from:

- Overfill
- Operator handling errors
- System failure
- Faulty product routing, which results in a mix-up between sterilised and non-sterilised product.

Chapter 6

Food Additives

Despite modern-day associations food additives have been used for centuries. Food preservation began when man first learned to safeguard food from one harvest to the next and by the salting and smoking of meat and fish. The Egyptians used colours and flavourings, and the Romans used saltpetre (potassium nitrate), spices and colours for preservation and to improve the appearance of foods. Cooks regularly used baking powder as a raising agent, thickeners for sauces and gravies, and colours, such as cochineal, to transform good-quality raw materials into foods that were safe, wholesome and enjoyable to eat. The overall aims of traditional home cooking remain the same as those prepared and preserved by today's food manufacturing methods.

Food additives play an important role in today's complex food supply also. Never before has the range and choice of foods been so wide either in supermarkets, specialist food shops or when eating out. Whilst a shrinking proportion of the population is engaged in primary food production, consumers are demanding more variety, choice and convenience alongside higher standards of safety and wholesomeness at affordable prices. Meeting these consumer expectations can only be achieved using modern food processing technologies which include the use of a variety of food additives proven effective and safe through long use and rigorous testing.

Additives carry out a variety of useful functions which we often take for granted. Foods are subjected to many environmental conditions, such as temperature changes, oxidation and exposure to microbes, which can change their original composition. Food additives play a key role in maintaining the food qualities and characteristics that consumers demand, keeping food safe, wholesome and appealing from farm to fork. Over the last 50 years, developments in food science and technology have led to the discovery of many new substances that can fulfill numerous functions in foods. These food additives are now readily available and include; emulsifiers in margarine, sweeteners in low-calorie products and a wider range of preservatives and antioxidants which slow product spoilage and rancidity whilst maintaining taste.

A food additive is defined as "any substance not normally consumed as a food in itself and not normally used as a characteristic ingredient of food whether or not it has nutritive value, the intentional addition of which to food for a technological purpose in the manufacture, processing, preparation, treatment, packaging, transport or storage of such food results, or may be reasonably expected to result, in it or its by-products becoming directly or indirectly a component of such foods" (Council Directive 89/107/EEC). Many food additives are naturally occurring and some are even essential nutrients; it is the technical purpose that leads to these being classified as food additives and given an E number.

FAO and WHO in 1956 defined food additives as 'non-nutritive substances added intentionally to food, generally in small quantities, to improve its appearance, flavour, texture or storage properties.' The Food and Drug Administration of the United States in collaboration with the Food Protection Committee has defined a food additive as 'a substance or a mixture of substances, other than a basic foodstuff, which is present in a food as a result of any aspect of production, processing, storage or packaging.' The term does not include chance contaminants. This definition includes intentional as well as unintentional additives. Intentional additives are substances added to perform specific functions while unintentional additives are substances, which have no intentional function in the finished products but become part of the food product during some phase of production or handling. Both kinds of additives in excessive amounts can be detrimental to health.

The current legal definition of food additive according to USA (FDA) food additives is any substance, the use of which results or may reasonably be expected to result, directly or indirectly, in its becoming a component of or otherwise affecting the characteristics of any food.

Functions and Uses of Food Additives

Food additives perform several functions such as drying, emulsifying, enhancing flavour, enriching, firming, flavouring, leavening, antidrying, antifoaming, antihardening, antisticking, bleaching, colour retaining, conditioning, texturing, thickening, preserving, refining, stabilizing, replacing air in food packages etc.

Major uses of food additives are:

1. Enhancement of the attractiveness of food by means of colouring and flavouring agents, emulsifiers, stabilizers, thickeners, clarifier and bleaching agents.
2. Maintenance of nutritional quality, such as by the use of antioxidants.
3. Facilitating food processing by means of acids, alkalis, buffers, sequestraints and various other chemicals.
4. Enhancement of keeping quality or stability by the use of antioxidants, antimicrobial agents etc.

Types of Food Additives

The intentional food additives may be classified under various groups:

Preservatives

Any substance which is capable of inhibiting, retarding or arresting the growth of microorganisms is known as preservative. These include chemical preservatives against bacterial, yeast and moulds such as sodium benzoate used in fruit drinks and sodium and calcium propionate used in breads and cakes as mould inhibitors, sorbic acid used in cheese and metabisulphite and SO_2 used as preservative in fruit products.

Preservatives limit, retard or arrest the growth of micro-organisms (*e.g.* bacteria, yeast, mould) that are present in or gain entry to the food, preventing spoilage or food poisoning. They are used in baked foods, wine, cheese, cured meats, fruit juices and margarine among others. Examples include:

1. Sulphur dioxide and sulphites (E 220-228)–these help to prevent colour changes in dried fruits and vegetables. Sulphites also inhibit the growth of bacteria in wine and fermented foods, some snack foods and baked goods. Sulphites also have antioxidant properties.
2. Calcium propionate (E 282)–prevents bread and baked foods from turning mouldy.
3. Nitrates and nitrites (sodium and potassium salts) (E 249-252)–are used as a preservative in processed meats such as ham and frankfurters to keep the products safe by preventing the growth of botulinum bacteria, *Clostridium botulinum*, which is highly pathogenic.

Antioxidant

An oxidant is a substance which when added to fats and fat containing foods prevents their oxidation and thus prolongs their shelf life, wholesomeness and palatability. Without them, fatty acids (*e.g.* potato chips, salted nuts, fat containing dehydrated foods) cannot be stored for any length of time without becoming rancid. Some important antioxidants which are used in foods are butylated hydroxyanisole (BHA), butylated hydroxytoluene (BHT), tertiarybutylhydroquinone (TBHQ) butylhydroquinone (TBHQ) and propylgallate (PG). The antioxidants also include such diverse materials as ascorbic acid, stannous chloride, and tocopherols (Vit E). Sulphur dioxide plays dual role as preservative and an antioxidant.

Naturally occurring substances that act as antioxidants are the tocopherols. It acts as biological antioxidant in plant and animal tissues, but rarely used as it is very expensive than synthetic antioxidants.

Antioxidants play important role in interrupting the free radical chain mechanism involved in lipid oxidation. They are effective in small concentrations (0.01-0.02 per cent). Browning of cut fruits and vegetables is due to enzymic oxidation of phenolic substances. Antioxidants prevent this discolouration. Ascorbic acid is used as an antioxidant in this case. Acids such as citric acid and phosphoric acid, increase the effectiveness of ascorbic acid in preventing browning.

Antioxidants prevent the oxidation of foods that causes rancidity or discoloration. They are used in baked foods, cereals, fats, oils and salad dressings. The major fat soluble antioxidants are:

1. Tocopherols (E 306-309), BHA (butylated hydroxyanisole or E 320) and BHT (butylated hydroxytoluene or E 321)–these protect edible fats, vegetable oils and salad dressings from turning rancid.
2. Ascorbic acid (E 300) and citric acid (E 330)–which preserve the colour of freshly cut fruits and vegetables.

Sequestrants

These are also known as chelating agents or metal scavengers since they combine with metals such as iron and copper present in foods and remove them from solution. The trace elements are active catalysts, catalyze oxidation and also cause discolouration, rancidity and flavour changes in foods and must be removed by chelating agents such as ethlenediamine tetra acetic acid (EDTA), polyphosphates, tartaric acid and citric acid are also examples of sequestrants. They are used to suppress the action of some objectionable and practically unavoidable ingredients *e.g.*, in the manufacture of soft drinks, iron in the water reacts with one of the flavouring substances to form an insoluble compound that makes the product cloudy. Citrates form a soluble complex with iron, whereas others render citric acid insoluble.

Emulsifier or Surface Active Agents

These are also known as emulsifiers and are used to stabilize oil-in-water, water-in-oil, gas-in-liquid and gas-in-solid emulsions. The purpose of these food additives is to maintain consistent texture and to prevent the separation of ingredients in such products as margarine, low-fat spreads, ice cream, salad dressings and mayonnaise. Many reduced-fat and low-fat versions of common foods are dependent on this technology. Any recipe that requires the mixing of ingredients that normally do not mix, such as fat and water, need emulsifiers and stabilisers to impart and maintain the desired consistency. Besides emulsifiers of natural origin such as lecithin and synthetic ones such as mono and diglycerides and their derivatives, certain fatty acids and their derivatives and bile acids can be used as emulsifying agents. Synthetic surface active agents, include defoaming compounds and detergents, *e.g.*, propylene glycol monostearate and monosodium phosphate. They also include several defoaming of detergent chemicals. Food emulsifiers or surface acting agents categorize as:

Naturally Occurring

Bile acids, phospholipids, lecithin, cholesterol, saponin etc.

Synthetic

Glycerol mono esters, propylene glycol mono esters, sorbital esters etc.

Applications in Foods

1. In cereal based products, emulsifier can function as dough conditioners which means greater resistance to mixing, better gas retention etc. Examples are acetic, citric, fumaric, lactic, succinic acids etc. They are used by the baking industry as dough conditioners.
2. Emulsifiers in ice-cream improve fat dispersion, facilitate fat-protein interaction, reduce whipping time etc.
3. In candy products, they can be used as crystal structure modifier in mixtures of triglycerides.
4. Certain emulsifiers can also be used to control in product viscosity in cream fillings in chocolates.
5. Egg yolk containing lecithin and phosphoproteins act as excellent foam stabilizing agents in cakes etc.
6. Liquid foods can easily be thickened by the use of surface acting agents.
7. The stickiness of caramel or peanut butter can be reached by adding a surface acting agents.

Colouring Agents

Colour is one of the first and most important sensory qualities and it helps us to accept or reject particular foods. Whilst adding colour may appear to some to be purely cosmetic, there is no doubt that colour is important in consumer perception of a food and it is often associated with a specific flavour and intensity of flavour. Colours are used to add or restore colour in a food in order to enhance its visual appeal and to match consumer expectations. The processing of peas and the preparation of jams can lead to loss of colour, and hence food colours can compensate for these losses. Some colours are used purely for visual decoration on cakes and confectionery items. Masking or disguising inferior quality, however, are unacceptable uses of colours.

The primary reasons for adding colours to foods include:

1. To offset colour loss due to exposure to light, air, extremes of temperature, moisture and storage conditions
2. To compensate for natural or seasonal variations in food raw materials or the effects of processing and storage to meet consumer expectations.
3. To enhance colours that occur naturally but at levels weaker than those usually associated with a given food.

The colouring agents include colour stabilizers, colour fixatives, colour retention agents etc. They consist of synthetic colours. Colours natural or synthetic are used to improve the aesthetic appeals of food. Colours naturally present in foods may be classified as water-soluble and water-insoluble, and may be of plant or animal origin. Natural colouring matters (pigments) present in vegetables and fruits can be classified on the basis of chemical structure as carotenoids (yellow-orange), chlorophylls (green), flavonoids and anthocynins (red, blue and purple) and anthoxanthins (green yellow).

The choice of natural colour for use in food is limited as their stability during processing and storage is influenced by pH, heat, light, oxidation etc. Even though synthetic colours add nothing to the nutritive value of foods. But they are frequently added to restore the natural ones lost in food processing or to give the preparation the natural colour we accept. Colours of synthetic origin, include the certified coal tar dyes (carbonated beverages, candies, gelatin desserts and bakery goods are among items coloured with certified coal tar dyes). In addition to the organic coal tar dyes and the natural organic colouring agents, food colours also include such materials as iron oxide to give redness and titanium dioxide to intensify whiteness. Extract of annatto, caramel, carotene and saffron are examples of food colours which are of natural origin. Only a few coal tar food colours were found to be safe and have received the approval of FDA, USA. The amount of any permitted coal tar dye which may be added to any fruit product should not exceed 0.2 g/kg of the final product.

Although colours add to the attractiveness of food products, it is better to avoid their use as far as possible. Recently, the use of some colours in foods has been banned. In India, Acid Magenta II and Blue V.R.S., which were used in tomato ketchup and canned peas, respectively have been banned. Both are triphenylmethane dyes, instead of Blue V.R.S., Fast Green S and Green FCF have been recommended. Red 6 B, Red FB and Brilliant Black have also been banned. These are added to a large number of food items to make them attractive and appetizing.

Buffers, Acids and Alkalis

These additives are primarily used to control or adjust the pH of foods and affect properties such as flavour, texture and cooking qualities, *e.g.* acetic acid, lactic acid, citric acid, tartaric acid, sodium bicarbonate, sodium hydroxide etc.

Stabilizers and Thickners

These substances help to improve the texture of foods, inhibit crystallization of sugar and formation of ice, stabilize emulsions and foams and reduce stickiness of icings on baked products. They combine with water to form gels and make the food viscous, *e.g.* agar-agar, starch, gum arabic, gelatin, pectin, amylose, carboxymethylcellulose (CMC) etc. Gravies, pie fillings, cake toppings, jellies, puddings and salad dressings are some foods that contain stabilizers and thickeners.

Nutrient Supplements

When foods are processed or stored there may be loss of some nutrients. In order to restore this loss to provide more nutritional value some nutrient supplements are added. These are mainly vitamins and minerals: Vitamin D (added to milk), vitamin B, iron and calcium (to cereal products), iodine (to salt), vitamin C (to fruit-juices) etc.

Non-nutritive and Special Dietary Sweeteners

Various sweetening agents have been used as substitutes for cane sugar (sucrose). Non-nutritive sweeteners are mainly used in the low-calorie soft drinks. They are also added to low-calorie liquid foods, canned fruit, frozen desserts, some baked products etc.

Both "bulk" and "intense" sweeteners impart a sweet taste to foodstuffs and are useful in low-calorie products and for special dietary products, such as those for diabetics. Intense sweeteners, such as acesulfam K (E 950), aspartame (E 951) and saccharin (E 954) are 130-200 times, 200 times and 300-500 times sweeter, respectively, than sugar-and they have zero calories. Thaumatin (E 957), a naturally sweet protein extracted from the fruit of the plant Thaumatococcus danielli, is 2500 times sweeter than sugar and is used at very low levels for its flavouring properties. Bulk sweeteners include sorbitol (E 420), isomalt (E 953) and maltitol (E 965) and these can be incorporated into "table-top" sweeteners and in energy-reduced foods, in which they provide volume and mouth feel. These substances have reduced caloric value, providing 2.4 kcal/gram compared with 4 kcal/gram for other carbohydrates.

Saccharin, cyclamates and aspartame are some of the well known non-nutritive dietary sweeteners. The first synthetic sweetening agent used was saccharine which is about 300 times sweeter than sucrose but it leaves a bitter and unpleasant after-taste.

Cyclamates are about 30 times sweeter than sucrose but have little after taste of saccharine. Earlier they were widely used in soft drinks and other low calorie liquid foods but now they have been banned due to the their carcinogenic effects.

Aspartame is about 200 times sweeter than sugar. It is good tasting, enhances citrus and other fruit flavours and does not contribute to tooth decay. Similarly, the high-intensity sweetener aspartame (another substance made from naturally occurring amino acids, aspartic acid and phenylalanine) has been blamed for a wide variety of adverse effects, none of which have been validated by scientific studies. It is being used in beverages, foods and other products like instant breakfast mixtures, carbonated soft drinks, fruit syrups, fruit spreads etc.

Sucralose is the only low-calorie sweetener made from sugar. It is safe for everyone to consume. Its acceptable daily intake is 5 mg/kg of body weight per day.

Flavouring Agents and Flavour Enhancers

Flavouring agents are those substances, naturally occurring or synthetic, which are responsible for the characteristic flavours of almost all the foods in our diet. Currently, a large number of flavouring agents are used as food additives. Among the flavour enhancers monosodium glutamate (MSG) and ribonucleotides (5-nucleotides) extracted from yeast are commonly used.

MSG is made up of sodium and glutamic acid. Glutamic acid is an amino acid found naturally in high protein foods such as meats and dairy products like Camembert cheese. MSG is also a flavour enhancer used in prepared meals, some Chinese food, certain sauces and soups. MSG has been "blamed" for a variety of side effects including headaches and body tingling, however scientific studies show no link between MSG and these reactions suggesting that some other component of the meal, or even psychological responses, may be responsible for any adverse effects. Recently, it was reported that monosodium glutamate caused damage to the brain when injected in mice. Therefore, FDA, USA prohibited the use of MSG in baby foods. Its use is permitted in food preparations meant for older children and adults.

Bleaching and Maturing Agents (Flour Improvers) and Starch Modifiers

Freshly milled flour has suboptimal baking qualities. Both the colour and baking properties improve slowly on normal storage. This improvement can be brought about rapidly by the use of certain oxidising agents such as oxides of nitrogen and chlorine dioxide. These have been used for bleaching and maturing flour. Bromate is added to flour to improve its baking quality.

Leavening Agents

Leavening agents produce CO_2 which helps to give characteristic structure to bread, cakes, biscuits etc. They produce light fluffy baked goods. The important groups are:

1. Yeast
2. Chemical leavening agents

Yeast

It is used in baking. It causes fermentation of sugar and produces CO_2 and alcohol. This process gives the characteristic structure of bread after baking.

Chemical Leavening Agents

Baking powder is used in the preparation of cake and biscuits. It generates CO_2 for leavening purposes.

Miscellaneous

Miscellaneous additives are acetic acid, caramel, glycerine, nitrogen, phosphoric acid, sodium carbonate and bicarbonate etc.

Chapter 7

HACCP: Approach to Food Safety

HACCP is now a widely used logical approach to safer food processing. By describing and analysing each processing operation, deciding what the hazards are, and identifying the critical control points, one is taking control of processing in a way which can provide reassurance to processors, control agencies and consumers alike.

HACCP systems, together with accurate, reliable measurements, and good record-keeping and documentation, provide safeguards as more food is eaten away from the home, and increasing quantities of food are distributed internationally. Food safety objectives (FSOs) provide targets for the control of hazards in foods, concentrating on the actual public health hazard rather than focussing on pathogen reduction as an endpoint. The most nutritious foodstuffs potentially provide the greatest hazards, and we all have seen great concerns about meat in recent years.

Fish and fish products present their own problems, particularly from high levels of amines which can be present, causing illness.

Fruits and vegetables are increasingly traded internationally, as we expect a wide range of tropical and temperate fruits, salads and vegetables to be available everywhere in all seasons; often there is no final consumer process or cooking as a safety measure. Fruit and vegetable surfaces can be the agents for direct transmission of food borne illness 'from field to mouth', several thousand miles away.

Can legal standards be set, and through HACCP and FSOs, human health be improved, rather than allowing increasing globalisation, in all directions, to help replace 'local' foodborne illness with more exotic imported challenges? Much has already been achieved through HACCP and international awareness and understanding. So many people travel these days, and consumers want to enjoy a wide international diet, at home and abroad.

What is HACCP

Traditionally, industry and regulators have depended on spot-checks of manufacturing conditions and random sampling of final products to ensure safe food. This approach, however, tends to be reactive, rather than preventive, and can be less efficient than the new system.

The new system is known as Hazard Analysis and Critical Control Point, or HACCP (pronounced hassip). HACCP has been endorsed by the National Academy of Sciences, the Codex Alimentarius Commission (an international food standard-setting organization), and the National Advisory Committee on Microbiological Criteria for Foods. A number of international food companies already use the system in their manufacturing processes.

HACCP is a management system in which food safety is addressed through the analysis and control of biological, chemical, and physical hazards from raw material production, procurement and handling, to manufacturing, distribution and consumption of the finished product. Hence, HACCP is an effective and rational means of assuring food safety from harvest to consumption. For successful implementation of a HACCP plan, management must be strongly committed to the HACCP concept.

HACCP is designed for use in all segments of the food industry from growing, harvesting, processing, manufacturing, distributing, and merchandising to preparing food for consumption. Food safety systems based on the HACCP principles have been successfully applied in food processing plants, retail food stores, and food service operations. The seven principles of HACCP have been universally accepted by government agencies, trade associations and the food industry around the world.

Principles and Application Guidelines of HACCP

HACCP is a systematic approach to the identification, evaluation, and control of food safety hazards based on the following seven principles:

Principle 1: Conduct a Hazard Analysis

Potential hazards associated with a food and measures to control those hazards are identified. The hazard could be biological, such as a microbe; chemical, such as a toxin; or physical, such as ground glass or metal fragments.

Principle 2: Identify the Critical Control Points (CCPs)

These are points in a food's production-from its raw state through processing and shipping to consumption by the consumer-at which the potential hazard can be controlled or eliminated. Examples are cooking, cooling, packaging, and metal detection.

Principle 3: Establish Critical Limits

For a cooked food, for example, this might include setting the minimum cooking temperature and time required to ensure the elimination of any harmful microbes.

Principle 4: Establish Procedures to Monitor the Critical Control Points

Such procedures might include determining how and by whom cooking time and temperature should be monitored.

Principle 5: Establish Corrective Actions

Establish corrective actions to be taken when monitoring shows that a critical limit has not been met, for example, reprocessing or disposing of food if the minimum cooking temperature is not met.

Principle 6: Establish Verification Procedures

Establish procedures to verify that the system is working properly, for example, testing time-and-temperature recording devices to verify that a cooking unit is working properly.

Principle 7: Establish Record-keeping and Documentation Procedures

Establish effective recordkeeping to document the HACCP system. This would include records of hazards and their control methods, the monitoring of safety requirements and action taken to correct potential problems.

Each of these principles must be backed by sound scientific knowledge, for example, published microbiological studies on time and temperature factors for controlling foodborne pathogens.

Guidelines for Application of HACCP Principles

The following guidelines will facilitate the development and implementation of effective HACCP plans. These guidelines should be applied as appropriate to each segment of the food industry under consideration.

Prerequisite Programs

The production of safe food products requires that the HACCP system be built upon a solid foundation of prerequisite programs. Each segment of the food industry must provide the conditions necessary to protect food while it is under their control. This has traditionally been accomplished through the application of cGMPs. These conditions and practices are now considered to be prerequisite to the development and implementation of effective HACCP plans.

1. Prerequisite programs provide the basic environmental and operating conditions that are necessary for the production of safe, wholesome food. The Codex Alimentarius General Principles of Food Hygiene describe the basic conditions and practices expected for foods intended for international trade.
2. In addition to the requirements specified in regulations, industry often adopts policies and procedures that are specific to their operations. Many of these are proprietary.
3. While prerequisite programs may impact upon the safety of a food, they also are concerned with ensuring that foods are wholesome and suitable for consumption. HACCP plans are narrower in scope, being limited to ensuring food is safe to consume.

4. The existence and effectiveness of prerequisite programs should be assessed during the design and implementation of each HACCP plan.
5. All prerequisite programs should be documented and regularly audited.
6. Prerequisite programs are established and managed separately from the HACCP plan.
7. Certain aspects, however, of a prerequisite program may be incorporated into a HACCP plan. For example, many establishments have preventive maintenance procedures for processing equipment to avoid unexpected equipment failure and loss of production. During the development of a HACCP plan, the HACCP team may decide that the routine maintenance and calibration of an oven should be included in the plan as an activity of verification. This would further ensure that all the food in the oven is cooked to the minimum internal temperature that is necessary for food safety.

Examples of Common Prerequisite Programs

The production of safe food products requires that the HACCP system be built upon a solid foundation of prerequisite programs. Each segment of the food industry must provide the conditions necessary to protect food while it is under their control. This has traditionally been accomplished through the application of cGMPs. These conditions and practices are now considered to be prerequisite to the development and implementation of effective HACCP plans. Prerequisite programs provide the basic environmental and operating conditions that are necessary for the production of safe, wholesome food. Common prerequisite programs may include, but are not limited to:

Facilities

The establishment should be located, constructed and maintained according to sanitary design principles. There should be linear product flow and traffic control to minimize cross-contamination from raw to cooked materials.

Supplier Control

Each facility should assure that its suppliers have in place effective GMP and food safety programs. These may be the subject of continuing supplier guarantee and supplier HACCP system verification.

Specifications

There should be written specifications for all ingredients, products, and packaging materials.

Production Equipment

All equipment should be constructed and installed according to sanitary design principles. Preventive maintenance and calibration schedules should be established and documented.

Cleaning and Sanitation

All procedures for cleaning and sanitation of the equipment and the facility should be written and followed. A master sanitation schedule should be in place.

Personal Hygiene

All employees and other persons who enter the manufacturing plant should follow the requirements for personal hygiene.

Training

All employees should receive documented training in personal hygiene, GMP, cleaning and sanitation procedures, personal safety, and their role in the HACCP program.

Chemical Control

Documented procedures must be in place to assure the segregation and proper use of non-food chemicals in the plant. These include cleaning chemicals, fumigants, and pesticides or baits used in or around the plant.

Receiving, Storage and Shipping

All raw materials and products should be stored under sanitary conditions and the proper environmental conditions such as temperature and humidity to assure their safety and wholesomeness.

Traceability and Recall

All raw materials and products should be lot-coded and a recall system in place so that rapid and complete traces and recalls can be done when product retrieval is necessary.

Pest Control

Effective pest control programs should be in place.

Other examples of prerequisite programs might include quality assurance procedures; standard operating procedures for sanitation, processes, product formulations and recipes; glass control; procedures for receiving, storage and shipping; labelling; and employee food and ingredient handling practices.

Education and Training

1. The success of a HACCP system depends on educating and training management and employees in the importance of their role in producing safe foods.
2. This should also include information about the control of food borne hazards related to all stages of the food chain.
3. Employees must first understand what HACCP is and then learn the skills necessary to make it function properly.
4. Specific training activities should include working instructions and procedures that outline the tasks of employees monitoring each CCP.
5. Management must provide adequate time for thorough education and training.
6. Personnel must be given the materials and equipment necessary to perform these tasks. Effective training is an important prerequisite to successful implementation of a HACCP plan.

Advantages of HACCP

HACCP offers a number of advantages over the current system. Most importantly, HACCP:

1. Focuses on identifying and preventing hazards from contaminating food
2. Is based on sound science

3. Permits more efficient and effective government oversight, primarily because the recordkeeping allows investigators to see how well a firm is complying with food safety laws over a period rather than how well it is doing on any given day
4. Places responsibility for ensuring food safety appropriately on the food manufacturer or distributor
5. Helps food companies compete more effectively in the world market
6. Reduces barriers to international trade.

Developing a HACCP Plan

The format of HACCP plans varies. In many cases the plans will be product and process specific. However, some plans may use a unit operations approach. Generic HACCP plans can serve as useful guides in the development of process and product HACCP plans; however, it is essential that the unique conditions within each facility be considered during the development of all components of the HACCP plan.

In the development of a HACCP plan, five preliminary tasks need to be accomplished before the application of the HACCP principles to a specific product and process. The five preliminary tasks are given in Figure 7.1.

Assemble the HACCP Team

1. The first task in developing a HACCP plan is to assemble a HACCP team consisting of individuals who have specific knowledge and expertise appropriate to the product and process. It is the team's responsibility to develop the HACCP plan.
2. The team should be multi-disciplinary and include individuals from areas such as engineering, production, sanitation, quality assurance, and food microbiology.
3. The team should also include local personnel who are involved in the operation as they are more familiar with the variability and limitations of the operation. In addition, this fosters a sense of ownership among those who must implement the plan.

Assemble the HACCP Team

↓

Describe the Food and its Distribution

↓

Describe the Intended Use and Consumers of the Food

↓

Develop a Flow Diagram Which Describes the Process

↓

Verify the Flow Diagram

Figure 7.1: Preliminary Tasks in the Development of the HACCP Plan

4. The HACCP team may need assistance from outside experts who are knowledgeable in the potential biological, chemical and/or physical hazards associated with the product and the process. However, a plan which is developed totally by outside sources may be erroneous, incomplete, and lacking in support at the local level.
5. Due to the technical nature of the information required for hazard analysis, it is recommended that experts who are knowledgeable in the food process should either participate in or verify the completeness of the hazard analysis and the HACCP plan.
6. Experts in the team should have the knowledge and experience to correctly: (a) conduct a hazard analysis; (b) identify potential hazards; (c) identify hazards which must be controlled; (d) recommend controls, critical limits, and procedures for monitoring and verification; (e) recommend appropriate corrective actions when a deviation occurs;

(f) recommend research related to the HACCP plan if important information is not known; and (g) validate the HACCP plan.

Describe the Food and its Distribution

The HACCP team first describes the food. This consists of a general description of the food, ingredients, and processing methods. The method of distribution should be described along with information on whether the food is to be distributed frozen, refrigerated, or at ambient temperature.

Describe the Intended Use and Consumers of the Food

Describe the normal expected use of the food. The intended consumers may be the general public or a particular segment of the population (*e.g.*, infants, immunocompromised individuals, the elderly, etc.).

Develop a Flow Diagram which Describes the Process

1. The purpose of a flow diagram is to provide a clear, simple outline of the steps involved in the process.
2. The scope of the flow diagram must cover all the steps in the process which are directly under the control of the establishment.
3. In addition, the flow diagram can include steps in the food chain which are before and after the processing that occurs in the establishment.
4. The flow diagram need not be as complex as engineering drawings.
5. A block type flow diagram is sufficiently descriptive. Also, a simple schematic of the facility is often useful in understanding and evaluating product and process flow.

Verify the Flow Diagram

The HACCP team should perform an on-site review of the operation to verify the accuracy and completeness of the flow diagram. Modifications should be made to the flow diagram as necessary and documented.

After these five preliminary tasks have been completed, the seven principles of HACCP are applied.

Implementation and Maintenance of the HACCP Plan

The successful implementation of a HACCP plan is facilitated by commitment from top management. The next step is to establish a plan that describes the individuals responsible for developing, implementing and maintaining the HACCP system. Initially, the HACCP coordinator and team are selected and trained as necessary. The team is then responsible for developing the initial plan and coordinating its implementation. Product teams can be appointed to develop HACCP plans for specific products. An important aspect in developing these teams is to assure that they have appropriate training. The workers who will be responsible for monitoring need to be adequately trained. Upon completion of the HACCP plan, operator procedures, forms and procedures for monitoring and corrective action are developed. Often it is a good idea to develop a timeline for the activities involved in the initial implementation of the HACCP plan. Implementation of the HACCP system involves the continual application of the monitoring, record-keeping, corrective action procedures and other activities as described in the HACCP plan.

Maintaining an effective HACCP system depends largely on regularly scheduled verification activities. The HACCP plan should be updated and revised as needed. An important aspect of maintaining the HACCP system is to assure that all individuals involved are properly trained so they understand their role and can effectively fulfill their responsibilities.

Need for HACCP in Future

New challenges to the world food supply have prompted us to consider adopting a HACCP-based food safety system on a wider basis. One of the most important challenges is the increasing number of new food pathogens. For example, between 1973 and 1988, bacteria not previously recognized as important causes of food-borne illness such as *Escherichia coli* O157:H7 and *Salmonella enteritidis* became more widespread.

There also is increasing public health concern about chemical contamination of food: for example, the effects of lead in food on the nervous system.

Another important factor is that the size of the food industry and the diversity of products and processes have grown tremendously in the amount of domestic food manufactured and the number and kinds of foods imported.

The need for HACCP in future is further fuelled by the growing trend in international trade for worldwide equivalence of food products and the Codex Alimentarius Commission's adoption of HACCP as the international standard for food safety

Terms Related to Hazard Analysis and Critical Control Point (HACCP)

CCP Decision Tree

A sequence of questions to assist in determining whether a control point is a CCP.

Control

1. To manage the conditions of an operation to maintain compliance with established criteria.
2. The state where correct procedures are being followed and criteria are being met.

Control Measure

Any action or activity that can be used to prevent, eliminate or reduce a significant hazard.

Control Point

Any step at which biological, chemical, or physical factors can be controlled.

Corrective Action

Procedures followed when a deviation occurs.

Criterion

A requirement on which a judgement or decision can be based.

Critical Control Point

A step at which control can be applied and is essential to prevent or eliminate a food safety hazard or reduce it to an acceptable level.

Critical Limit

A maximum and/or minimum value to which a biological, chemical or physical parameter must be controlled at a CCP to prevent, eliminate or reduce to an acceptable level the occurrence of a food safety hazard.

Deviation

Failure to meet a critical limit.

HACCP

A systematic approach to the identification, evaluation, and control of food safety hazards.

HACCP Plan

The written document which is based upon the principles of HACCP and which delineates the procedures to be followed.

HACCP System

The result of the implementation of the HACCP Plan.

HACCP Team

The group of people who are responsible for developing, implementing and maintaining the HACCP system.

Hazard

A biological, chemical, or physical agent that is reasonably likely to cause illness or injury in the absence of its control.

Hazard Analysis

The process of collecting and evaluating information on hazards associated with the food under consideration to decide which are significant and must be addressed in the HACCP plan.

Monitor

To conduct a planned sequence of observations or measurements to assess whether a CCP is under control and to produce an accurate record for future use in verification.

Prerequisite Programs

Procedures, including Good Manufacturing Practices, that address operational conditions providing the foundation for the HACCP system.

Severity

The seriousness of the effect(s) of a hazard.

Step

A point, procedure, operation or stage in the food system from primary production to final consumption.

Validation

That element of verification focused on collecting and evaluating scientific and technical information to determine if the HACCP plan, when properly implemented, will effectively control the hazards.

Verification

Those activities, other than monitoring, that determine the validity of the HACCP plan and that the system is operating according to the plan.

Principle 1: Conduct a Hazard Analysis

After addressing the preliminary tasks discussed above, the HACCP team conducts a hazard analysis and identifies appropriate control measures. The purpose of the hazard analysis is to develop a list of hazards which are of such significance that they are reasonably likely to cause injury or illness if not effectively controlled. Hazards that are not reasonably likely to occur would not require further consideration within a HACCP plan. It is important to consider in the hazard analysis the ingredients and raw materials, each step in the process, product storage and distribution, and final preparation and use by the consumer. When conducting a hazard analysis, safety concerns must be differentiated from quality concerns. A hazard is defined as a biological, chemical or physical agent that is reasonably likely to cause illness or injury in the absence of its control. Thus, the word hazard as used in this chapter is limited to safety.

A thorough hazard analysis is the key to preparing an effective HACCP plan. If the hazard analysis is not done correctly and the hazards warranting control within the HACCP system are not identified, the plan will not be effective regardless of how well it is followed.

The hazard analysis and identification of associated control measures accomplish three objectives: Those hazards and associated control measures are identified. The analysis may identify needed modifications to a process or product so that product safety is further assured or improved. The analysis provides a basis for determining CCPs in Principle 2.

The process of conducting a hazard analysis involves two stages:

1. The first, hazard identification, can be regarded as a brain storming session. During this stage, the HACCP team reviews the ingredients used in the product, the activities conducted at each step in the process and the equipment used, the final product and its method of storage and distribution, and the intended use and consumers of the product. Based on this review, the team develops a list of potential biological, chemical or physical hazards which may be introduced, increased, or controlled at each step in the production process. Hazard identification focuses on developing a list of potential hazards associated with each process step under direct control of the food operation.
2. In stage two of the hazard analysis, the HACCP team decides which potential hazards must be addressed in the HACCP plan. During this stage, each potential hazard is evaluated based on the severity of the potential hazard and its likely occurrence. Severity is the seriousness of the consequences of exposure to the hazard. Considerations of severity (*e.g.*, impact of sequelae, and magnitude and duration of illness or injury) can be helpful in understanding the public health impact of the hazard. Consideration of the likely occurrence is usually based upon a combination of experience, epidemiological data, and information in the technical literature. When conducting the hazard evaluation, it is helpful to consider the likelihood of exposure and severity of the potential consequences if the hazard is not properly controlled. In addition, consideration should be given to the effects of short term as well as long term exposure to the potential hazard. Such considerations do not include common dietary choices which lie outside of HACCP.
3. During the evaluation of each potential hazard, the food, its method of preparation, transportation, storage and persons likely to consume the product should be considered to determine how each of these factors may

influence the likely occurrence and severity of the hazard being controlled. The team must consider the influence of likely procedures for food preparation and storage and whether the intended consumers are susceptible to a potential hazard. However, there may be differences of opinion, even among experts, as to the likely occurrence and severity of a hazard. The HACCP team may have to rely upon the opinion of experts who assist in the development of the HACCP plan.

Hazards identified in one operation or facility may not be significant in another operation producing the same or a similar product. For example, due to differences in equipment and/or an effective maintenance program, the probability of metal contamination may be significant in one facility but not in another. A summary of the HACCP team deliberations and the rationale developed during the hazard analysis should be kept for future reference. This information will be useful during future reviews and updates of the hazard analysis and the HACCP plan.

In Table 7.1 three examples of using a logic sequence in conducting a hazard analysis have been illustrated. While these examples relate to biological hazards, chemical and physical hazards are equally important to consider.

This table is for illustration purposes to further explain the stages of hazard analysis for identifying hazards. Hazard identification and evaluation as outlined in Table 7.1 may eventually be assisted by biological risk assessments as they become available. While the process and output of a risk assessment is significantly different from a hazard analysis, the identification of hazards of concern and the hazard evaluation may be facilitated by information from risk assessments

Thus, as risk assessments addressing specific hazards or control factors become available, the HACCP team should take these into consideration.

Upon completion of the hazard analysis, the hazards associated with each step in the production of the food should be listed along with any measure(s) that are used to control the hazard(s). The term control measure is used because not all hazards can be prevented, but virtually all can be controlled. More than one control measure

Table 7.1: Examples of how the Stages of Hazard Analysis are Used to Identify and Evaluate Hazards

Hazard Analysis Stage	Frozen cooked beef patties produced in a manufacturing plant	Product containing eggs prepared for foodservice	Commercial frozen pre-cooked, boned chicken for further processing
Stage 1 Determine potential Hazard hazards associated Identification with product	Enteric pathogens (*i.e.*, E. coli O157:H7 and Salmonella)	Salmonella in finished product.	*Staphylococcus aureus* in finished product.
Stage 2 Hazard Evaluation Assess severity of health consequences if potential hazard is not properly controlled.	Epidemiological evidence indicates that these pathogens cause severe health effects including death among children and elderly. Undercooked beef patties have been linked to disease from these pathogens.	Salmonellosis is a food borne infection causing a moderate to severe illness that can be caused by ingestion of only a few cells of *Salmonella*.	Certain strains of *S. aureus* produce an enterotoxin which can cause a moderate foodborne illness.
Determine likelihood of occurrence of potential hazard if not properly controlled.	*E. coli* O157:H7 is of very low probability and salmonellae is of moderate probability in raw meat.	Product is made with liquid eggs which have been associated with past outbreaks of salmonellosis. Recent problems with *Salmonella* serotype Enteritidis in eggs cause increased concern. Probability of Salmonella in raw eggs cannot be ruled out. If not effectively controlled, some consumers are likely to be exposed to Salmonella from this food.	Product may be contaminated with *S. aureus* due to human handling during boning of cooked chicken. Enterotoxin capable of causing illness will only occur as S. aureus multiplies to about 1,000,000/g Operating procedures during boning and subsequent freezing prevent growth of S. aureus, thus the potential for enterotoxin formation is very low.

Contd...

Table 7.1–Contd...

Using information above, determine if this potential hazard is to be addressed in the HACCP plan.	The HACCP team decides that enteric pathogens are hazards for this product. *Hazards must be addressed in the plan.*	HACCP team determines that if the potential hazard is not properly controlled, consumption of product is likely to result in an unacceptable health risk. *Hazard must be addressed in the plan.*	The HACCP team determines that the potential for enterotoxin formation is very low. However, it is still desirable to keep the initial number of S. aureus organisms low. Employee practices that minimize contamination, rapid carbon dioxide freezing and handling instructions have been adequate to control this potential hazard. *Potential hazard does not need to be addressed in plan.*

may be required for a specific hazard. On the other hand, more than one hazard may be addressed by a specific control measure (*e.g.* pasteurization of milk).

For example, if a HACCP team were to conduct a hazard analysis for the production of frozen cooked beef patties, enteric pathogens (*e.g., Salmonella* and verotoxin-producing *Escherichia coli*) in the raw meat would be identified as hazards. Cooking is a control measure which can be used to eliminate these hazards.

Examples of Questions to be considered when Conducting a Hazard Analysis

The hazard analysis consists of asking a series of questions which are appropriate to the process under consideration. The purpose of the questions is to assist in identifying potential hazards.

Ingredients

1. Does the food contain any sensitive ingredients that may present microbiological hazards (*e.g.*, Salmonella, Staphylococcus aureus); chemical hazards (*e.g.*, aflatoxin, antibiotic or pesticide residues); or physical hazards (stones, glass, metal)?
2. Are potable water, ice and steam used in formulating or in handling the food?
3. What are the sources (*e.g.*, geographical region, specific supplier)

Intrinsic Factors

Physical characteristics and composition (*e.g.*, pH, type of acidulants, fermentable carbohydrate, water activity, preservatives) of the food during and after processing.

1. What hazards may result if the food composition is not controlled?
2. Does the food permit survival or multiplication of pathogens and/or toxin formation in the food during processing?
3. Will the food permit survival or multiplication of pathogens and/or toxin formation during subsequent steps in the food chain?

4. Are there other similar products in the market place? What has been the safety record for these products? What hazards have been associated with the products?

Procedures used for processing

1. Does the process include a controllable processing step that destroys pathogens? If so, which pathogens? Consider both vegetative cells and spores.
2. If the product is subject to recontamination between processing (*e.g.*, cooking, pasteurizing) and packaging which biological, chemical or physical hazards are likely to occur?

Microbial content of the Food

1. What is the normal microbial content of the food?
2. Does the microbial population change during the normal time the food is stored prior to consumption?
3. Does the subsequent change in microbial population alter the safety of the food?
4. Do the answers to the above questions indicate a high likelihood of certain biological hazards?

Facility Design

1. Does the layout of the facility provide an adequate separation of raw materials from ready-to-eat (RTE) foods if this is important to food safety? If not, what hazards should be considered as possible contaminants of the RTE products?
2. Is positive air pressure maintained in product packaging areas? Is this essential for product safety?
3. Is the traffic pattern for people and moving equipment a significant source of contamination?

Equipment Design and Use

1. Will the equipment provide the time-temperature control that is necessary for safe food?
2. Is the equipment properly sized for the volume of food that will be processed?

3. Can the equipment be sufficiently controlled so that the variation in performance will be within the tolerances required to produce a safe food?
4. Is the equipment reliable or is it prone to frequent breakdowns?
5. Is the equipment designed so that it can be easily cleaned and sanitized?
6. Is there a chance for product contamination with hazardous substances; *e.g.*, glass?
7. What product safety devices are used to enhance consumer safety?
   (*a*) metal detectors
   (*b*) magnets
   (*c*) sifters
   (*d*) filters
   (*e*) screens
   (*f*) thermometers
   (*g*) bone removal devices
   (*h*) dud detectors
8. To what degree will normal equipment wear affect the likely occurrence of a physical hazard (*e.g.*, metal) in the product?
9. Are allergen protocols needed in using equipment for different products?

Packaging

1. Does the method of packaging affect the multiplication of microbial pathogens and/or the formation of toxins?
2. Is the package clearly labeled "Keep Refrigerated" if this is required for safety?
3. Does the package include instructions for the safe handling and preparation of the food by the end user?
4. Is the packaging material resistant to damage thereby preventing the entrance of microbial contamination?
5. Are tamper-evident packaging features used?

6. Is each package and case legibly and accurately coded?
7. Does each package contain the proper label?
8. Are potential allergens in the ingredients included in the list of ingredients on the label?

Sanitation

1. Can sanitation have an impact upon the safety of the food that is being processed?
2. Can the facility and equipment be easily cleaned and sanitized to permit the safe handling of food?
3. Is it possible to provide sanitary conditions consistently and adequately to assure safe foods?

Employee Health, Hygiene and Education

1. Can employee health or personal hygiene practices impact upon the safety of the food being processed?
2. Do the employees understand the process and the factors they must control to assure the preparation of safe foods?
3. Will the employees inform management of a problem which could impact upon safety of food?

Conditions of Storage Between Packaging and the End User

1. What is the likelihood that the food will be improperly stored at the wrong temperature?
2. Would an error in improper storage lead to a microbiologically unsafe food?

Intended Use

1. Will the food be heated by the consumer?
2. Will there likely be leftovers?

Intended Consumer

1. Is the food intended for the general public?

---

* For illustrative purposes only. The potential hazards identified may not be the only hazards associated with the products listed. The responses may be different for different establishments.

1. Receiving (Beef)

↓

2. Grinding

↓

3. Mixing

↓

4. Forming

↓

5. Cooking

↓

6. Freezing

↓

7. Boxing

↓

8. Distributing

↓

9. Reheating

↓

10. Serving

Figure 7.2: Example of a Flow Diagram for the Production of Frozen Cooked Beef Patties

2. Is the food intended for consumption by a population with increased susceptibility to illness (*e.g.*, infants, the aged, the infirmed, immunocompromised individuals)?
3. Is the food to be used for institutional feeding or the home?

The following is an excerpt from a hazard analysis summary table for this product.

Step	*Potential Hazard(s)*	*Justification*	*Hazard to be Addressed in plan? Y/N*	*Control Measure(s)*
5. Cooking	Enteric pathogens: *e.g., Salmonella*, verotoxigenic-*E. coli*	Enteric pathogens have been associated with outbreaks of foodborne illness from undercooked ground beef	Y	Cooking

The hazard analysis summary could be presented in several different ways. One format is a table such as the one given above. Another could be a narrative summary of the HACCP team's hazard analysis considerations and a summary table listing only the hazards and associated control measures.

Principle 2: Determine Critical Control Points (CCPs)

A critical control point is defined as a step at which control can be applied and is essential to prevent or eliminate a food safety hazard or reduce it to an acceptable level. The potential hazards that are reasonably likely to cause illness or injury in the absence of their control must be addressed in determining CCPs.

Complete and accurate identification of CCPs is fundamental to controlling food safety hazards. The information developed during the hazard analysis is essential for the HACCP team in identifying which steps in the process are CCPs. One strategy to facilitate the identification of each CCP is the use of a CCP decision tree (Examples of decision trees are given in Tables 7.2 and 7.3. Although application of the CCP decision tree can be useful in determining if a particular step is a CCP for a previously identified hazard, it is merely a tool and not a mandatory element of HACCP. A CCP decision tree is not a substitute for expert knowledge.

Table 7.2: Example I of a CCP Decision Tree

Important considerations when using the decision tree:

1. The decision tree is used after the hazard analysis.
2. The decision tree then is used at the steps where a hazard that must be addressed in the HACCP plan has been identified.
3. A subsequent step in the process may be more effective for controlling a hazard and may be the preferred CCP.
4. More than one step in a process may be involved in controlling a hazard.
5. More than one hazard may be controlled by a specific control measure.

Q 1. Does this step involve a hazard of sufficient likelihood of occurence and severity to warrant its control?

↓ YES ↓

↓ NO → Not a CCP

Q 2. Does a control measure for the hazard exist at this step?

↓ YES ↓

↓ NO ↓ Is control at this step necessary for safety? → YES ↑ Modify the step, process or product ↑

Is control at this step necessary for safety? ↓ NO → Not a CCP → STOP*

Q 3. Is control at this step necessary to prevent, eliminate, or reduce the risk of the hazard to consumers?

↓ YES ↓ CCP

↓ NO → Not a CCP → STOP*

* Proceed to next step in the process.

Table 7.3: Example II of a CCP Decision Tree

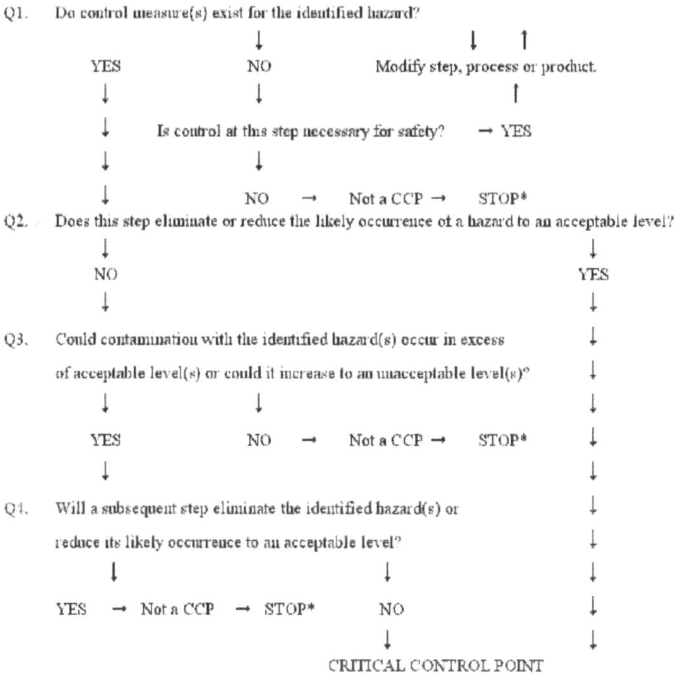

* Proceed to next step in the process.

Critical control points are located at any step where hazards can be either prevented, eliminated, or reduced to acceptable levels. Examples of CCPs may include: thermal processing, chilling, testing ingredients for chemical residues, product formulation control, and testing product for metal contaminants. CCPs must be carefully developed and documented. In addition, they must be used only for purposes of product safety. For example, a specified heat process, at a given time and temperature designed to destroy a specific microbiological pathogen, could be a CCP. Likewise, refrigeration of a precooked food to prevent hazardous microorganisms from multiplying, or the adjustment of a food to a pH necessary to prevent toxin formation could also be CCPs. Different facilities preparing similar food items can differ in the hazards identified and the steps which are CCPs. This can be due to differences in each facility's layout, equipment, selection of ingredients, processes employed, etc.

Principle 3: Establish critical limits

A critical limit is a maximum and/or minimum value to which a biological, chemical or physical parameter must be controlled at a CCP to prevent, eliminate or reduce to an acceptable level the occurrence of a food safety hazard. A critical limit is used to distinguish between safe and unsafe operating conditions at a CCP. Critical limits should not be confused with operational limits which are established for reasons other than food safety.

Each CCP will have one or more control measures to assure that the identified hazards are prevented, eliminated or reduced to acceptable levels. Each control measure has one or more associated critical limits. Critical limits may be based upon factors such as:

1. Temperature
2. Time
3. Physical dimensions
4. Humidity
5. Moisture level
6. Water activity (a_w)
7. Ph
8. Titratable acidity
9. Salt concentration
10. Available chlorine
11. Viscosity
12. Preservatives
13. Sensory information such as aroma and visual appearance.

Critical limits must be scientifically based. For each CCP, there is at least one criterion for food safety that is to be met. An example of a criterion is a specific lethality of a cooking process such as a 5D reduction in *Salmonella*. The critical limits and criteria for food safety may be derived from sources such as regulatory standards and guidelines, literature surveys, experimental results, and experts.

An example is the cooking of beef patties (Figure 7.2). The process should be designed to ensure the production of a safe product. The hazard analysis for cooked meat patties identified enteric pathogens (*e.g.*, verotoxigenic *E. coli* such as *E. coli* O157:H7, and salmonellae) as significant biological hazards. Furthermore, cooking is the step in the process at which control can be applied to reduce the enteric pathogens to an acceptable level. To ensure that an acceptable level is consistently achieved, accurate information is needed on the probable number of the pathogens in the raw patties, their heat resistance, the factors that influence the heating of the patties, and the area of the patty which heats the slowest. Collectively, this information forms the scientific basis for the critical limits that are established. Some of the factors that may affect the thermal destruction of enteric pathogens are listed in the following table. In this example, the HACCP team concluded that a thermal process equivalent to 155°F for 16 seconds would be necessary to assure the safety of this product. To ensure that this time and temperature are attained, the HACCP team for one facility determined that it would be necessary to establish critical limits for the oven temperature and humidity, belt speed (time in oven), patty thickness and composition (*e.g.*, all beef, beef and other ingredients). Control of these factors enables the facility to produce a wide variety of cooked patties, all of which will be processed to a minimum internal temperature of 155°F for 16 seconds. In another facility, the HACCP team may conclude that the best approach is to use the internal patty temperature of 155°F and hold for 16 seconds as critical limits. In this second facility the internal temperature and hold time of the patties are monitored at a frequency to ensure that the critical limits are constantly met as they exit the oven. The example given below applies to the first facility.

Process Step	*CCP*	*Critical Limits*
5. Cooking	YES	Oven temperature:___°F Time; rate of heating and cooling (belt speed in ft/min): ____ft/min Patty thickness:____in. Patty composition: *e.g.* all beef Oven humidity: ____ per cent RH

Principle 4: Establish monitoring procedures

Monitoring is a planned sequence of observations or measurements to assess whether a CCP is under control and to produce an accurate record for future use in verification. Monitoring serves three main purposes.

1. First, monitoring is essential to food safety management in that it facilitates tracking of the operation. If monitoring indicates that there is a trend towards loss of control, then action can be taken to bring the process back into control before a deviation from a critical limit occurs.
2. Second, monitoring is used to determine when there is loss of control and a deviation occurs at a CCP, *i.e.*, exceeding or not meeting a critical limit. When a deviation occurs, an appropriate corrective action must be taken.
3. Third, it provides written documentation for use in verification.

An unsafe food may result if a process is not properly controlled and a deviation occurs. Because of the potentially serious consequences of a critical limit deviation, monitoring procedures must be effective. Ideally, monitoring should be continuous, which is possible with many types of physical and chemical methods. For example, the temperature and time for the scheduled thermal process of low-acid canned foods is recorded continuously on temperature recording charts. If the temperature falls below the scheduled temperature or the time is insufficient, as recorded on the chart, the product from the retort is retained and the disposition determined as in Principle 5. Likewise, pH measurement may be performed continually in fluids or by testing each batch before processing. There are many ways to monitor critical limits on a continuous or batch basis and record the data on charts. Continuous monitoring is always preferred when feasible. Monitoring equipment must be carefully calibrated for accuracy.

Assignment of the responsibility for monitoring is an important consideration for each CCP. Specific assignments will depend on the number of CCPs and control measures and the complexity of monitoring. Personnel who monitor CCPs are often associated with production (*e.g.*, line supervisors, selected line workers and maintenance personnel) and, as required, quality control personnel. Those individuals must be trained in the monitoring technique for which they are responsible, fully understand the purpose and importance of monitoring, be unbiased in monitoring and reporting, and accurately report the results of monitoring. In addition, employees should be trained in procedures to follow when there is a trend towards loss of control so that adjustments can be made in a timely manner to assure that the process remains under control. The person responsible for monitoring must also immediately report a process or product that does not meet critical limits.

All records and documents associated with CCP monitoring should be dated and signed or initialed by the person doing the monitoring.

When it is not possible to monitor a CCP on a continuous basis, it is necessary to establish a monitoring frequency and procedure that will be reliable enough to indicate that the CCP is under control. Statistically designed data collection or sampling systems lend themselves to this purpose.

Most monitoring procedures need to be rapid because they relate to on-line, "real-time" processes and there will not be time for lengthy analytical testing. Examples of monitoring activities include: visual observations and measurement of temperature, time, pH, and moisture level.

Microbiological tests are seldom effective for monitoring due to their time-consuming nature and problems with assuring detection of contaminants. Physical and chemical measurements are often preferred because they are rapid and usually more effective for assuring control of microbiological hazards. For example, the safety of pasteurized milk is based upon measurements of time and temperature of heating rather than testing the heated milk to assure the absence of surviving pathogens.

With certain foods, processes, ingredients, or imports, there may be no alternative to microbiological testing. However, it is important to recognize that a sampling protocol that is adequate to reliably detect low levels of pathogens is seldom possible because of the large number of samples needed. This sampling limitation could result in a false sense of security by those who use an inadequate sampling protocol. In addition, there are technical limitations in many laboratory procedures for detecting and quantitating pathogens and/or their toxins.

Principle 5: Establish Corrective Actions

The HACCP system for food safety management is designed to identify health hazards and to establish strategies to prevent, eliminate, or reduce their occurrence. However, ideal circumstances do not always prevail and deviations from established processes may occur. An important purpose of corrective actions is to prevent foods which may be hazardous from reaching consumers. Where there is a deviation from established critical limits, corrective actions are necessary. Therefore, corrective actions should include the following elements:

1. Determine and correct the cause of non-compliance
2. Determine the disposition of non-compliant product and
3. Record the corrective actions that have been taken.

Specific corrective actions should be developed in advance for each CCP and included in the HACCP plan. As a minimum, the HACCP plan should specify what is done when a deviation occurs, who is responsible for implementing the corrective actions, and that a record will be developed and maintained of the actions taken. Individuals who have a thorough understanding of the process, product and HACCP plan should be assigned the responsibility for oversight of corrective actions. As appropriate, experts may be consulted to review the information available and to assist in determining disposition of non-compliant product.

Principle 6: Establish Verification Procedures

Verification is defined as those activities, other than monitoring, that determine the validity of the HACCP plan and that the system is operating according to the plan. The major infusion of science in a HACCP system centers on proper identification of the hazards, critical control points, critical limits, and instituting proper verification procedures. These processes should take place during the development and implementation of the HACCP plans and maintenance of the HACCP system. An example of a verification schedule is given in Figure 7.3.

One aspect of verification is evaluating whether the facility's HACCP system is functioning according to the HACCP plan. An effective HACCP system requires little end-product testing, since sufficient validated safeguards are built in early in the process. Therefore, rather than relying on end-product testing, firms should rely on frequent reviews of their HACCP plan, verification that the HACCP plan is being correctly followed, and review of CCP monitoring and corrective action records.

Figure 7.3: Example of a Company Established HACCP Verification Schedule

Activity	*Frequency*	*Responsibility*	*Reviewer*
Verification Activities Scheduling	Yearly or Upon HACCP System Change	HACCP Coordinator	Plant Manager
Initial Validation of HACCP Plan	Prior to and During Initial Implementation of Plan	Independent Expert(s)*	HACCP Team
Subsequent validation of HACCP Plan	When Critical Limits Changed, Significant Changes in Process, Equipment Changed, After System Failure, etc.	Independent Expert(s)*	HACCP Team
Verification of CCP Monitoring as Described in the Plan (*e.g.*, monitoring of patty cooking temperature)	According to HACCP Plan (*e.g.*, once	According to HACCP Plan (*e.g.*, Line Supervisor)	According to HACCP Plan (*e.g.*, Quality Control)
Review of Monitoring, Corrective Action Records to Show Compliance with the Plan	Monthly	Quality Assurance	HACCP Team
Comprehensive HACCP System Verification	Yearly	Independent Expert(s)*	Plant Manager

* Done by others than the team writing and implementing the plan. May require additional technical expertise as well as laboratory and plant test studies.

Another important aspect of verification is the initial validation of the HACCP plan to determine that the plan is scientifically and technically sound, that all hazards have been identified and that if the HACCP plan is properly implemented these hazards will be effectively controlled. Information needed to validate the HACCP plan often include (1) expert advice and scientific studies and (2) in-plant observations, measurements, and evaluations. For example, validation of the cooking process for beef patties should include the scientific justification of the heating times and temperatures needed to obtain an appropriate destruction of pathogenic microorganisms (*i.e.*, enteric pathogens) and studies to confirm that the conditions of cooking will deliver the required time and temperature to each beef patty.

Subsequent validations are performed and documented by a HACCP team or an independent expert as needed. For example, validations are conducted when there is an unexplained system failure; a significant product, process or packaging change occurs; or new hazards are recognized.

In addition, a periodic comprehensive verification of the HACCP system should be conducted by an unbiased, independent authority. Such authorities can be internal or external to the food operation. This should include a technical evaluation of the hazard analysis and each element of the HACCP plan as well as on-site review of all flow diagrams and appropriate records from operation of the plan. A comprehensive verification is independent of other verification procedures and must be performed to ensure that the HACCP plan is resulting in the control of the hazards. If the results of the comprehensive verification identifies deficiencies, the HACCP team modifies the HACCP plan as necessary.

Verification activities are carried out by individuals within a company, third party experts, and regulatory agencies. It is important that individuals doing verification have appropriate technical expertise to perform this function. The role of regulatory and industry in HACCP was further described by the NACMCF (1994).

Examples of verification activities are included as Appendix I.

APPENDIX I
Examples of Verification Activities

Verification procedures may include

1. Establishment of appropriate verification schedules.
2. Review of the HACCP plan for completeness.
3. Confirmation of the accuracy of the flow diagram.
4. Review of the HACCP system to determine if the facility is operating according to the HACCP plan.
5. Review of CCP monitoring records.

6. Review of records for deviations and corrective actions.
7. Validation of critical limits to confirm that they are adequate to control significant hazards.
8. Validation of HACCP plan, including on-site review.
9. Review of modifications of the HACCP plan.
10. Sampling and testing to verify CCPs.

Verification should be conducted

1. Routinely, or on an unannounced basis, to assure CCPs are under control.
2. When there are emerging concerns about the safety of the product.
3. When foods have been implicated as a vehicle of foodborne disease.
4. To confirm that changes have been implemented correctly after a HACCP plan has been modified.
5. To assess whether a HACCP plan should be modified due to a change in the process, equipment, ingredients, etc.

Verification reports may include information on the presence and adequacy of.

1. The HACCP plan and the person(s) responsible for administering and updating the HACCP plan.
2. The records associated with CCP monitoring.
3. Direct recording of monitoring data of the CCP while in operation.
4. Certification that monitoring equipment is properly calibrated and in working order.
5. Corrective actions for deviations.
6. Sampling and testing methods used to verify that CCPs are under control.
7. Modifications to the HACCP plan.
8. Training and knowledge of individuals responsible for monitoring CCPs.
9. Validation activities.

Principle 7: Establish Record-keeping and Documentation Procedures

Generally, the records maintained for the HACCP System should include the following:

1. A summary of the hazard analysis, including the rationale for determining hazards and control measures.
2. The HACCP Plan:
   (*a*) Listing of the HACCP team and assigned responsibilities.
   (*b*) Description of the food, its distribution, intended use, and consumer.
   (*c*) Verified flow diagram.
   (*d*) HACCP Plan Summary Table that includes information for:
   (*e*) Steps in the process that are CCPs
   (*f*) The hazard(s) of concern.
   (*g*) Critical limits
   (*h*) Monitoring*
   (*i*) Corrective actions*
   (*j*) Verification procedures and schedule*
   (*k*) Record-keeping procedures*

* A brief summary of position responsible for performing the activity and the procedures and frequency should be provided

The following is an example of a HACCP plan summary table:

CCP	*Hazards*	*Critical Limit(s)*	*Monitoring*	*Corrective Actions*	*Verification*	*Records*

1. Support documentation such as validation records.
2. Records that are generated during the operation of the plan.

Examples of HACCP records are given in Appendix II.

APPENDIX II
Examples of HACCP Records

Ingredients for which critical limits have been established.

1. Supplier certification records documenting compliance of an ingredient with a critical limit.
2. Processor audit records verifying supplier compliance.
3. Storage records (*e.g.*, time, temperature) for when ingredient storage is a CCP.

Processing, storage and distribution records

1. Information that establishes the efficacy of a CCP to maintain product safety.
2. Data establishing the safe shelf life of the product; if age of product can affect safety.
3. Records indicating compliance with critical limits when packaging materials, labeling or sealing specifications are necessary for food safety.
4. Monitoring records.
5. Verification records.

Deviation and corrective action records.

Employee training records that are pertinent to CCPs and the HACCP plan.

Documentation of the adequacy of the HACCP plan from a knowledgeable HACCP expert.

Hazard Analysis Critical Control Points (HACC)/ISO: 9000

The Food Industry plays an important role in the national economy. In today's global market, quality and food safety have become competitive edges for the 48CODEX Codex Alimentarius Commission is an International Body constituted by Food and Agriculture Organization (FAO) and World Health Organization (WHO) of the United Nations with the objective to protect health of consumers and to ensure fair practices in the food trade. The term "Codex Alimentarius" is taken from the Latin language and means "food code". The Codex Alimentarius Commission is a collection of International Standards for the safety and quality of food as well as codes of good manufacturing practice and other guidelines to protect the health of the consumers and remove unfair practices in international trade. These standards, guidelines and recommendations are recognized worldwide for international trade negotiations and also for settlings of disputes by WTO.

The Codex Contact Point in India is the Directorate General of Health Services (DGHS) in the M/o Health and Family Welfare.

However, the Ministry of Food Processing Industries is closely associated with the activities of Codex Alimentarious in the country. Ministry of Food Processing Industries is the Chairman of five Shadow Committees of Codex Committee which are as under:

1. Shadow Committee on Food Additives and Contaminants.
2. Shadow Committee on Food Labelling.
3. Shadow Committee on Processed Fruits and Vegetables. Enterprises producing foods and providing services. Therefore, the installation of ISO:9000 Quality Management Systems and Hazard Analysis Critical Control Points (HACCP) based food safety system is extremely desirable in view of the changing scenario in the international trade. The M/o Food Processing Industries under its Plan Scheme is providing financial assistance to all agencies engaged in the field of food processing for implementation of Total Quality Management (TQM) including HACCP, ISO 9000 etc. Under this Scheme, assistance in the form of grant-in-aid is provided to all agencies to the extent of 50 per cent subject to a max. limit of Rs. 10 lakhs towards the cost of implementing Total Quality Management including obtaining ISO:9000 Certification, HACCP etc.

During the period, a grant-in-aid of Rs. 10.00 lakhs has been provided to M/S Jumbo International, New Delhi towards implementation of ISO 9000-HACCP system in their Fruit and Vegetable Processing Unit. A grant-in-aid of Rs. 9.90 lakhs has been sanctioned to M/S Avon Agro Industries Pvt. Ltd, New Delhi for implementation of ISO 9000-HACCP System in their Spice processing unit.

A grant-in-aid of Rs. 10.00 lakhs has also been sanctioned to M/S Haldiram Foods Ltd, Gurgaon towards implementation of ISO 9000-HACCP system in their Food Processing Unit.

A grant-in-aid of Rs. 10.00 lakhs has also been sanctioned to M/S Kayem Food Industries Pvt. Ltd, Gurgaon towards implementation of ISO 9000-HACCP system in their Cereal Processing Unit.

National Food Safety Board

Ministry of Food Processing Industries has a proposal to establish an autonomous apex body in the form of National Food Safety Board which would not only accredit various certification bodies but also control the certification activity. In addition, the objectives of the Board include:

1. To promote food safety concepts in Indian Food Industries for developing confidence among domestic and foreign customers
2. To establish certification and accreditation mechanism in the area of food safety systems.
3. Organisation of training of auditors for food safety systems
4. Registration of auditors of food safety systems
5. Advising Government and Industry on food safety verification matters.

This Ministry has assigned work to FICCI Quality Forum for development of sector specific guidelines/booklets on GHP/GMP/HACCP for the use of food processing industries, Training material for trainers, Generic check list for audit of GHP/GMP/HACCP system and constitution of accreditation Board etc.

HACCP systems, to introduce Good Laboratory Practices, to set up/upgrade the infrastructure such as laboratories / codex cells etc. the Ministry of Food Processing has formulated Schemes under 10th Plan to provide financial assistance to various stakeholders relating to food processing to keep pace with the developments in international market. The key objectives of these Schemes are as under:

Total Quality Management

Objectives

1. To motivate the food processing industries for adoption of food safety and quality assurance mechanisms such as Total Quality Management (TQM) including ISO-9000, ISO14000, Hazard Analysis and Critical Control Points (HACCP), Good Manufacturing Practices (GMP), Good Hygienic Practices (GHP)
2. To prepare them to face the global competition in International trade in post WTO regime
3. To enable adherence to stringent quality in hygiene norms
4. To enhance product acceptance by overseas buyers
5. To keep Indian industry technologically abreast of international best practices..

Bar-Coding

Objectives

1. To popularise the concept of bar coding

Networking of Food Networking of Food Testing Laboratories

The M/o Food Processing Industries in association with CIFTI organized a national seminar in October 2002 on networking of food testing laboratories which was attended by various concerned stakeholders such as Ministries/Departments, research institutions, industry associations and other organizations etc. The workshop concluded highlighting the following basic recommendations:

1. To set up a network of laboratories so as to have at least one sophisticated lab in every state preferably the capital unless exclusive concentration of units are observed in locations other than the capital or two of them in zones like North- East, hill areas etc.
2. To set up a Committee to identify the laboratory in each State to form part of this network from amongst the laboratories already existing.

3. To evolve a system of gathering test data from all strata of laboratories and also work on a system of compiling them in centralized manner.

Plan Schemes for Quality Assurance

Scheme for total quality management, codex cell, research and development and quality control laboratories to popularise the concept of bar coding, to encourage adoption of GHP, GMP and system following National/International Standards and to encourage food processors to affix bar codes on their processed food packages in order to keep abrest with the developments in the international markets

1. To avoid any non-tariff barriers in future, as bar coding has become an important requirement in the context of stringent food safety standards.

Strengthening the Codex Cell

Objectives

To strengthen / setting up of Codex cell in the Ministry as well as at the point of various stakeholders such as State level Nodal Agencies for food processing industries, designated by the State Government, Industry Associations, National Research Institutions for enabling pro-active participation in Codex deliberations and adequate projection of Indian view point in Codex system. This includes computerisation, compilation and maintenance of Codex documents, study/survey for scientific data generation, engagement of consultants for preparing Discussion/ Position papers, code of practices etc. especially for Codex purposes only.

Setting up of Quality Control Laboratory

Objectives

1. To ensure compliance with National food standards
2. To assist industries in the food sector to develop and implement quality management systems such as ISO 9000, HACCP etc
3. To analyse the samples received from food processing industries, and other stakeholders.

4. To impart training in the areas relating to quality improvement through own expertise.
5. To provide information on quality standards and requirements of various markets on quality of products.
6. To reduce the time of analysis of samples by reducing transportation time of samples.
7. To generate scientific data on levels of contaminants, pesticide residue etc.

Assistance for Setting Up/Modernisation of Laboratories

For successful food processing sector in India, various aspects of the Total Quality Management such as quality control, quality system and quality assurance should function in a horizontal fashion for total success. These are vital today, if one has to reach the world market or avoid being swamped by imported food items. MFPI aims at setting up a network of laboratories to help implement quality regime forprocessed food. A number of prestigious laboratories have been assisted in upgrading facilities for the quality revolution in the country. Some of the important ones are:

1. Mahratta Chamber of Commerce, Industry and Agriculture, Pune, was granted Rs. 37.56 lakhs for setting up of Food Testing Laboratory.
2. Pradeshik Cooperative Dairy Federation Ltd., Lucknow, was granted Rs. 90 lakhs for for strengthening of Quality Control Laboratory.
3. Raja Dinesh Singh Krishi Vigyan Kendra, Kalakankar, Pratapgarh, was granted Rs. 20 lakhs to set up a Quality Control Laboratory.
4. Food Research and Analysis Centre, New Delhi, was granted Rs. 50 lakhs for setting up an Organic Food Testing facility at New Delhi.
5. Food Research and Analysis Centre, New Delhi, was granted Rs.5.48 lakhs for Upgradation of Quality Control Laboratory.

Part III
Postharvest Processing and Utilization of Fruits and Vegetables

Chapter 8

Physiological Study of Perishable Foods

Introduction

As we know almost all agricultural produce is stored in a fresh state for part of its postharvest life. The reactions and processes that occur in such produce are many and complex and their overall effect manifests itself as quality change (usually quality loss) during storage. The rates and equilibria of these reactions are determined by many variables, particularly by temperature and atmospheric composition and humidity. Of these temperature has by far the greatest effect and fortunately it is the easiest to control.

It is not possible to consider the effect of storage temperature on the rate and equilibrium of each and every biochemical reaction. Such a model would be far too complex, some reactions are not completely understood and the mechanisms of some are unknown. It is necessary, therefore, to use gross simplification and even the nebulous concept of 'quality' itself in order to produce working models of storage changes. In spite of this, many of the models proposed and evaluated in the literature can be used to predict, with adequate precision, the effects of temperature on storage changes in agricultural produce.

Produce is rarely stored at constant temperature. Methods are described which enable the effects of fluctuating temperatures on storage changes to be assessed.

Most food products, no matter how they are processed, stored or preserved, deteriorate with time. For some products, notably these that are sterilised or dehydrated and properly packaged, the storage life is so long that it is often not of primary interest. But for other products, especially those that are frozen or preserved by cool storage in a fresh condition, storage life is a major consideration in the process. Of all the parameters that influence storage life, temperature is perhaps the most important and it is with the effects of temperature that we are concerned here.

It has been estimated that between 25 and 80 per cent of fresh fruit and vegetables are lost after harvest, the most serious losses occurring in developing countries where they can least be tolerated. As Lorentzen puts it; no matter how productivity is achieved, it will always be necessary to take good care of the harvest and bring it to the consumers with a minimum of quality loss.

Delays between the harvesting and utilisation of crops are inevitable and fresh fruits and vegetables are particularly prone to quality loss during the intervening period. Quality loss in produce destined for further processing can often be reduced by synchronisation of harvest, transport to the factory and processing. But the world's major food preservation process is fresh storage–90 per cent of agricultural production is processed in no other way and the distribution chain that ends with domestic consumption of 'fresh' produce is usually of unpredictable duration and subject to considerable changes in environmental conditions.

Individual products differ greatly in the length of time for which they can be maintained in an acceptable condition, in their responses to various treatments and to adjustments in environmental conditions. In general, the rate of deterioration of vegetables is related to the botanical function of the tissue. The shortest storage lives, and the highest metabolic rates, are found in tissues that were growing rapidly at the time of harvest, such as mushrooms, asparagus and other shoots. Next come leaves and other aerial parts of mature plants (cabbage, spinach, celery), followed by storage roots (carrots, turnips) and finally by specialised storage organs such as tubers (potatoes) and bulbs (onions).

Table 8.1 indicates the typical storage lives and metabolic rates for a selection of vegetables related to botanical function.

Table 8.1: Typical Storage Lives and Metabolic Rates for a Selection of Vegetables Related to Botanical Function

Product	*Relative Respiration Rate*	*Botanical Function*	*Typical Storage Life (Weeks at 2°C)*
Asparagus	40	Actively growing shoots	0.2–0.5
Mushrooms	21		
Artichokes	17		
Spinach	13	Aerial parts of plants	1–2
Lettuce	11		
Cabbage	6		
Carrots	5	Storage roots	5–20
Turnips	4		
Beetroot	3		
Potatoes	2	Specialised storage organs	25–50
Garlic	2		
Onions	1		

Physiological deterioration in vegetables occurs primarily because of the stress induced by the removal of the tissue from the parent plant. The tissue is unable to replenish either the water or the metabolic substrates that it loses and may be unable to dispose of metabolic products. The moribund condition that results, mainly from excessive water loss, often predisposes the tissue to invasion by facultatively parasitic microorganisms and this condition is often exacerbated by mechanical damage during handling of the produce.

Deterioration in fruits is rather different in that it is closely related to the natural ripening and senescence of the fruit. For it is the natural function of the fleshy parts of fruits to deteriorate after the fruit leaves the tree and to release the seeds.

Considerations that apply to the effects of temperature on fresh vegetables can often be applied to frozen produce, although the temperatures involved are typically 25°C lower. Frozen foods in general present a simpler system than fresh fruits and vegetables.

They are often blanched before freezing so that enzyme-catalysed reactions are of secondary importance, if they occur at all, and the main cause of deterioration may be comparatively simple reactions such as oxidation.

Important features of the environment which influence the longevity of fresh produce and which are amenable to control are temperature, humidity and the composition of the atmosphere surrounding the produce. Of these, temperature has the greatest effect on storage life and is, conveniently, the most readily controlled.

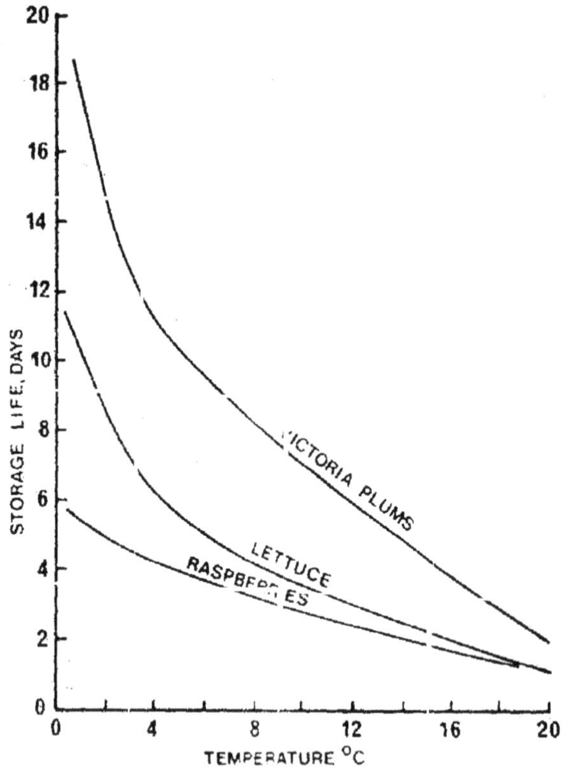

Figure 8.1: The Effect of Temperature on the Storage Life of Some Produce

Humidity is not readily controlled, but can have a considerable effect on deterioration; not only does it regulate the rate of water loss, but it has also been implicated in physiological storage disorders such as scald in apples. Regulation of respiration by the control of atmospheric composition has been practised commercially for half a century since the pioneering work of Kidd and West. Today, controlled atmosphere storage for apples and pears in the United States amounts to almost 400000 tonnes capacity and in Britain to 250000 tonnes.

Acceptability of a Food

The criterion by which the acceptability of a food is judged by the consumer is that of quality. Although an individual consumer will have no qualms about describing the quality of a food, quality is a complex function of many attributes of a food. For a fruit or vegetable, it will, typically, involve flavour and aroma, texture and firmness, size, shape and appearance. Many of these attributes are extremely difficult to measure quantitatively and, even if measurement is possible, the part that each plays in the overall concept of quality will vary from consumer to consumer. It is not, therefore, possible to develop a universal measurement of quality for any product. Not only does the relative importance of individual attributes vary, but many of them have limiting values that render the product unacceptable whatever the values of the other attributes. For example, a tomato could be unacceptable because it was too soft, even if its flavour and aroma, size and shape were perfect. Because of the difficulty of specifying quality, the concept of a quality indicator is often used. This is the application of one or more attributes of a product to indicate storage changes. Ideally, a quality indicator should be a major quality attribute, but, even more important, it must be readily measurable with sufficient precision and reproducibility. With tomatoes, for example, firmness, surface colour and sugar, acid ratio have been used as quality indicators. Other quality indicators proposed have included changes in vitamins and internal colour of fruits. The use of quality indicators is not satisfactory in that it does not give a real assessment of quality, but it is necessary in experimental work to utilise some objective and reproducible measurement rather than the nebulous and subjective concept of overall quality. The inherent variability of quality makes hypotheses based on it unreproducible. And the scientific acceptability of a hypothesis requires that it be reproducible at all places and at all times. So there is a strong case for the use of quality indicators although they are less realistic in practical terms. Nevertheless, direct measurement of quality–usually be sensory analysis–can give valuable practical data providing its limitations are realised. And much published information purporting to be based on quality measurement turns out, on close scrutiny, to be based on a quality indicator.

Now let us try to understand, how the temperature influences the quality of food.

Because temperature is the major factor influencing quality, it is very desirable to be able to predict the quality change of produce

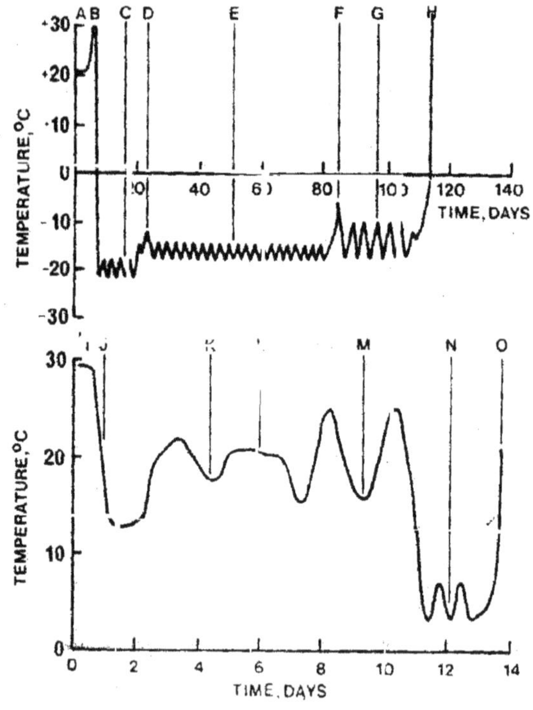

Figure 8.2: Typical Temperature Histories of Fresh and Frozen Produce

from known time-temperature data as a function of storage time. If the temperature is constant this can be done by experience, but transport and storage between the region of production and that of consumption usually involves great changes in temperature. Typical temperature histories of fresh and frozen produce are most variable and often subject to unpredictable delays, especially fresh produce which has an indeterminate sojourn at the wholesale market and retail outlet.

Different temperature histories induce different degrees of quality change, depending on the nature of the time-temperature pattern. Therefore, it is necessary to develop a simple procedure to give an estimation of the accumulated quality change in a product at any time in the distribution system, no matter how variable or irregular the time-temperature pattern might be. This procedure should provide information on which to act for the protection of quality at any stage in the commercial life of the food. It should also provide information on the effects of individual operations in the distribution chain on quality; for example, on the effect of a non-refrigerated journey or on the effect of leaving a pallet exposed to the sun for a short time.

Every analytical calculation needs a mathematical model. Various models have been proposed for studying the effects of temperature fluctuations on the storage of fruits and vegetables, both fresh and frozen. The 'effective temperature' theory, computer aided simulation models and the 'time-temperature-tolerance' hypothesis (TTT) are the most significant. Each of these models has partially fulfilled the prediction requirements already discussed. In particular, the effects of frozen storage on the quality of foodstuffs have quite successfully been predicted by the time-temperature-tolerance hypothesis.

The Change in Metabolic Indicators

As is true of most chemical reactions, the rate of enzyme-controlled reactions usually increases with temperature within the temperature range over which the enzyme is stable and retains full activity. Enzyme-catalysed reactions have an optimum temperature. The peak in a plot of catalytic activity against temperature occurs because enzymes, being proteins, are denatured by heat and become inactive as the temperature is raised beyond a certain point, usually between 55 and 60°C. The apparent optimum temperature is thus the result of two opposing processes; (1) the usual increase in reaction rate following the Arrhenius equation and (2) the increasing rate of thermal inactivation above a critical temperature. So the effect of temperature on any of the individual enzyme-catalysed reactions which, together, affect storage deterioration is as illustrated in Figure 8.3.

Physiological processes such as respiration, which are enzymic and can be used as general metabolic indicators, are affected by temperature in a similar way. Temperature coefficients of respiration,

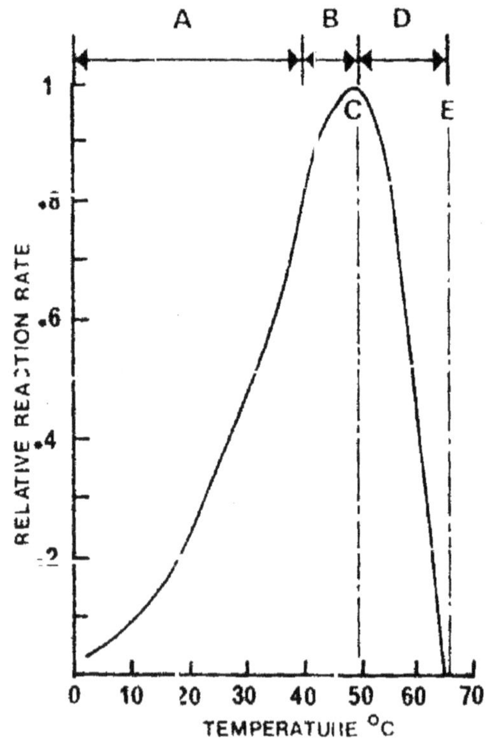

Figure 8.3: The Effect of Temperature on the Rate of Enzyme-Catalysed Reactions

Q_{10} (defined as the ratio of respiration rate at temperature T°C to that at $T + 10$°C) increase less rapidly with increasing temperature than would be expected from the Arrhenius model. It has been suggested that at low temperatures respiration rate is controlled by an enzyme reaction with a high Q_{10} value, whereas at higher temperatures it is limited by the rate of oxygen diffusion through the tissue, a process with a low Q_{10} value. Temperature coefficients of respiration also depend on the tissue age. But James points out that, at higher temperatures, differences between temperature coefficients of similar tissues at different ages tend to disappear: evidence for oxygen diffusion limitation at these temperatures.

During post-harvest storage of vegetables or ripening and senescence of fruits, a great many metabolic changes occur simultaneously, for example, softening of the cell walls, synthesis and destruction of pigments resulting in colour changes, changes in carbohydrate composition; and concentration resulting in changes in sweetness and destruction or polymerisation of phenolics resulting in reduction of astringency. Some of these reactions are metabolically essential to the tissue. Some are secondary or by-products of the mainstream reactions. All have different temperature coefficients. If the fruits or vegetables remain attached to the parent plant or ripen in their natural environment, any possible imbalance of relationships or metabolic reactions resulting from fluctuations of temperature will be corrected by the tissue's hormonal system and by the interdependence of individual reaction rates on the product and substrate concentrations of other reactions. In other words, fruits will mature and senesce normally and vegetables will change only as required by the normal growth of the plant, free of physiological stress despite the metabolic changes taking place. But when the produce is separated from its natural environment, this ability to compensate for temperature change is severely impaired. Temperature changes can now result in the accumulation or depletion of metabolic substrates or products, resulting in considerable changes in those attributes that constitute quality.

When temperature fluctuation occurs, both equilibrium constants and the concentrations of the substrates involved are altered. For instance, the equilibrium constant for the hydrolysis of starch to glucose, a vital process, rises with temperature, *i.e.* high temperatures displace the equilibrium towards starch and low temperatures towards the sugars. Low temperatures also lead to the conversion of lipids into sugars and the accumulation of organic acids. It has been shown that when potatoes are transferred from room temperature to temperatures below 8°C, reducing and nonreducing sugars accumulate, which explains the respiration drift that they suffer when transferred again to room temperature. Upward changes of temperature are always accompanied by a temporary peak in carbon dioxide production when, presumably, these sugars are oxidised. Such effects are particularly important if the potatoes are to be used for frying for, in addition to flavour change, the sugar/ starch ratio influences the degree of browning during subsequent cooking.

Work with apples and pears has shown that a temperature transition from 1 to 20 °C is closely followed by an increase in carbon dioxide production, while production of ethylene–the hormone involved in ripening–rises to a peak after 5 or 6 h (before complete temperature equilibrium is achieved) and then falls. It has been suggested that this increase in ethylene production is related either to the increased activity of an ethylene-producing system (which has developed at low temperatures due to the availability of substrates) or to tissue stress.

When produce, particularly fruit, is exposed to extremes of temperature, either high or low, physical and chemical disorders may occur; membrane permeabilities may be increased or reduced or some reactions may be accelerated or retarded to excess, resulting in permanent disruption of the normal ripening pattern. Due to under-concentration or essential substrates or to excessive accumulation of toxic products, physiological disorders obvious to sight or taste can ensue. A great variety of physiological injuries has been reported to occur in fruits subjected to chilling temperatures, for example internal browning in apples, skin spotting in citrus fruits, damage to vascular bundles in avocados, wooliness in peaches, and flesh breakdown in plums.

Chilling disrupts the entire physiology of some sensitive fruits, particularly those of tropical or sub-tropical origin, and a number of mechanisms have been proposed to account for the effects. It seems likely that cell membranes undergo a physical phase transition at the critical chilling temperature from a normal, flexible liquid crystal to a solid gel structure. This change of state would be expected to bring about a concentration of the membrane components, causing the formation of holes and increased membrane permeability. The phase transition may also increase the activation energy of membrane-bound enzymes, leading to interference with metabolic processes. Reduction in the supply of ATP coupled with the increased rigidity of membranes would account for the cessation of protoplasmic streaming which is one of the immediate symptoms of chilling injury.

Exposure to chilling temperatures, unlike freezing which effects immediate damage, often needs to be prolonged before injury occurs. However, there are exceptions and bananas are damaged by even brief exposures to temperatures below 10°C. In general, the degree of chilling injury increases as temperature is lowered or exposure time is extended at any temperature below the critical one. In some cases, it has been observed that the effects of chilling are ameliorated if the tissue is subsequently transferred to a higher temperature. Analysis has revealed that there is a higher proportion of saturated fatty acids in the lipids of chill-sensitive than of chill-resistant varieties. It is also believed that membranes containing a high proportion of saturated fatty acids are less stable and more liable to undergo phase transition at chilling temperatures.

Time Temperature-Tolerance Hypothesis

The rate at which any individual chemical reaction proceeds varies exponentially with temperature. Although the reactions and processes that affect quality changes in fruits and vegetables are many and varied, an exponential relationship to describe the overall effect of these reactions is still approximately valid, at least over moderate temperature ranges. Such a relationship can be used to relate storage life and temperature. Thorne and Meffert fitted exponential equations to published temperature/storage life data for fresh and frozen fruits and vegetables and obtained very high correlation coefficients. Previously, several authors had reported similar relationships for a number of individual products.

If storage life at any temperature is defined as the time required for various physiological changes to accumulate to such an extent that quality is just unacceptable, then the reciprocal of storage life is the rate at which these changes take place at that temperature. As an exponential curve can often be fitted to observed temperature/storage life data it should be possible to construct a curve of the rate of change (1/storage life) against time for any time-temperature history. The area under this curve up to any time would be a measure of the total change suffered by the produce up to this time. This is the basis of the method known as the time-temperature-tolerance hypothesis, first proposed by Van Arsdel. It is analogous to the method first suggested by Bigelow *et al.* and later modified by Ball for determining the effectiveness of sterilising processes for canned foods: the universal method today. The 'lethal rate' of the Bigelow *et al.* and Ball method is analogous to the reciprocal of storage life in the TTT hypothesis.

The term 'storage life' requires acceptance of that dubious parameter 'quality' in its definition and therefore, lacks nearly all the elements of a satisfactory quantitative measure. Nevertheless, it is possible to derive meaningful relationships between temperature and storage life, especially if a valid quality indicator can be used. Using such data, a graphical procedure for summing the effect of an irregular temperature history on quality or on a quality indicator can be used.

Figure 8.4 indicates a graphical solution of the time-temperature-tolerance hypothesis.

It is unlikely that a quality indicator will be related linearly to temperature (or to time) and it must be linearised by a suitable

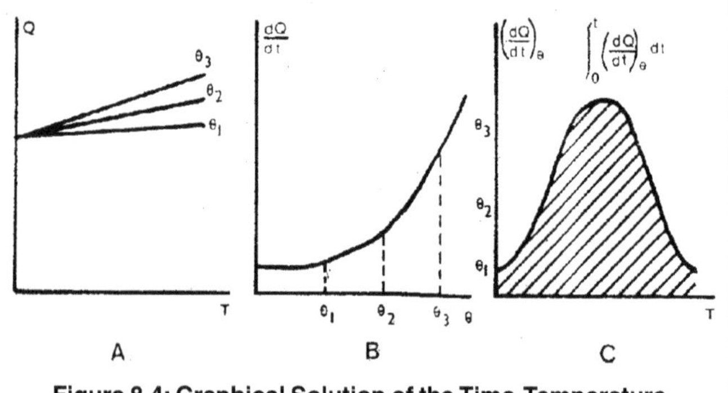

Figure 8.4: Graphical Solution of the Time-Temperature-Tolerance Hypothesis

mathematical function (exponential, logarithmic, etc.) so that the change of the property with time can be calculated from the slopes of these lines at different temperatures. In other words, the quality indicator, or a function of it, must depend only on temperature and never on the quality indicator itself: it must be explicitly dependent on temperature.

Working on the time-temperature-tolerance of frozen fruit and vegetables. Guadagni *et al.* followed the rate of change of ascorbic acid in sliced strawberries and found that this decreased linearly with time at –12°C and could, therefore, be used as an objective measure of deterioration during storage. Similarly, the logarithm of the chlorophyll content of green snap beans has been used as a quality indicator. Segurajauregui Alvarez has used surface colour and firmness to indicate quality in fresh tomatoes.

Approaches to a graphical solution of the time temperature-tolerance hypothesis can be explained as follows:

$$(1/L)_{\text{remaining}} = 1 - \int_0^t (1/L).dt \quad (1)$$

$$Q = Q_0 + \int_0^t (dQ/dt).dt \quad (2)$$

where Q is the value of the quality indicators at any time, Q_0 is its initial value and $1/L$ is the fraction of the storage life in the production. In the second of these approaches, a correlation has to be found between the quality indicator and the point at which the produce becomes unacceptable and there is a difficulty in identifying any quality attribute that might be a liming factor for the produce. Complications in this procedure can arise if two or more deterioration processes are involved; and there usually are at least two. For instance, if the reactions resulting in texture change in a commodity have a greater temperature coefficients than reactions causing colour change, then texture changes could limit storage life at higher storage temperatures whilst discolorations could be the limiting factor at low temperatures.

An example of the calculation will illustrate the time-temperature-tolerance hypothesis and its utility. Storage life (L) of green mature tomatoes can be related to temperature (T) by equation (3):

$$L = 97.\exp(-0.13T) \quad (3)$$

The effect on the quality of tomatoes of the temperature history illustrated in Figure 8.5 is to be investigated. For small time increments along the curve of Figure 8.5, the rate of quality change in the tomatoes ($1/L$) can be calculated from equation (3). A curve of $1/L$ against time corresponding to Figure 8.5 can be constructed. The area under this latter curve is a measure of the total loss of storage life in the tomatoes. Integrating graphically between 0 and 10 days we get:

$$\int_0^{10} (1/L).dt = 0.87$$

i.e. 87 per cent of the tomatoes storage life has been lost and 13 per cent remains. From equation (3) it can be calculated that, for example, the time-temperature history for 10 days was equivalent to 6.7 days storage at a constant 20°C and that the tomatoes could be stored for a further 1 day at 20°C without becoming unacceptable. [From equation (3), $L = 7.7$ days at 20°C, $1/L = 0.13$ day^{-1}].

As already mentioned, the TTT hypothesis rests on three important assumptions: (1) additivity, (ii) commutativity and (iii) non-existence of additional effects due to temperature change itself. Additivity means that the total alteration in properties of a product, produced by a succession of experiences at various temperatures, is the simple sum of the separate amounts of change at each temperature

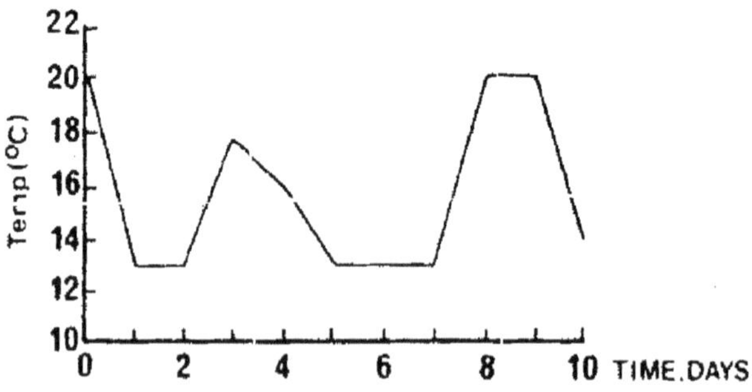

Figure 8.5: Temperature History of Tomatoes

and remains the same regardless of the number of breaks of continuity. Essentially, it means that changes or quality once produced in the food persist and are added to by all other changes that may be produced subsequently; that is, quality changes are irreversible. Commutativity means that the total alteration in properties is independent of the order of presentation of the various temperature experiences. Figure 8.6 illustrates these assumptions. Lines T_1 and T_2 represent quality changes with time at two different temperatures. Produce (I) is stored for 3 days at T_1 and then for 1 day at T_2; produce (II) is stored for 1 day at T_2 and then for 3 days at T_1. If the assumptions of additivity and commutativity are both satisfied, both lots will end up with the same quality after 4 days. Produce (I) will travel along the line T_1 for 3 days, cross to line T_2 without further

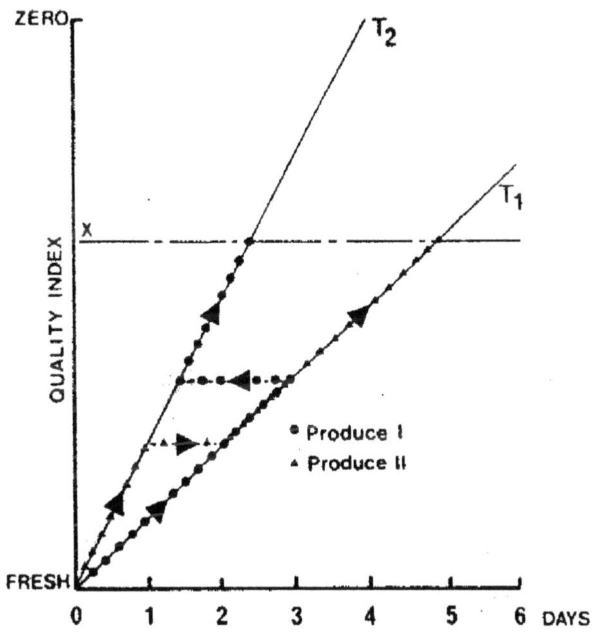

Figure 8.6: Quality Changes for Produce I and II with Different Patterns of Storage Temperature

quality change and proceed along this line for a further day. Produce (II) will proceed along line T_2 for 1 day, cross to T_1 and continue along this line for 3 days. If additivity and commutativity are valid assumptions, both produce (I) and (II) will end up with the same residual quality, as they do in Figure 8.6. The non-existence of additional effects due to temperature change itself implies, in particular, that temperature gradients within the produce created when the ambient temperature changes have no effect.

These three assumptions have been tested in frozen food for several commodities, for example strawberries, sliced peaches, raspberries, orange juice concentrate, peas and green snap beans, spinach, cauliflower, turkey, and chicken. Charm *et al.* have proposed a graphical method for estimating the shelf life of fish at any given set of storage temperatures based on these assumptions.

Because of the greater biochemical and metabolic complexity of fresh produce, application of the temperature-time-tolerance hypothesis is not easy and is often subject to limitations for individual products. Chilling injury is usually outside the scope of the hypothesis as it is neither additive nor commutative. But the method can often be applied within specified limits. Baxter *et al.* have developed exponential equations to predict ripening of pears during and after cool storage by selecting firmness as a quality indicator. Segurajauregui Alvarez has successfully used firmness and surface colour as quality indicators for ripening tomatoes.

The Sprenger Institute presents temperature-quality data for fresh produce in such a way that it can be used as an approximate indicator of the effect of storage at more than one temperature. Superimposed on their quality diagram for lettuce is the effect of storing lettuce for 2 days at 8°C followed by 2 days at 1°C. The remaining quality (*X*) can be calculated from the diagram as being equal to 2½ days storage at 0°C, 1¾ days at 2°C or 1¼ days at 4°C.

Additivity and commutativity are implicit in the use of these diagrams and, although these have not formally been demonstrated for many products, such diagrams have been used successfully commercially. They are, of course, incapable of being used for predicting the effects of frequent or cyclic changes in temperature.

In general, it would appear that the assumptions of additivity, commutativity and the absence of temperature effects are valid for

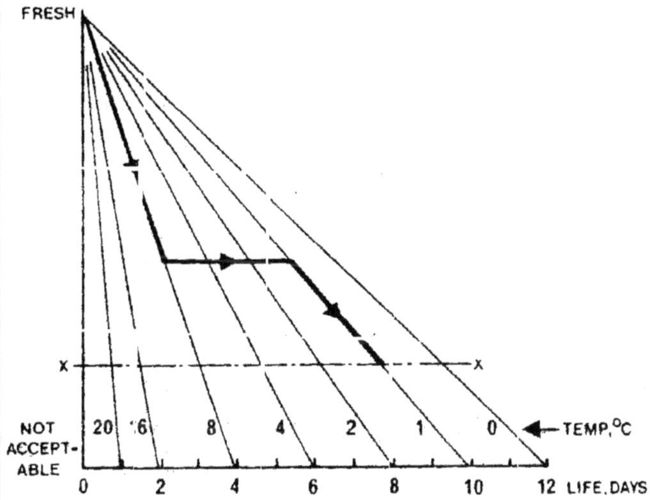

Figure 8.7: Quality Diagram for Lettuce

fresh fruits and vegetables, even during the ripening of climacteric fruits, unless drastic effects such as chilling injury occur.

Although the time-temperature-tolerance hypothesis, in its various guises, is the most widely applied theory to predict loss of quality in produce, a number of alternative theories have been proposed. The most important of these are the 'effective temperature' theory and various computer-aided simulation models.

Hicks Theory

Hicks proposed his theory depending on temperature variation, which is later known as effective temperature theory.

The effective temperature theory proposes that if temperature varies with time according to some regular periodic wave then the extent of a particular reaction or change after a period of time will be the same as it would have been if the material had been held at some constant temperature for the same time. This theory was first proposed by Hicks, who developed an equation for the estimation of the effect of sinusoidal diurnal temperature fluctuation on reaction rates in foods stored at ambient temperatures. It was further developed by Schwimmer *el al.* who analysed different models of regular temperature fluctuation, for example saw-tooth, square and sine waves. Equations relating effective temperature to mean temperature and the wave amplitude temperature coefficient were developed for each model based on a zero-order reaction rate and on the following assumptions:

1. That the entire system fluctuated homogeneously so that analysis of one cycle would give data for the entire process, and
2. That the wave amplitude temperature coefficient was constant throughout the temperature range of fluctuation and throughout the entire time.

The effective temperature theory has been widely tested for a number of frozen products in which deterioration is controlled by specific reactions, for example, rancidity in turkey, acid loss in strawberries, oxidative browning in peaches, migration of acidity from berries to syrup in raspberries, rancidity in chicken, retention of chlorophyll in snap beans, retention of chlorophyll in peas, and ascorbic acid retention and colour in cauliflower. Repeated oscillations of as much as ± 5.6°C (± 10°F) within the range of –23 to –1°C (–10 to + 30°F) and cycle periods of 24–72 h were considered. In all cases studied, deteriorative effects on observed quality were, within the limits of experimental error, the same as they would have been after storage for the same length of time at the predicted steady temperature. Deteriorative reactions were independent of cycle frequency.

The limitations of the theory proposed by Schwimmer *et al.* were discussed by Singh and Wang and these limitations are even more important when considering the possible application of the model to the cool storage of fruits and vegetables. The effective temperature theory assumes a single kind of deteriorative reaction with a constant temperature coefficient. If two or more reactions with different temperature coefficients proceed simultaneously then the overall effect will not be predictable by the theory. Furthermore, it has been shown that temperature coefficients depend on the temperature interval considered, their value decreasing as the temperature interval increases. In the case of fresh fruits and vegetables, temperature coefficients might also depend on the state of maturity and change during storage. Schwimmer *et al.* suggested that when Q_{10} varies with temperature, it should be possible to calculate the effective temperature by stepwise integration over the pertinent temperature ranges. They did not select the Arrhenius model for interpreting the relationship between temperature and reaction rate in frozen foods but considered an exponential model instead. If the logarithm of reaction rate constant is plotted against temperature for both the exponential and Arrhenius model, both follow approximately the same profile if the temperature range considered is fairly small.

Another major assumption of the effective temperature theory is that the reaction rate is constant at constant temperature: this suggests a zero-order deteriorative reaction. The validity of this assumption is supported by Labuza who points out that although many foods deteriorate by first-order reactions, the difference between zero- and first-order is small except for very high degrees of deterioration. A limitation of the theory itself is that it is applicable only to cyclic temperature fluctuations. Cyclic fluctuations may occur in cool storage, but they are usually more complex than the simple functions which may be explained by the effective temperature theory. And the theory is unable to explain the non-cyclic changes that occur during transport and transfer to and from different stores, trucks, etc.

Although the validity of the effective temperature theory has been confirmed for a variety of frozen foods, its application to fresh produce has yet to be investigated.

Browning of Frozen Food

Many workers examined the food quality for frozen and cool storage.

The main objective of these studies has been to incorporate information on laboratory experiments on:

1. The properties of foodstuffs,
2. The kinetics of food deterioration reactions and
3. The properties of packaging protection into a computer model of the entire system.

The extent of browning of frozen and dried cabbage stored in packages made of materials with different water vapour permeabilities was studied by Mizrahi *et al.* The model that they developed was used in accelerated tests of browning in dehydrated cabbage. Quality simulation models have also been proposed for products where two or more simultaneous deteriorative changes are important. Quast *et al.* investigated the storage life of potato chips (crisps), used as a model of a food that deteriorates by two simultaneous mechanisms: oxidation by atmospheric oxygen and textural changes due to water gain or loss. Kinetic information on ascorbic acid oxidation has been used in a computer simulation to predict quality changes in infant-feeding formulae during storage. Factors such as light, dissolved oxygen concentration and package characteristics were included in the model to simulate quality.

Normally in frozen storage, a decrease in temperature is accompanied by an increase in stability of the product. However, some products show an opposite trend below a certain critical temperature. For example, bacon and liver accumulated more free fatty acids when stored at –30°C than at –12°C. Singh developed a computer-aided simulation model that incorporates the effects of storage temperature variations on the stability of frozen food products showing this reversed stability.

An important pre-requisite in computer-aided simulation of stored foods is the determination of kinetic information about deteriorative reactions. This kinetic data can then be incorporated into a program for mathematical model that accounts for any variation in storage conditions. When several quality attributes are considered, the simulation should allow prediction of any one factor that limits quality; the consumer might, for example, reject a product with excessive colour deterioration even though all other quality attributes might still be at acceptable levels.

Analogue computer models for predicting physical and chemical properties of biological materials have also been investigated. Jabbari *et al.* developed a model for apple fruit which predicts the change with time of starch, sucrose, glucose, fructose, malic acid, pyruvic acid, protopectin, soluble pectin, respiration rate and the total carbon dioxide output in storage under various atmospheric compositions and temperatures.

The utility and precision of computer models is limited by two major factors:

1. The model has to be constructed for an individual, cultivar and
2. A very considerable quantity of kinetic information is required to simulate an entire biological system.

Food systems are extremely complex and those involving fresh produce in particular involve many thousands of individual reactions, all with different temperature coefficients of rate and equilibrium. Furthermore, the rate of each reaction depends on the concentrations of its substrates and products, which in turn depend on the rate and equilibrium of other reactions. Not only would this approach involve a prodigious quantity of information, but detailed mechanisms of many reactions and even the reactions themselves are unknown. So practical information must be derived from simplified models based on such information that it is within human ability to obtain. Usually this means that, instead of considering individual reactions, it is necessary to consider the results of many reactions as if they were a single reaction; to consider colour changes or changes in texture for example. Such simplifications mean that there is no universal theory to explain the effects of temperature on deterioration and that there is a considerable empirical element in all theories. Nevertheless, such theories do provide usable information provided that their limitations are appreciated. These limitations usually involve a limited temperature range and restriction of a method to one specified humidity and atmospheric composition. Even when the limitations are observed, unpredicted failures of previously workable methods do occur occasionally because of lack of understanding of the kinetic detail of processes. But generally, if based on adequate experimental information, prediction is sufficiently precise for practical utility.

The most universally useful method for predicting the effect of various temperature histories on produce is the time-temperature-tolerance hypothesis. It is the simplest method, mainly derived from fundamental changes in the food. Great care must be taken to ensure that time-temperature-tolerance data is applied only to the cultivar for which it was derived, that conditions of atmospheric composition, humidity, etc. are not changed and that the data are not extrapolated to temperatures beyond the original experimental range. Given these limitations, the time-temperature-tolerance hypothesis, in its various guises, is a workable theory.

It is doubtful that even a theory based on fundamental data would be completely foolproof, for it could never account for all variations that occur from batch to batch even with the same cultivar of any one fruit or vegetable. Behaviour depends on, among other considerations, time of planting and harvest, soil and weather conditions. Post-harvest storage behaviour can also vary significantly from one end of a field to the other.

Effective temperature theory is one step nearer to considering fundamental changes, but it is limited to conditions of regularly cycling temperature. Computer simulations can come nearest to a fundamental approach, provided that adequate and sufficiently accurate information can be provided: if such information is inaccurate it can easily induce greater inaccuracies than the gross assumptions of the time-temperature-tolerance hypothesis.

Prediction of the effect of temperature on stored produce is imperfect. Doubtless methods will improve and doubtless they will never be perfect. But provided the limitations of methods are appreciated, useful practical information can be obtained.

Chapter 9

Preservation of Fruits and Vegetables

The use of modified atmospheres for preservation of fresh fruits and vegetables can probably be traced back even earlier in history.

An ancient Chinese poem describes the fondness of an empress for lychees in the Tang Dynasty in the 8th century. She demanded that freshly picked lychees be transported from southern China to her in Chang-An, now Xi-An, approximately 1000 km away by continuous horseback riding, the fastest possible way through rugged terrain at that time. It was said that the carriers discovered that lychees would keep better if the fruit were sealed inside the hollow centers of bamboo stems along with some fresh leaves. However, the poem did not explain why. We know today that the atmospheres within the bamboo centers, modified by the respiration of lychees and the fresh leaves, were probably the key factor responsible for keeping the freshness of those lychees.

Tremendous progress has been made during the past 60 years on the technology of controlled atmosphere (CA) storage. CA storage has withstood the test of time to prove that it is effective for maintaining quality and extending storage life of a number of horticultural crops. Each type of fruit or vegetable has its own specific requirement and tolerance for atmosphere modification. The optimum CA condition for each commodity is continually being revised and improved.

The beneficial effects of CA on specific crops will be described in other chapters. This chapter will be confined to discussion of the physiological and biochemical effects of CA. There have been several reviews and conferences on various aspects of CA during the past few years. For additional discussion of the physiological and biochemical basis for CA effects, reference should be made also to Burton, Do and Salunkhe, Kader, Knee and Sharples, Metlitskii *et al.*, and Ulrich.

It has been well documented that the rate of respiration of fresh fruits and vegetables is markedly affected by low O_2 or high CO_2 concentrations in the atmospheres. The pioneer work of Kidd and West, circa 1930, has established that the onset of the climacteric rise in apples and pears was retarded by lowering the O_2 concentration or raising the CO_2 level. Subsequently, reduction in respiration rate and delay of climacteric peak by CA has been demonstrated in many commodities. Solomos reported that the rate of respiration of preclimacteric bananas was reduced by about 30 per cent in 3 per cent O_2 and respiration of sweet potatoes gradually decreased with decreasing O_2 concentration in the atmosphere down to 2 per cent. He suggested that the decrease in respiration rate in response to low O_2 may be the result of a diminution of activity of an oxidase which has a five to six times lower oxygen affinity than cytochrome oxidase.

The respiration quotient (RQ), which measures the ratio of the volume of CO_2 produced to that of the O_2 consumed, usually gives an indication of the type of substrate being metabolized as the main source of respiration. The RQ generally increases with the ripening and senescence of most fruits and vegetables. An increase in RQ indicates an increased use of organic acids rather than carbohydrates or fatty acids as the major substrate for respiration. Organic acids have more O_2 per carbon atom than sugars or fatty acids, therefore requiring less O_2 consumption for the production of CO_2.

Addition of organic acids to the tissues of postclimacteric fruit induced a large increase in CO_2 output, but induced little change in

O_2 uptake, thus increasing the RQ values. This phenomenon was not observed in preclimacteric fruit, indicating an increase of oxidative, decarboxylation of organic acids with senescence in fruit. It has been found that the RQ in apples decreased markedly in gas mixtures containing high CO_2 and low O_2 or high CO_2 alone compared to that in the air storage. This may possibly be due to an increase in the fixation of CO_2 into organic acids in fruit. A reduction in RQ value by CA also indicates a retardation of the ripening and senescence of fruit and a reduced catabolism of organic acids, resulting in higher acid retention under CA conditions.

Closely related to the respiration rate is heat generation. Since the respiratory rate is generally suppressed by CA, it is conceivable that crops stored under CA evolve less heat. Data collected by Toledo *et al.* on peas, lima beans, sweet corn, and apples suggest that heat of respiration of fresh produce under optimum CA condition was about 30 per cent of that in air. This information is useful in computing refrigeration load requirements for CA storage.

The Role of O_2 and CO_2 in Ethylene Biosynthesis

Recent researches have shown that the importance of steps in the biosynthetic pathway which require presence of oxygen.

When oxygen is absent or when plant tissue is under anaerobic atmosphere, ethylene biosynthesis ceases. This phenomenon was first observed by Gane and Hansen in apples and pears. In 1942, when Hansen transferred Bartlett pears from air to N_2, H_2, or He atmospheres, ethylene production was either greatly retarded or totally inhibited. However, when the fruit were transferred back to air, ethylene production showed a rapid increase to above the maximum rate for the control fruit. It was later suggested by Burg and Thimann that anaerobic treatment caused the accumulation of an ethylene precursor which could be converted to ethylene very rapidly in the presence of O_2. In 1979, Adams and Yang examined the metabolism of methionine in the absence and presence of O_2 and discovered that 1-aminocyclopropane-1-carboxylic acid (ACC) was the immediate precursor of ethylene. They demonstrated that methionine was first converted to S-adenosylmethionine which in turn fragmented to ACC and methylthioadenosine. ACC accumulated in tissue under nitrogen, but was efficiently converted to ethylene when the tissue was incubated in air. Concurrently, but independently, Lurssen *et al.* also confirmed that ACC was the intermediate between S-adenosylmethionine and ethylene. Incubation of apple tissue slices in 6 per cent or lower O_2 caused accumulation of ACC and reduction of ethylene production.

Another step in the ethylene' biosynthetic pathway which is O_2 dependent is in the recycling of 5-methylthioadenosine (MTA) to methionine between 5-methylthioribose-1-phosphate (MTR-1-P) and 2-keto-4-methylthiobutyrate (KMB). Figure 9.1 shows the two steps where O_2 is required in the methionine cycle and between ACC and ethylene in the ethylene biosynthetic pathway.

The effect of low O_2 on the capacity of tissue for ethylene biosynthesis is dependent on the O_2 concentration. The time of onset of ethylene production in apples was inversely related to O_2 concentration and the maximum rate of production was directly related to O_2. Apples kept in 3 per cent O_2 produced two to three times less ethylene than those kept in air. Mapson *et al.* showed that at O_2 tensions of 10 to 13 per cent ethylene biosynthesis in bananas was delayed. When O_2 tension was reduced to below 7.5 per cent, bananas did not synthesize significant amounts of ethylene. Hesselman and Freebairn found that reducing O_2 levels below 10 per cent caused similar proportional reductions in rate of banana ripening down to 2.5 per cent O_2. The delay in the onset of ripening of banana fruit in these studies was attributed to the inability of fruit tissue to synthesize ethylene in low O_2 atmospheres. By transfer of the fruit from 5 per cent O_2 to air, the normal rate of ethylene production was resumed. However, after 11 d at 18°C in 1 per cent O_2, bananas lost their capacity to produce ethylene or acceptable flavor and sweetness, even after transfer to air.

Figure 9.1 indicates about the steps, where oxygen is required in ethylene biosynthetic pathway.

Now let us try to understand the role of CO_2 on ethylene production.

Elevated concentrations of CO_2 can also suppress ethylene production, though it is not clear as to the mechanism involved. Reduction in ethylene production by high CO_2 has been reported in apples, pears, avocados, tomatoes and sweet peppers. A close association was also found between CO_2 level in the storage chambers and the amount of ethylene evolved by the fruits in the

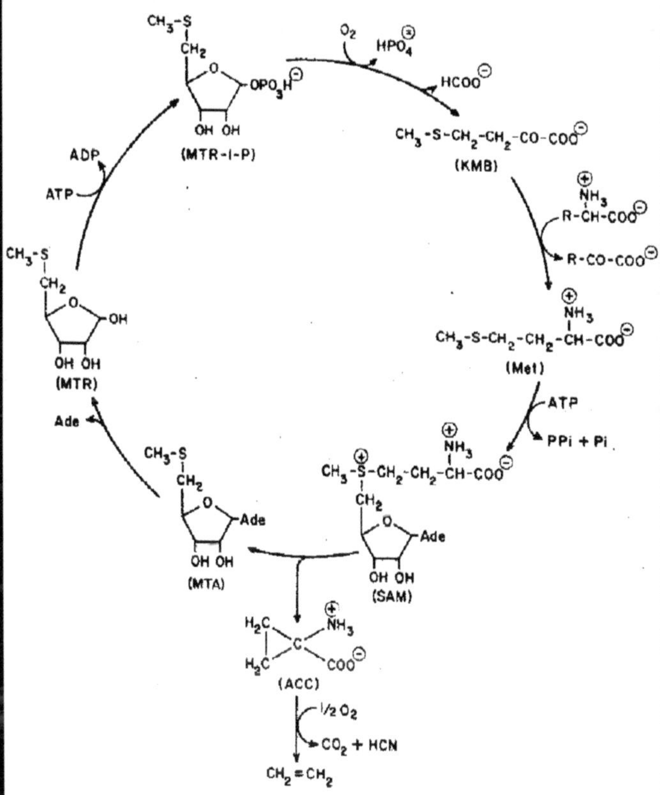

Figure 9.1: The Steps, where Oxygen is required in Ethylene Biosynthetic Pathway

chambers. Ethylene production in sweet peppers was greatly suppressed by CO_2 but rapidly recovered after transfer to air. Intermittent exposure of tomato fruit to high CO_2 and air caused an oscillating pattern of ethylene production. Ethylene declined in the presence of CO_2 but increased in air. Intermittent treatment of Spartan and McIntosh apples with high CO_2 also reduced internal ethylene level, maintained juiciness and flavor, and delayed softening.

Li *et al.* found that when preclimacteric apples were treated continuously in elevated CO_2 (20 to 30 per cent) or low O_2 (1 to 3 per cent) at 15 to 20°C, there was an initial inhibition of ACC synthesis, which was then followed by a substantial accumulation of ACC. Delay of the opening process by low O_2 and high CO_2 might have caused the initial inhibition on ACC formation, but the inhibitory effects of these CA treatments on the conversion of ACC to ethylene eventually resulted in the buildup of ACC in the tissue. Similarly, Chaves and Tomas observed an inhibition of ethylene production and an accumulation of ACC in apple tissue under high CO_2 atmosphere and concluded that CO_2 reduced ethylene production by inhibiting the enzyme system responsible for the conversion of ACC to ethylene. These results are not consistent with the view that CO_2 may activate the enzyme involved in the ACC conversion to ethylene and stimulate ACC-dependent ethylene production.

These seemingly contradictory observations can be explained by the differences in the response of different types of tissues to CO_2. Inhibition of ethylene production by CO_2 primarily occurred in fruits, while stimulation of ethylene production was mostly observed in photosynthetic vegetative tissues or leaf disks. In addition, the concentrations of CO_2 producing an inhibitory effect on ethylene production were much higher than those inducing a stimulatory effect. Carbon dioxide may be required to enhance the enzyme system for the conversion of ACC to ethylene, but concentrations higher than the saturation point (about 0.1 per cent) are no longer necessary for stimulation. In the ripening fruits or other climacteric-type tissues, endogenous CO_2 concentrations produced by respiration often exceed what is required for the activation of the enzyme; therefore, addition of CO_2 exogenously may not have any stimulatory effect on the activity of the existing enzyme.

Burg and Burg postulated that ethylene exerts its action by attaching to a metal-containing receptor through a noncovalent bond and this attachment is competitively inhibited by CO_2. The antagonism between CO_2 and ethylene has been demonstrated in several reactions. Oxygen is also required for ethylene action. Oxygen concentration affects binding of ethylene to the receptor site. The receptor has an O_2 affinity closely similar to that of cytochrome oxidase.

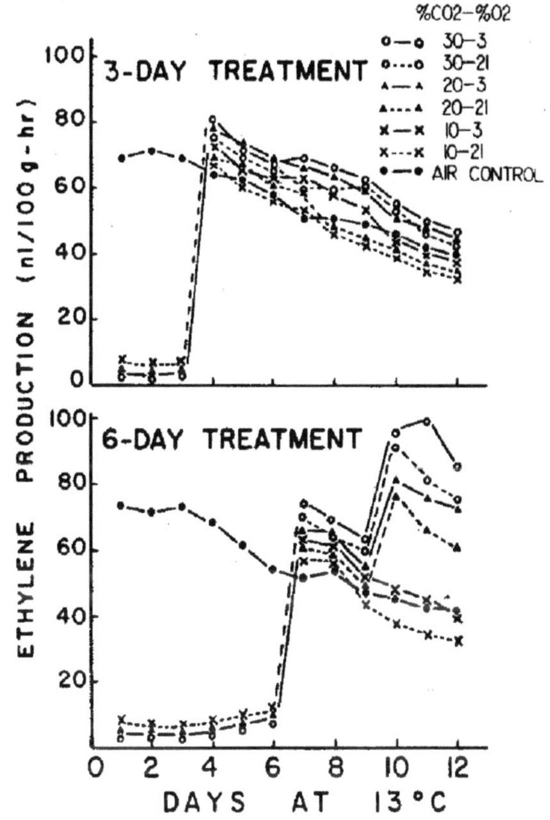

Figure 9.2: Suppression of Ethylene Production by CO_2 and Recovery of Ethylene Evolution after Removal of Sweet Peppers from Elevated Levels of CO_2

The autocatalytic production of enzymes involved in ethylene biosynthesis, including, ACC synthase and ethylene forming enzyme, may be inhibited by the antagonistic effect of high CO_2 atmospheres. This indirect inhibitory effect of CO_2 on ethylene biosynthesis through the retardation of ethylene action requires time and does not occur as instantly as the stimulatory effect of CO_2 on enzyme activity.

Lau *et al.* treated Golden Delicious apples with a "prestorage high CO_2" and found that treated fruit had much less accumulation of ACC and lower internal ethylene level than air-stored fruit. This slower rate of ethylene biosynthesis and accumulation of ACC was closely associated with firmness retention of the fruit. High CO_2 undoubtedly inhibits ethylene action, suppresses general metabolism, and delays onset of the ripening process. Consequently, all reactions associated with ripening are delayed, including rise in respiration, autocatalytic ethylene production, rapid acid catabolism, synthesis of ripening enzymes, softening, and changes of pectic substances in the cell wall.

Higher organic acid retention in CA-stored apple and pear fruits than in air-stored fruits has been reported by Kollas and Li and Hansen. Numerous other reports have also shown that the decrease in titratable acidity during cold storage can be markedly inhibited by CA. Reduction in the loss of acidity in CA-stored apples and pears is mainly due to the reduced loss of malic acid. However, other organic acids can also be affected by CA storage, including citric, succinic, citramalic, shikimic, quinic, oxalacetic, pyruvic, tartaric, and α-ketoglutaric acids. This decreased loss of organic acids in CA fruit has been suggested to be due to an increase in CO_2 fixation, an inhibition of the respiratory metabolism, and a lower consumption of the acid under CA.

Allentoff *et al.* showed that CO_2 was fixed mainly into malic acid in apples. Young and Biale suggested that the first product of CO_2 fixation by lemon fruit in the dark was oxalacetic acid, but it is converted rapidly to malic, citric, and aspartic acids. Citric acid was found to be higher in fruit from air storage than those from CA storage. The production, of citric acid may be oxygen dependent and quinic acid may convert to citric acid upon oxidation. Sweet cherry fruit stored at high CO_2 concentration had a lower amount of tyrosine and higher amounts of α-aminobutyric and malic acids than fruit stored in air.

Oxidation of malate to oxalacetate is catalyzed by malate dehydrogenase which is a pyridine-linked enzyme regulated by NAD^+. The hydrogen atom removed from malate by the enzyme is transferred as a hydride ion which attaches to the pyridine ring and yields the reduced form of the coenzyme NADH. The rate of this oxidation-reduction process, involving malate dehydrogenase is important because malic acid is one of the main organic acids in most fruits and vegetables and malic enzyme activity has been demonstrated to regulate both the oxidative decarboxylation of malate and the fixation of CO_2 into malic acid. Murata and Minamide showed that the incorporation of atmospheric CO_2 into organic constituents, including malic, citric, and pyruvic acids, was greater in apples stored in high CO_2 than in air. The highest C activity was observed to be in the outermost part of the fruit.

Increase in succinic acid levels in the fruit tissue has been correlated to CO_2 injury and CO_2-induced core breakdown in apples and pears in CA storage. A small increase in succinic acid in Bartlett pears was observed at a CO_2 concentration of 2.5 or 5 per cent, while a large amount of the acid along with CO_2 injury symptoms was detected in fruit exposed to 10 or 15 per cent CO_2. The accumulation of succinic acid is thought to be caused by an inhibition of the activity of succinate dehydrogenase by CO_2 and a lower conversion of succinic acid into fumaric acid. Ranson *et al.* showed that succinate and aminobutyrate accumulated, whereas malate, aspartate, glutamate, and alanine were depleted; the conversion of succinate into malate and fumarate was markedly inhibited by CO_2 at concentration above 10 per cent. However, the conversion of fumarate into malate was not affected. At higher CO_2 concentrations, conversion of pyruvate into citrate and other tricarboxylic acid cycle acids was also depressed, suggesting an inhibitory effect of CO_2 on pyruvic oxidase or condensing , enzyme. Similarly, McGlasson and Wills found that low O_2 limited the operation of the tricarboxylic acid cycle between pyruvate and citrate, and 2-oxoglutarate and succinate in green bananas.

Shipway and Bramlage reported that CO_2 can have a strong controlling effect on apple mitochondrial activity. They found that CO_2 suppressed succinate, fumarate, pyruvate, α-ketoglutarate, citrate, and NADH oxidations, but stimulated malate oxidation. The wide range of effects by CO_2 was attributed not to a single enzyme system, such as succinate dehydrogenase, but to structural or conformational changes of the mitochondria.

Organic acids may be utilized as a substrate and as a source of energy for respiratory metabolism. Some organic acids can be replenished by amino acids or fatty acids. For example, α-ketoglutaric acid and oxalacetic acid may be generated from glutamate or aspartate by transaminase reactions. Acetyl-CoA from fatty acids may combine with glyoxylate or oxalacetate to form malate or citrate, respectively, through the action of malate synthase and condensing enzymes. Therefore, study of organic acid metabolism as affected by CA requires a thorough investigation of the interaction between various metabolites in order to have a better understanding of this dynamic system of the living tissue.

Biochemical Changes Under CA Storage

Storage in CA apparently has a profound effect on the entire metabolism of harvested fruits and vegetables.

Total amino acids in head lettuce decreased after 1 month of storage at 0.5°C in air, while no change in concentration occurred in CA storage (1 to 10 per cent O_2 and 0 to 6 per cent CO_2). In Chinese cabbage, total amino acid content increased considerably after 5 months of storage at 0.5°C under 3 per cent O_2, whereas the content remained constant or decreased in air storage. After 19 d of storage at 8°C, cucumbers kept at 5 per cent O_2 or 10 per cent O_2 and 10 per cent CO_2 had higher content of amino acids, particularly glutamine, the predominant amino acid, than those kept in air.

Reduction in loss of ascorbic acid in CA storage has been reported in many fruits and vegetables. Figure 9.3 shows higher retention of ascorbic in Chinese cabbage in 1 per cent O_2 than in air. However, in some cases, CA treatment results in lower ascorbic acid content. For example, high CO_2 treatment of sweet peppers remarked the increase of ascorbic acid which was associated with ripening

Retardation of changes of protopectin to soluble pectin by CA was shown in Cox's Orange Pippin, Golden Delicious, Jonathan, and Roter Boskoop apples. A higher concentration of total pectins and tannins was retained in CA-stored apricot and peach fruits than in conventional, refrigerated stored fruits.

Bartlye showed that phospholipid content in apples increased more slowly in low O_2 storage than in air storage. However, a rapid accumulation of phospholipid occurred following transfer of the fruits from low O_2 to air. An increase in phosphatidylcholine concentration accounted for most of the observed phospholipid changes. These data indicate that phospholipid metabolism is influenced by oxygen concentration. It was suggested that the O_2

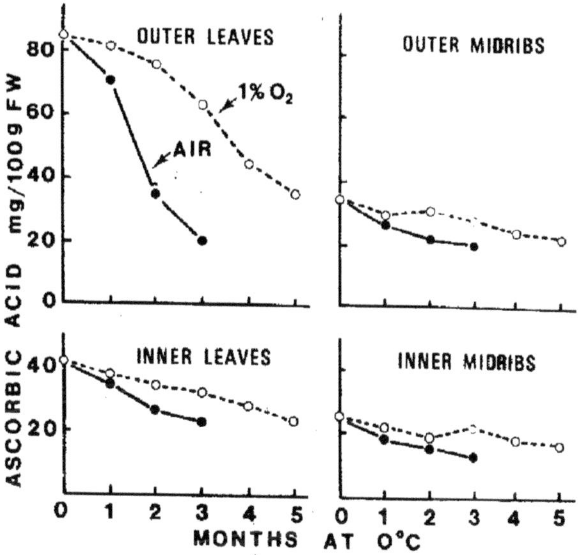

Figure 9.3: Changes of Ascorbic Acid in Various Parts of Chinese Cabbage During Storage in 1 per cent O_2 and in Air at 0°C

effect may be related to the adenylate energy charge which was proposed to be a regulator of cellular metabolism.

Now let us try to understand about the influence of carbohydrate retention on Chinese cabbage.

In Chinese cabbage, the loss of sugars, including fructose, glucose, and sucrose after 3 months of storage at 0°C in 1 per cent O_2 atmosphere was found to be significantly lower than those in air storage. In sugarbeet, the decreased loss of sucrose was attributed principally to the inhibition of respiratory by the high CO_2, while the increase in reducing sugars in CA condition was thought to be due to their lowered utilization in the biosynthesis of raffinose. A slower decline in total sugars was also reported in CA-stored apricots and peaches than in air-stored fruits. In tomatoes, the inhibition of starch degradation and subsequent sugar formation by low O_2 was more distinct in the early than in the late stages. In carrots, the ratio of sucrose to monosaccharides was found to decrease with decreasing atmospheric oxygen concentration. Since this ratio was considered to be positively correlated to the storage ability of carrots, a lower value may indicate a negative effect of CA on storage quality.

Influence of Many Factors on Food Storage

The activities of lipase, ribonuclease, and acid phosphatase in apple peel tissue rise during the climacteric period. A positive correlation was also found between chlorophyllase activity and the climacteric rise in respiration in apples and bananas. The increase in activities of cellulase, acid phosphatase, and polygalacturonase during ripening of avocado fruit in 2.5 per cent O_2 was suppressed 50 to 60 per cent as compared to those ripened in air. Introduction of 100 µl/l of ethylene into the 2.5 per cent O_2 atmosphere did not reverse the inhibition of the activities of these enzymes. Softening of avocados generally occurs within 2 to 3 d after polygalacturonase reaches its maximum activity and pectinesterase reaches its minimal activity. Under ideal CA storage conditions, there was very little change in either pectinesterase activity or firmness during storage. An increase in β-galactosidase activity coincided with the loss of galactose residue from the conical cell walls of apples during storage in air, whereas this increase in enzyme activity was delayed in 2 per cent O_2.

Failure of pears to ripen after prolonged cold storage was postulated to result from the loss of capacity of the tissue to synthesize ripening enzymes. Storage of Anjou and Bartlett pears under 2 to 2.5 per cent O_2 and 2 to 3 per cent CO_2 suppressed the increase of protein nitrogen during storage at 0°C and preserved the ability of fruit tissue to synthesize additional protein nitrogen upon transfer of the fruit from CA storage to a ripening temperature. Anjou pears kept in conventional refrigerated air storage did not show an increase in protein nitrogen and did not ripen normally after transfer to ripening temperature following 5 months of storage. The protein nitrogen induced at 21°C following storage at –1°C in CA-stored pears was probably related to enzymes necessary for normal ripening. Increases in free amino acids in CA-stored fruit were also reported by Littlefield.

The inhibition of polyphenol oxidase activity by high CO_2 or low O_2 has been found in lettuce, mushrooms, and snap beans. Since

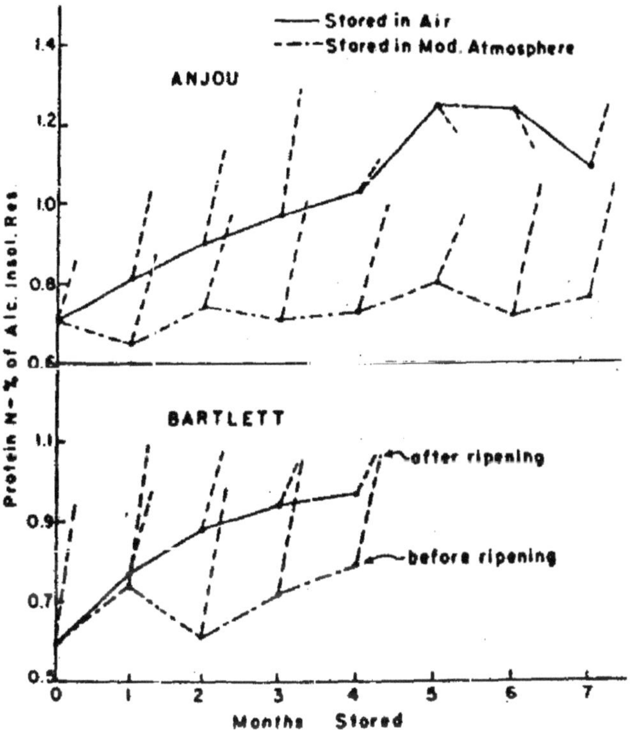

Figure 9.4: Changes of Protein N of Anjou and Bartlett Pears During Ripening at 21°C (Dotted Lines) After Storage at –1°C in Air and in CA

polyphenol oxidase may oxidize phenolic compounds to form quinones, a reduction in activity of this enzyme by CO_2 also prevents browning of these tissues. Carbon dioxide also prevented development of cinnamic acid-4-hydroxylase in lettuce and potato tissues. This enzyme, however, does not appear to be, involved in the browning process.

The inhibition of succinate dehydrogenase and the accumulation of succinic acid by high CO_2 have been found in a number of plants and isolated mitochondria. This effect has been associated with CO_2 injury as mentioned earlier.

Goodenought and Thomas studied isoenzymes involved in the glycolysis and tricarboxylic acid cycle in tomato fruit and found that hexokinase, gucose-6-phosphate dehydrogenase, NADH, and NADPH dehydrogenase were affected during long-term storage in CA. Other enzymes activities which have been reported to be inhibited by CA include the specific catalytic activity of esterase and phosphoenolpyruvic carboxylase.

Storage in CA generally maintains better flavor for most commodities. Prolonged exposure to certain CA conditions, however, may lead to off-flavor. Although alcohols and aldehydes contribute part of a complex volatile mixture to a typical aroma of many fruits, excessive accumulation of these substances, as often occurs after prolonged storage or in anaerobic atmospheres, may induce off-flavor or physiological disorders and lead to an unacceptable quality. Various studies have indicated that production of volatiles by fruits can be affected by different storage conditions and the fruits may lose their characteristic flavor after prolonged storage in certain CA conditions. It has been suggested by Bangerth that decrease in the sensitivity of fruit to ethylene after long storage may contribute to the poor flavor. Others have shown that lack of flavor and loss of aroma probably were due to abnormality in the synthesis of volatile precursors.

Deficiencies in volatile ester production have been noted in fruit from CA storage. The deficiency in ester formation may be an important factor influencing the eating quality of fruit. Volatile compounds important to apple flavor including butyl acetate, hexyl acetate, and estragole formed slowly and reached a lower maximum amount during ripening after prolonged CA storage as compared to a shorter CA storage period. Storage of McIntosh apples in atmospheres of 1.5 per cent O_2 and 1.5 per cent CO_2 or 1 per cent O_2 and 1.5 per cent CO_2 at 2.8°C significantly suppressed production of hexanol, ethyl butyrate, acetaldehyde, and ethanol from intact fruit. This reduction in the volatile production contributed to the lower intensity of the characteristic McIntosh flavor which was detected in the low-O_2-stored fruit by sensory taste panelists. The capacity of apples to produce ethyl butyrate and hexanal can be regenerated upon returning the fruit to air. However, after prolonged storage in low O_2 atmospheres, the capacity of fruit for subsequent regeneration or volatiles in air may be blocked.

When Cox's orange Pippin apples were stored in 2 per cent O_2 at 3.5°C, they gradually lost their capacity to ripen normally and when these apples were transferred to air at 20°C, they produced much smaller amounts of volatiles with the exception of ethylene than air-stored fruits. The diminished volatile production was thought to be due to loss of substrates or enzymes essential to the formation of esters during CA storage. When exogenous ethanol and butanol were supplied as substrate to fruit slices, aldehydes, acetate esters, and butyl butyrate were formed, indicating that these alcohols might be the precursors for apple flavor. Knee and Hatfield reported that the butanol content of Cox's Orange Pippin apples stored in 2 per cent O_2 was lower, but esterification of butanol proceeded at rates similar to those of air-stored fruit. They concluded that the lack of esters in low O_2-stored fruit was a result of lack of alcohol supply and not because of absence of an esterifying system.

One of the hazards of CA storage is injury caused by insufficient oxygen in the atmosphere. In an atmosphere lacking molecular oxygen, glycolysis serves as an important emergency mechanism to provide energy. Fresh fruits and vegetables shift to this anaerobic degradation of glucose as a major pathway for generating ATP for energy source when oxygen in the atmosphere becomes critical. In aerobic respiration, fresh fruits and vegetables obtain their energy from oxidation-reduction reactions and oxygen serves as the final electron acceptor in respiration. Under anaerobic conditions, however, oxygen is not available and, electrons are passed from one organic intermediate of sugar breakdown to other organic, metabolites and ultimately yields products of fermentation, such as lactate, acetaldehyde, and ethanol. As a result, fruits and vegetables produce off-odor and off-flavor or physiological disorders, particularly when these anaerobic products accumulate in large quantities in the tissue. Off-odor and off-flavor induced by low O_2 or high CO_2 can dissipate if the commodity is returned to air after short exposure and aerobic respiration is resumed. However, off-odor and off-flavor will persist even after aeration, if the damage caused by anaerobic products becomes irreversible after prolonged exposure.

Figure 9.5 explains how to identify the anaerobic enzymes.

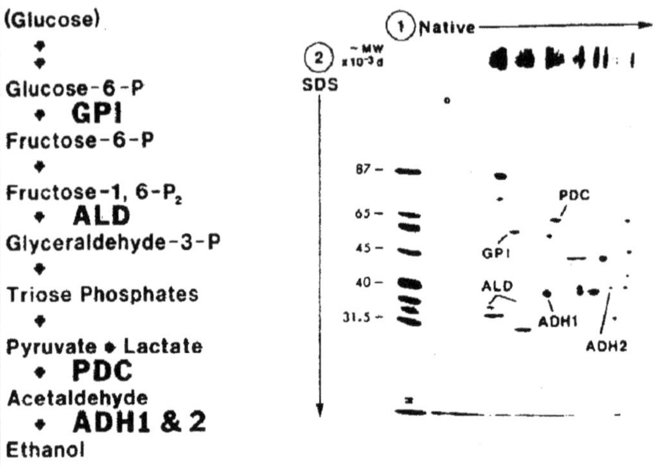

Figure 9.5: Identify the Anaerobic Enzymes

One of the most disturbing physiological disorders which develops in apples, pears, and certain other fruits during and after storage is superficial scald. Various contributing factors during the growing season predispose the fruit to develop this disorder. Preharvest weather conditions and fruit maturity have a direct bearing on susceptibility of fruit to superficial scald. Post-harvest handling and storage conditions also can affect the severity of scald. Storage of fruit in CA, particularly in a low O_2 atmosphere, has been shown to reduce the susceptibility of fruit to scald.

An unsaturated sesquiterpene hydrocarbon, α-farnesene, has been suggested to play an important role in causing scald. Higher concentrations of α-farnesene in the cuticle and adjacent cells of apple fruit have been detected in affected fruit. Huelin and co-workers found that earlier-picked apples had more α-farnesene and had higher susceptibility to scald. Also, varieties which were more liable to scald contained more farnesene. They also showed that diphenylamine delayed oxidation of α-farnesene and was an effective inhibitor of scald. Subsequently, it was shown that the accumulation of oxidation products of α-farnesene was responsible for the appearance of this disorder.

It has been reported that the dynamics of α-farnesene was greatly affected by storage conditions. CA retarded scald through its reduction of α-farnesene concentration and its, oxidation production. The rate and level of accumulation of α-farnesene and its products in an atmosphere containing 3 per cent O_2 and 5 per cent CO_2 were lower in comparison to that in air. However, some investigators have observed that although scald developed more rapidly in air than in CA, changes of α-farnesene content and the accumulation of α-farnesene oxidation products in apple skins proceeded at similar rates in air and in CA storage. Thus, CA may also reduce the sensitivity of fruit tissue to α-farnesene and its oxidation products.

The effects of CA on chilling injury vary with different commodities. CA reduces chilling injury in certain sensitive crops, while it aggravates or has no effect on the chilling symptoms in others.

Holding zucchini squash in 4, 2, or 1 per cent O_2 ameliorated chilling injury during storage at 2.5°C. Conditioning grapefruit with short-term prestorage treatment of high CO_2 (40 per cent), at 21°C has been shown to reduce brown staining and rind pitting, two types of chilling injury, at 1°C. Intermittent exposure of unripe avocados to 20 per cent CO_2 reduced chilling injury at 4°C. Addition of CO_2 to the storage atmospheres was also effective in reducing chilling-induced internal breakdown and retaining ripening ability of peaches.

An atmosphere of 2 per cent O_2 and 10 per cent CO_2 prevented development of anthracnose (*Colletotrichum gloeosporioides*) at 7.5°C and reduced chilling injury at 4.4°C in Fuchs avocados. Reduction of chilling injury by increased CO_2 levels before or during chilling also has been reported for okra. It was suggested that CO_2 may reduce chilling injury of grapefruit and avocados by limiting the breakdown of the ATP/ADP energy transport system during chilling.

CA is not always beneficial in reducing chilling injury. In some crops, metabolic imbalance induced by chilling stress can probably be aggravated by CA-induced alteration of normal functions. Increasing levels of CO_2 accentuated symptoms of chilling injury in cucumbers. Various combinations of O_2 and CO_2 concentrations also did not reduce chilling injury of cucumbers. The severity of chilling injury in tomatoes held at 0 or 5°C for 10 or 15 d under 2.5 to 50 per cent O_2 or under 5 to 70 per cent CO_2 was comparable to that exhibited by tomatoes held in air, and the elevated concentrations of CO_2 induced CO_2-injury symptoms similar to those of chilling injury.

The use of CA to control chilling injury needs to be evaluated for each commodity under a given circumstance. It appears that the efficiency of CA in ameliorating chilling injury symptoms is dependent upon the variety, the concentrations of O_2 and CO_2, the timing and duration of treatment, and the storage temperature.

The development of a brown discoloration and a dry breakdown in and around the core and adjacent cortex tissue of fruits has been associated with certain CA conditions during cold storage. Kidd and West first described the brown-heart symptom in apples as the result of high CO_2 injury at low temperature. Subsequently, brown discoloration in the core has been observed in Sturmer Pippin apples, McIntosh apples, Jonathan apples, Conference pears, Bartlett pears, and Anjou pears in CA storage, all related to the excessive concentration of CO_2. It was reported that Anjou pears packed in sealed polyethylene box liners developed brown core after the buildup of CO_2 in the polyethylene bags. However, the disorder did not develop under the low O_2 condition. Hydrated lime packaged within polyethylene box liners maintained CO_2 concentration below 3 per cent and reduced brown core injury in pears.

Combating Insect Infestation

Many chemical treatments, such as ethylene dibromide are used to combat insect infestation, on fresh fruits and vegetables.

In addition to its capacity to disinfect the produce of insects and microorganisms, ionizing radiation has been shown to delay fruit ripening, to extend shelf life, and to inhibit sprouting of potatoes and onions.

Treatment of fresh produce with doses of up to 100 krads of ionizing radiation has been approved as a commercial process. With growing government and industry interest in the use of radiation for food preservation, it is timely to review the compatibility and possible interaction of irradiation with established post-harvest practices. We are particularly concerned with the effects of irradiation on the quality of CA-stored produce and the effects of CA on irradiated produce.

The effects of γ-irradiation on biochemical changes in apples during conventional refrigerated storage and CA storage have been investigated. Okamoto *et al.* irradiated apples with γ-rays at the dose of 5 krad prior to CA and air storage and found that the respiration rate was substantially suppressed. Some effects of γ-irradiation on chemical properties of CA-stored apples, such as decrease in acidity, were detected by Saito and (Igarashi at a dose as low as 50 krad. However, the changes of other components, such as vitamin C and nonprotein nitrogen contents, in fruit during 6 months of storage in 3 to 15 per cent O_2 and 2 to 2.5 per cent CO_2 at 0°C were not affected by the irradiation up to 200 krad. Saito and Igarashi concluded that γ-irradiation resulted in only slight or limited improvement, of quality for CA-stored apples and was not considered to be an effective treatment for apples intended for CA storage.

Shrikhande and Kaewubon studied the effects of CA at different levels of O_2 and CO_2 during and following irradiation on fruit quality and reported that some beneficial effects could be obtained by irradiating fruits and vegetables under CA conditions. They found that ethylene production by irradiated lemons was oxygen dependent, and low O_2 suppressed the radiation-induced ethylene production and its subsequent action in degreening of lemon fruit. The sensitivity of the irradiated fruit to ethylene decreased at lower O_2 levels and in the presence of high CO_2. A higher amount of ethylene was required in irradiated fruit to induce the same action on chlorophyll degradation in 2.5 per cent O_2 and 5 per cent CO_2 than in air. In addition, the conversion of oxygen to ozone during irradiation can be injurious to fruits and vegetables. Removal of oxygen from the surrounding atmosphere during irradiation treatment, however, can minimize injuries caused by ozone.

With the issuance of the irradiation regulations by the US government and the approval of the commercial use of irradiation, the technology is expected to have a major impact on international trade. Additional studies are necessary to investigate the relation of irradiation and CA for different maturities, temperatures, O_2 and CO_2 combinations, and the feasibility for various crops.

Aspects of Ripening

Ripening consists number of processes.

Separation of individual reactions during ripening has been demonstrated between softening and the climacteric rise in respiration, ethylene production and changes in organic acids, protein synthesis and the respiratory climacteric, amino acid incorporation and rate of O_2 uptake, and color development and CO_2 evolution.

Storage of fruits under different gas atmospheres has also produced differential effects on various aspects of ripening. At oxygen concentrations between 1 and 15 per cent, banana fruits ripened, while the peel remained green, creating a "green ripe" condition. Looney showed that 10 per cent CO_2 suppressed ethylene production and softening, but acid catabolism was not affected in McIntosh apples. Softening of apples may require continuing involvement of ethylene and is therefore subject to competitive inhibition by CO_2, while rapid acid catabolism tends to proceed after being initiated regardless of the presence or absence of ethylene.

Goodenough and Thomas showed that CA (5 per cent O_2 and 5 per cent CO_2 repressed chlorophyll loss and synthesis of lycopene, carotenoids, and xanthophylls, but did not restrict changes in organic acids, starch degradation, and sugar accumulation in tomato fruit. They also demonstrated that CA prevented synthesis of polygalacturonase and red pigment but did not alter monosaccharide and organic acid concentrations in tomato fruit. Goodenough *et al.* postulated that expression of genes for various ripening reactions may be separated by storing fruit in CA and the reduced rate of respiration may supply less ATP to the biosynthetic systems of fruit. Storage of mature green tomatoes in 6 per cent O_2 and 6 per cent CO_2 at 12°C retarded synthesis of ethylene, polygalacturonase, and lycopene, while starch degradation and changes in tricarboxylic acid cycle enzymes, citrate synthase, and malate dehydrogenase proceeded with little inhibition. These results not only suggest that different ripening events responded differently to CA treatment, but also indicate that some biochemical changes in ripening fruit can occur independent of ethylene. Based on these findings, Goodenough and co-workers further postulated that CA may provide a greater retardation on systems which require transcription and translation processing for *de novo* enzyme synthesis, such as synthesis of pigments and hydrolytic enzymes, than systems which require activation of existing enzymes or regulation of enzyme activities, such as changes in sugar and acid concentrations.

The increase of CO_2 and lowering of O_2 also influences the quality of food stuffs.

Many processes are influenced by both lowering of O_2 concentration and elevating of CO_2 level. However, certain processes are more sensitive to alteration of O_2 concentration, while others are affected more by the addition of CO_2. Blanpied and Hansen found a significant linear relationship between firmness loss and CO_2.

The effectiveness of high CO_2 and low O_2 also varies with different crops. Examples observed by various investigators are listed below. High CO_2 has been reported to be more effective in retarding chlorophyll degradation in broccoli and cabbage, decreasing the acidity in asparagus, beans, broccoli, and spinach, and enhancing tenderness and reducing the incidence of bacterial soft rot in asparagus. Low O_2 is more effective than elevated CO_2 in reducing the respiration rate of various crops, delaying ripening of bananas, retarding decline in fruit quality during storage of pears, reducing losses and extending storage life of Chinese cabbage, and reducing russet spotting in lettuce.

Different results have been obtained concerning the residual effects of CA treatment. High CO_2 lost its effect on suppression of ethylene production immediately following transfer of sweet peppers or tomatoes to air. The protection of zucchini squashes against chilling injury by low O_2 vanished after 2 d of aeration. However, long-term residual effects were observed in carrots, potatoes, and zucchini squashes on respiratory inhibition following exposure to either high CO_2 or low O_2 at non-chilling temperatures. The residual effects were also observed in grapefruit on prevention of chilling injury after conditioning of fruit with 40 per cent CO_2, in apples on delaying softening by short-term treatment with 15 to 20 per cent CO_2 and on suppressing volatile production by 2 per cent O_2, in pears on retaining ripening capacity by 2-week treatment with 12 per cent CO_2, in broccoli on inhibition of ethylene production by 20 per cent CO_2, and in lettuce on retardation of respiratory rate by 2.5 per cent O_2 and 2.5 per cent CO_2.

Rapid improvements have been made in CA technology during recent years. Several new CA strategies, including short-term high-CO_2 treatment, low-O_2 storage, and rapid CA storage, have been developed in response to increased production of apples and pears and the need for lengthening the marketing period for these fruits. In comparison, progress in understanding the physiological and biochemical basis of CA effects has been relatively slow. Although many studies have elucidated various aspects of CA effects as described in this review, the basic mechanism of how low O_2 and high CO_2 work is still not fully understood.

It was generally assumed that CA delayed senescence through its suppression of respiratory activity. Recent evidence, however, shows that several other processes, which are important during ripening and senescence, can be influenced by CA separately and independently of respiration. These same processes can also lead to the deterioration of fruits and vegetables. Therefore, the specific mechanisms involved in the observed effects of high CO_2 and low O_2 need to be examined.

One aspect which is important to the process of senescence but which has received little attention is the changes of growth substances during CA storage. Growth substances regulate cellular functions and participate in the control of senescence. Their dynamic changes are closely related to rate of metabolism. The interaction and balance between promotors and inhibitors also affect maturation, ripening, and deterioration. Studies by Isenberg showed that a CA of 2 to 3 per cent O_2 and 5 per cent CO_2 can affect gibberellins and auxins in cabbage. An atmosphere of 1 per cent O_2 was also found to reduce the level of abscisic acid in Chinese cabbage. Further studies are needed to determine the CA-induced changes of different growth regulators and their effect on delaying senescence.

Development of CA Storage Method

Today the CA storage is improved very much, by which the fresh perishable commodities, such as, vegetables, cut flowers and seeds can be stored without spoilage. The specific environmental conditions may vary according to the commodity being considered and may include:

1. Storage at the lowest safe temperature to avoid freezing or chilling injury.
2. The lowest safe oxygen level to attenuate respiration rate and arrest development of ripening or senescence.
3. The highest safe carbon dioxide level to reduce respiration rate and slow developmental processes of ripening and senescence.

4. The lowest practical ethylene levels to suppress initiation or development of ripening or senescence.
5. The highest humidity levels to attenuate moisture loss from the commodity.

The absolute or desirable levels of these environmental variables differ according to commodity and stage of development. Moreover, tissue response may vary because of interactions among these variables as influenced by variety and preharvest conditions and climatic factors peculiar to the region where the commodity is grown. The objective in applying CA technology is to extend the useful marketing period for the commodity during storage, transport, and distribution to maintain quality, nutritive value, or market value of the product for eventual consumption over that achievable by the use of temperature only. CA should therefore be considered as a supplement to refrigeration for product maintenance after harvest. Economic principles of supply and demand for the commodity in relation to cost, benefit, and practicality will ultimately determine the use of CA storage technology for post-harvest maintenance of the product for ultimate consumption or use.

Commercial interest in the development and implementation of CA or modified atmospheres (MAs) for transportation and storage of fresh fruits and vegetables is expanding. This is comprised of a worldwide expansion of improved CA technology application for fruits such as apples and pears which began more than 50 years ago, coupled with a growing interest and application to extend CA technology for an expanding list of fresh fruits and vegetables. This demands extending the post-harvest period for distributing and marketing these fresh commodities, produced in established production centers, beyond current practices. Development of processes to artificially establish and maintain controlled levels of temperature, humidity, oxygen, carbon dioxide, and ethylene has, in some instances, gone beyond the research base which defines the limits within which these parameters may be applied with predictable success in distributing fresh and wholesome produce to the consumer.

There are basic biologically defined limits to the environment that can be employed for the post-harvest preservation of fresh fruits and vegetables. The basic processes of transpiration, respiration, and biochemical transformations may continue in much the same manner in plant tissues after harvest as before harvest. However, this is done without replenishment of reserves or metabolic control from the parent plant. The nature of these cellular processes can be altered, and the chemical reaction rate can be attenuated or stimulated by manipulating the temperature and concentration of the biologically active gases such as water vapor, oxygen, carbon dioxide, and ethylene among others. Beyond certain limits of temperature and gas atmosphere composition and according to the nature and stage of development of the commodity, physiological disorders may develop which terminate the useful life of the commodity by destroying appearance, flavor, nutritive value, or wholesomeness.

Development and application or successful post-harvest preservation techniques for fresh fruits and vegetables depends largely on understanding certain fundamental aspects of biology, engineering, and economics that are important in the maintenance and distribution of these perishables. The purpose of this section is to provide a historical account of how CA storage or fruits and vegetables came into being and how these developments relate to contemporary application of this important technology. Past and current day application of CA storage has dealt mainly with storage of fruits and so explains the emphasis on fruit storage.

CA storage as a technique for post-harvest preservation or certain fresh fruits and vegetables originated several centuries ago. This was long before the role of oxygen and carbon dioxide in the basic respiration of plants and plant organs was understood. Dairymple wrote an interesting and amusing history of the development of CA storage of fruits. Some of the early trials and experiences deserve mentioning as they provide insight and perspective for preservation of fresh perishables.

A perusal of ancient literature reveals attempts at preservation of some fresh fruits and vegetables by burying them with departed souls in tombs and crypts. Ostensibly this was done to reduce the oxygen level within the tomb to an anaerobic level to aid in the preservation of the remains of the departed and for sustenance and enjoyment of the soul in the life hereafter. There is some evidence that this technique combined with gas-tight construction of the inner vaults was successful in creating an anaerobic environment within the tomb. Recently opened Ming tombs in China have been found to be anaerobic. While it is impossible to prove what exactly took place within the inner vault, but some of the objectives were obviously achieved. The vegetable matter had disappeared, the tomb atmosphere was anaerobic, and the remains of the departed were somewhat preserved. The relative contribution of respiration of the produce and microorganisms along with inorganic oxidations is difficult to assess in terms of establishing the anaerobic atmosphere. Other Ming tombs remain unopened for future generations of Chinese and other scholars and tourists to study and ponder these questions.

The centuries-old custom of burying certain fruits and vegetables in the ground after enclosing them in various wrappings and protecting them from freezing by covering them with soil and insulating materials is another example of success in extending the useful life of perishables. This crude practice is used to a limited extent in modern times realizing the beneficial effects on preservation to be gained by reducing the oxygen level, elevating the carbon dioxide concentration, and lowering the product temperature in the storage environment, there is reason to believe that this practice will continue in the future.

There are numerous other examples of the practice of subterranean storage on a larger scale to gain benefits from reducing product temperature and gas atmosphere modification. These include the practice of cave storage of fruits and vegetables as done in China and trench storage of grains, root crops, and silage. Again, these practices had their origins centuries before the physiological bases for the benefits that were obtained were fully understood. Today, cave storage of apples and pears is widely practiced in China and the results of the relatively, unsophisticated storage technology rivals results obtained by modern means and at only a fraction of the cost; current-day research is being conducted in China and elsewhere to improve and extend this economical means of produce preservation. Recent innovations include the use of polymeric film liners, controlled ventilation, and adsorbers to control the levels of carbon dioxide and ethylene.

According to Dalrymple, the first recorded scientific study of CA storage was conducted in the early 1800s in France by Jacques Etenne Berard, a chemist in the School of Pharmacy at Montpellier.

His studies revealed that harvested fruits utilized oxygen and produced carbon dioxide. He discovered that fruits kept in an atmosphere devoid of oxygen did not ripen. He extended these investigations which were apparently conducted at room temperature to several fruits. In the case of peaches, prunes, and apricots, he found a useful storage life of up to 1 month, and with apples and pears, he found useful storage life of about 3 months. Moreover, he found that the fruits would ripen after returning them to air, providing that they were not kept too long in storage without oxygen. The interesting perspective to be gained from Professor Berard's pioneering investigations of nearly 2 centuries ago is relevant to current research and observations by commercial CA storage operators that apples stored beyond 6 months at ultra low oxygen levels of 0.5 to 1.0 per cent ripen poorly or, if they ripen at all, do so with inferior dessert quality. Berard's studies published in 1821 won him acclaim and the Grand Prix de Physique got recognition from the French Academy of Science. These studies apparently stimulated other scientists to conduct similar investigations, some of which were reviewed in a 1905 publication by the US Department of Agriculture.

The first recorded attempts at CA of fruits in the US were apparently made in northern Ohio by Benjamin Nyce in about 1865. He was aware of the fundamentals of fruit respiration and constructed an airtight sheet metal-lined storage room of about 4000-bushel capacity that was cooled with ice to maintain a storage temperature of just above the freezing point of water. Apples were stored commercially at the Cleveland storage in saleable condition for periods up to 9 months following harvest. Nyce obtained several US patents on his storage technology. A subsequent CA storage facility built by Nyce in New York did not prove to be commercially acceptable. The patent protection afforded to Nyce on this technology may have discouraged commercial development of this innovative preservation system in the US during the late 19th century since the practice was not adopted commercially during this era. Alternatively, it may be conjectured that incidences of what may have been carbon dioxide injury to the fruit impacted negatively on commercial development of this CA storage technology. Continuing with exerpts from Dalrymple's narrative reveals that in the 1890s the San Jose Fruit Company in California conducted tests with rail car shipments of peaches, nectarines, pears, quinces, persimmons, and grapes in which non-refrigerated rail cars were sealed and gassed with carbon dioxide during shipment of produce from California to Chicago. The tests met with partial success and may be considered the forerunner of what subsequently has become known as MA transport of perishables. This practice is in current and widespread commercial use. Again, U.S. patents were issued on this process in the late 1900s; although additional shipments of peaches, plums, and pears were made for a couple of years and met with some success with the pears, the practice of carbon dioxide enrichment of the transport hold of perishables was abandoned for several decades and was reinstated as a supplement to refrigeration in the 1960s.

These early trials with CA or MA storage arid transport of perishables, although somewhat encouraging in terms of extending the useful life of some commodities, indicated a high level of risk of produce loss which probably was a factor in slow commercial acceptance of these practices. Moreover, a scientific explanation of understanding of either successful or unsuccessful results was lacking.

In the early 1900s, Hill at Cornell University made the observation that ripening of peaches could be slowed by storing them in carbon dioxide and, to a lesser extent, in other inert gases. The beneficial effects were ostensibly achieved by restricting the availability of oxygen. This confirmed, to some extent, the observations made at San lose Fruit Company noted earlier. Hill also made the interesting observation that the respiration rate of peaches exposed to carbon dioxide did not return to the prestorage level. The US Department of Agriculture, renewed investigations on the effects of carbon dioxide on storage life and respiration of apples in conjunction with studies on superficial scald of apples.

The basic groundwork that led to the widespread commercial application of CA storage was laid by the researchers Franklin Kidd and Cyril West, working at the Low Temperature Research Station at Cambridge, England, shortly after World War I. They began a systematic study of fruit respiration and ripening as influenced by temperature, carbon dioxide, and oxygen under the auspices of the Food Investigation Board of the UK Department of Scientific and Industrial Research. These studies, published over a period spanning nearly 40 years, provided the basis for most of the CA storage technology in use today and researchers of CA storage aspects would be remiss not to review these published works, particularly the bulletin, published in 1927. Their initial studies concerned both pome fruits and berries of several sorts. These studies have been reviewed in an excellent book by Fidler and co-workers. Emphasis was directed on apples, but later investigations included work on bananas, pears, and oranges. Their research was continued at the Ditton Laboratory which was constructed in 1930 on the grounds of the East Mailing Research Station in Kent, England, an important fruit-growing center southeast of London. A portion of the main building of Ditton Laboratory was constructed as a replica of the hold of a ship. It was reportedly built in this manner to address the problems of temperature and gas atmosphere maintenance encountered during maritime transport of fruits from the Commonwealth countries to England. This close attention to detail in research and problem solving is indicative of the creativity and thoroughness exemplified by these pioneering researchers. While problems of a physiological and pathological nature affecting fruit inspected upon receipt at Covent Gardens in nearby London focused Kidd and West's attention on practical research, they made major contributions of a fundamental nature on the role of temperature, oxygen, carbon dioxide, and ethylene in the control of fruit ripening.

J.C. Fidler is another noteworthy English scientist who conducted CA storage research at the Ditton and Covent Gardens Laboratories. Fidler and colleagues contributed significantly toward current-day recommendations for the CA storage of various varieties of apples and pears grown in the UK. These studies are also reviewed in the publication noted earlier.

The first commercial apple storage to employ the "gas storage" technique in England was constructed in 1929 near Canterbury in Kent County. The commercial success of this venture was realized in 1930 and led to rapid expansion in the adoption of gas storage technology for apples and pears in the decade of the 1930s. As an interesting aside, the first commercial CA storage in Michigan was put into operation in Kent County, MI, in the early 1950s.

The practice of sealing apple storage rooms provided most of the early expansion in CA storage volume in the US and Canada. It was not until the mid-1950s, when the practice of using bulk bins and lift trucks was introduced, that new construction of CA storages with high ceilings began. This rapidly replaced the practice of storing apples and pears in 1-bushel field crates. The increased efficiency in fruit handling and labor savings afforded by the use of bulk bins for harvesting, handling, and storing of fruit were important factors in the growth of CA storage capacity worldwide since 1950.

Returning to Dalrymple's historical account of CA storage development, it is clear that the pioneering research of Kidd and West in England in the 1920s stimulated CA storage research not only in the U.S. and Canada but elsewhere in the world in regions of fruit production. Overholser, at the University of California at Davis, initiated CA storage research including initial studies with avocados. Several years later, Biale also in California, began reinvestigating CA effects on avocados and over a long career made major contributions to fundamental knowledge on the biochemistry of tropical and subtropical fruit ripening.

In 1930, Allen and McKinnon joined the CA research effort in California and found, in 1935 that CA storage of Yellow Newton apples at 3 to 4°C allowed this variety to be stored successfully, avoiding low-temperature disorders found when the fruits were stored in air at 0 to 1°C. This confirmed the earlier observations of Kidd and West with several important English apple varieties where storage life was found to be limited by low temperature or chilling injury disorders and these could be largely eliminated by gas storage at 3 to 4°C. This was an important revelation in the 1930s and was responsible for much of the early growth in the adoption of CA storage technology. Spinoff from these investigations included energy savings in reducing refrigeration costs and this is still being realized in modern times. Moreover, the aspect of low-temperature injury is the subject of contemporary CA storage research worldwide as new apple varieties and other sorts of fruits and other perishables are tested for application of CA storage and transport technology.

Smock, recently deceased, joined Allen in California and extended the CA storage investigations on apples and broadened the studies with peaches, pears, and plums. Smock continued his investigations upon moving to Cornell University in the late 1930s. He made many significant contributions to CA storage technology and post-harvest physiology over a distinguished research and teaching career spanning more than 50 years. Smock and Van Doren authored a comprehensive treatise on CA storage in 1941 that did much to foster commercial acceptance of CA storage of apples in the

Eastern US and Canada. Nearly 20 years would pass before CA storage of apples was accepted technology in the Pacific Northwest and Archie Van Doren was largely responsible for its commercialisation.

The commercial development of CA storage of apples and pears in Canada can be mainly attributed to the research of Eaves of the Canada Department of Agriculture. Eaves culminated a distinguished research career at the Agriculture Canada Research Station at Kentville, Nova Scotia, where he made numerous significant and innovative contributions to the development of practical CA storage technologies over a 30-year period beginning in 1934. Among these was the development of a practical means to rapidly establish low-oxygen levels in the CA storage by using an open-flame propane-fueled oxygen burner. Various iterations of this device are still used to a limited extent today. Eaves was also instrumental in developing the use of dry lime (hydrated calcium oxide) as a practical means to scrub carbon dioxide from the storage atmosphere. Dry lime is still widely used today as a simple means to scrub carbon dioxide in CA storage rooms, particularly for some varieties of apples and pears which must be stored at levels of carbon dioxide at 1 per cent or less.

Apart from the early and aborted attempts at commercial CA storage of apples in the US in 1860s and the initiation of commercial CA in England in 1929, the first commercial CA storages outside of England were constructed in Elgin, South Africa, by the Elgin Fruit Packers Cooperative in 1935 and 1936. However, these CA rooms were only operated successfully for a short period of time before being converted for conventional refrigerated air storage. Nova Scotia, Canada, was the site of the first commercial CA storage in modern times in the Western Hemisphere in 1939. This was followed in 1940 by equipping existing refrigerated fruit storage rooms for operation as CA storages at Lockport and at Sodus, NY. This was shortly followed by construction of new CA storage enterprises in the Hudson Valley in New York State. Gradual expansion of CA storage capacity continued in New York and eastern Canada during the 1940s, and in New England in the 1950s. CA storage was largely dedicated to McIntosh, the major dessert apple variety in these regions. The McIntosh variety, highly prized for its dessert quality, like the Cox's Orange Pippen in England, has a relatively short storage life when kept under conventional refrigerated storage. This explains the primary attention in research and commercial storage given to the McIntosh variety. This tradition continues wherever McIntosh is important in the commercial sector.

The development and application of new CA storage technologies can be largely traced to research concentrated on extending the availability of McIntosh and Cox's Orange Pippen of prime dessert quality even after 6 to 9 months in CA storage, and, to a lesser extent, this can be said of other important apple and pear varieties.

A primary period of expansion in CA storage in the US and Canada began in the 1950s. As mentioned earlier, CA storage in Michigan began in the mid-1950s prompted by interest in not only storing McIntosh but also for storage of the Jonathan variety. The storage life of this variety was limited by excessive softening during prolonged refrigerated storage at 2 to 3°C necessary to attenuate the incidence of two physiological disorders, Jonathan spot and soft scald found during storage at 0°C. Investigations by Dewey revealed that both disorders could be largely if not completely controlled by storing this variety in CA storage at 3 per cent oxygen with 5 per cent carbon dioxide at 0°C which also delayed softening.

The most significant and accelerated expansion of CA storage in the US began in the 1960s when Washington State apple storage operators and packers and shippers realized the improvements to be gained in dessert quality and marketability of Red and Golden Delicious apples by storing them under CA. By 1960, the total CA holdings of apples and pears in the US amounted to over 4 million bushels. Expansion in CA storage capacity for apples and pears in the US and Canada and around the world has continued virtually unabated since 1960. For example, in US, CA storage capacity for apples and pears was nearly 100 million bushels by the late 1980s.

Attention will now be focused on means of establishing and maintaining the gaseous atmosphere. The early work of Kidd and West was conducted to determine if modification of the atmosphere could provide an alternative to refrigeration as a means to extend the storability of fruits. Gas mixtures containing 5 and 10 per cent carbon dioxide with 5 or 10 per cent oxygen were employed at relatively warm storage temperatures of 10 to 15°C. Temperature was not controlled, depending rather on natural cooling as the ambient temperature decreased with the onset of winter. Later trials were conducted under mechanical refrigeration to maintain the desired lower temperature commonly used with storage in air. As a consequence of these trials, they discovered that some varieties of English apples were subject to low temperature breakdown during gas storage at 0.5 to 1°C and this was not evident at warmer temperatures of 3 to 4°C. Fidler has reviewed the scientific and technical developments of gas storage relating to choice of atmospheres, construction, refrigeration, and atmosphere maintenance.

The gas composition in some of the early gas storages was maintained by ventilation with air to keep the carbon dioxide level in the range of 10 to 15 per cent and the oxygen level would then range accordingly from 11 to 6 per cent, respectively. The sum of the carbon dioxide and oxygen level would be 21 per cent as the air admitted was diluted with roughly four times as much nitrogen, and oxygen was consumed by the fruit. The accumulation of carbon dioxide and depletion of oxygen in the storage room were accomplished by respiration of the fruit. As the benefits of storing apples at lower oxygen levels than 6 per cent became apparent, it was found necessary to scrub the excess carbon dioxide by caustic soda or other means in order to achieve the desired lower oxygen levels. Moreover, it was found that physiological disorders such as brown core were prevalent when the level of carbon dioxide was too high in the storage atmosphere. Varieties were found to differ in their susceptibility to the carbon dioxide-linked disorders.

With few exceptions, the oxygen levels in the early CA storage rooms and those operated throughout the commercial development of CA into the 1950s were established by consumption of oxygen and production of carbon dioxide by the fruit. The time required to achieve the desired low oxygen level of about 3 per cent would vary from 1 to 3 weeks or longer, depending upon such factors as fruit temperature, fruit ripeness, and room tightness, among others. This length of time to establish the desired low oxygen level was not a critical factor in relation to quality maintenance for the maximum duration of storage commonly employed to market the fruit during this period. However, data began accumulating in the 1960s from various research centers, indicating that fruit ripening could be delayed further by reducing the oxygen level in the storage room more quickly by some means of flushing.

In the early 1960s, an alternative means became commercially available for quickly establishing and, maintaining the desired levels of oxygen and carbon dioxide in the CA room. This was introduced by the Whirlpool Corporation in Michigan as the Tectrol process. The process entailed combustion of propane or natural gas with fresh air using a catalytic burner to ensure clean combustion. The effluent gas from the burner was cooled and passed through a bed of activated carbon which removed carbon dioxide and residual hydrocarbons from the combustion process. The scrubbed gas, containing 1 to 2 per cent oxygen with the balance made up mostly of nitrogen. was used to purge the CA room by displacing the air in the void volume and to reduce the oxygen level to less than 5 per cent in a matter of 1 or 2 d. The actual time required achieve this, at a purge rate of about 1200 SCFM, was dependent upon the room size, void volume, room tightness, and respiration rate of the fruit. Subsequent control of the carbon dioxide level was achieved by recirculating the CA room atmosphere through an activated carbon scrubber. This system came into common usage for CA storage operation. Various iterations of this type of inert generator are in common use today throughout the world.

Also introduced in the early 1960s was a CA generator of a different design than the Tectrol system which was developed by the Atlantic Research Corporation. It was a recirculating catalytic oxygen burner known as Arcat and it also used propane. In this system, the atmosphere from the CA room was recirculated over a hot alumina catalyst coated with a trace of platinum. The flameless oxidation process consisted of a blower, preheater, catalyst, and cooler. Air from the CA room was passed through the catalytic oxygen burner reducing the oxygen content by 3 to 4 per cent with each pass. Carbon dioxide, produced in the process was adsorbed by a molecular sieve scrubber or other means. This recirculating system was slightly more efficient in oxygen reduction than the inert generator type and also came into widespread commercial application as a CA generator for CA storage operation. Again, several iterations of this type of atmosphere generator have evolved over the past 30 years and are in common use worldwide.

As mentioned earlier in reference to Eaves in Canada, an open-flame burner was also introduced in the 1960s as a CA generator. This system is known variously as the Eaves, Anderson, or Wilde

CA generator. In this system, propane or natural gas is combusted with fresh air in a flame confined in an open pipe without the aid of a catalyst. The combustion products containing a low level of oxygen are used to purge air from the CA room. The excess carbon dioxide is generally scrubbed with dry lime or other means. As noted above, several iterations of this type of CA generator have been introduced and some are still in common usage.

There have been numerous instances of explosions caused from the use of all three types of fossil fuel-based CA generators. Moreover, there have been cases of carbon monoxide poisonings associated with the use of some of these CA generators. Safety and operating concerns in the use of fossil fuel-based CA generators have been expressed.

A CA generator based on cracking of ammonia to nitrogen has also been developed. This system, known as the Smit or Oxy-drain nitrogen generator, was introduced in the 1980s. The process converts ammonia to nitrogen and hydrogen at a high temperature and the hydrogen is converted to water vapor. The nitrogen produced is used to purge the CA room as with the other inert generators.

Liquid nitrogen is also used as a means to establish and in some cases maintain CA atmospheres. In this system, liquid nitrogen from a bulk tank is converted to nitrogen gas which is then used to purge or displace air from the void volume in the CA room. This system is a simple, rapid, clean, and safe means to establish and maintain the CA atmosphere. Liquid nitrogen CA systems gained its popularity in the late 1980s, particularly in the Pacific Northwest, where the magnitude of scale in the size of the CA facilities and liquid nitrogen costs made the systems more competitive with alternative technologies.

A new era of nitrogen generators for CA storage application began in 1986. These are known as air separators which separate air into oxygen and nitrogen. One system is based on hollow-fiber membrane technology, and the other is based on pressure swing adsorption (PSA) technology. While the application of these technologies to CA storage was new, both of these systems had been widely employed in numerous industrial gas separation and inert blanketing applications for about a decade. Systems based on the hollow-fiber membrane principles were available from Permea, Inc., a Monsanto Company, and from Dow USA. PSA systems were available from Kaldair, Inc., a British Petroleum Company; Airco, Inc., and also from the Kemp Division of Monsanto. The new CA generator technology was quickly accepted by the CA industry in the US, Canada, and worldwide. Since these technologies are new in application for CA storage.

Chapter 10
CA Storage for Apples and Pears

Flesh Softening of Apples

Tests conducted more than 40 years ago indicated that flesh softening of apples could be retarded in air storage if they received an initial treatment of high CO_2 for a short period of time. Those early observations prompted Couey and Olsen to investigate the effect of prestorage CO_2 treatment on the softening of Golden Delicious apples in CA storage. The first commercial tests showed that CO_2 injury occurred on less than 1 per cent of the total quantity of CO_2-treated apples and that CO_2 pretreatment significantly increased flesh firmness which resulted in significantly less bruising and a longer shelf life for Washington State Golden Delicious. The following recommendations were made in 1977 and 1978 for prestorage CO_2 treatment of Golden Delicious in Washington State. Avoid fruit from trees of high or low vigor, trees with a light crop, trees in frost pockets, and mite-infested trees. The fruit should be in an early stage of maturity at harvest: 76 to 85 Newton flesh firmness, 11 per cent or higher total soluble solids, good fruit finish, and rounded calyxes: the skin color should have changed from light green to greenish white or greenish yellow. The total elapsed time to pick, fill, and seal the room should not exceed 5 d. Not more than 24 to 36 h should be taken to generate the high CO_2 atmosphere: 15 to 17 per cent CO_2 with 4 to 5 per cent O_2.

Table 10.1: US Five Year Average Storage Holdings of Fresh Apples on the First Day of Each Month

Month	*Crop Year**					
	1956–1960	*1961–1965*	*1966–1970*	*1971–1975*	*1976–1980*	*1981–1985*
December	868	1008	1047	1187	1445	1561
February	490	578	632	744	901	1002
April	198	244	294	359	454	538
June	38	59	73	94	153	198

* Holdings in 1000 t.

That atmosphere should be maintained for 8 to 9 d, after which the room should be quickly flushed with air and then the normal CA environment should be quickly generated.

Generation of the high CO_2 atmosphere was accomplished by burning the CA room atmosphere in a fossil-fueled atmosphere generator and/or addition of CO_2 in the gaseous, liquid, or solid state. The amount of CO_2 to be added can be calculated:

$$A = \frac{0.144\,V(P_f - P)}{100 - P_f}$$

where A = CO_2 required (pounds), V = room volume (cubic feet), P_t = percent CO_2 required, and P = percent CO_2 in the room. Although beneficial effects without significant CO_2 injury were reported by Mattus for Golden Delicious apples grown in Virginia, prestorage CO_2 treatment of Golden Delicious apples grown in British Columbia caused intolerably high instances of CO_2 injury. A large cooperative test by researchers in several McIntosh areas of North America showed that CO_2 prestorage treatment significantly improved flesh firmness, but also caused intolerable incidences of internal and/or external CO_2 injury. Mattus reported that in Virginia prestorage CO_2

treatment was of no benefit to Rome and increased the rate of flesh softening in Stayman apples.

Recommendations for prestorage CO_2 treatment were changed in 1980 to reduce the likelihood of CO_2 injury to Washington State Golden Delicious apples treated with high CO_2 before storage. The recommended period between harvest and sealing the CA room was reduced from 5 to 3 d and the length of treatment period at 15 to 17 per cent CO_2 was reduced from 8 or 9 to 4 d.

CO_2 prestorage treatment of Golden Delicious apples is not recommended in Washington State because occasional storage lots have suffered from high incidences of CO_2 skin burning. However, a few Washington State CA storage facilities continue to use prestorage CO_2 treatments for Golden Delicious. At these facilities, the likelihood of CO_2 skin burn is reduced by taking steps to be sure that the fruit is dry before the CO_2 treatment is initiated. These steps include not using a postharvest drench treatment for control of storage scald or drying the fruit after such treatment and using gaseous CO_2, rather than an atmosphere generator, to establish the high CO_2 environment. Water that is added to the air by atmosphere generators may condense on the skin of apples, rendering them more susceptible to CO_2 skin burn. Most storage operators, however, have changed their Golden Delicious CA strategy from prestorage CO_2 treatment to rapid CA.

Table 10.2: Total US Crop, December 1 Total US Controlled Atmosphere (CA) Storage Holdings, and Storage Holdings of Delicious Apples in Washington

Crop (Year)	*December 1 Storage Holdings of Apples*				
	US Crop (1000 t)	*Total CA (1000 t)*	*US Crop in CA (% of Total)*	*Washington Delicious*	
				Total (1000 t)	*CA (% of Total)*
1951–1955	1928	11	0.6	126	0.0
1956–1960	2157	57	2.7	149	1.6
1961–1965	2501	193	7.7	183	18.4
1966–1970	2654	290	10.9	216	35.4
1971–1975	2959	430	14.5	284	52.2
1976–1980	3436	680	19.8	414	68.4
1981–1985	3685	844	22.9	493	71.4

RC Atmosphere

On the west coast of North America, Golden Delicious trees are concentrated in a geographical area which extends northward from central Washington State, up the Columbia and Okanogan valleys, into south-central British Columbia. In the late 1970s, prestorage CO_2 treatment, which was commonly used in Washington, was not used in British Columbia because observations in that province indicated that British Columbia Golden Delicious were more susceptible to CO_2 injury and showed less firmness retention response to the Washington program for prestorage CO_2 treatment. An attempt was made to explain the difference in response of Washington and British Columbia Golden Delicious by comparing 32 lots of British Columbia Golden Delicious with 25 lots of Washington Golden Delicious, all lots having been subjected to the Washington program for prestorage CO_2 treatment. The study confirmed the results of the previous studies, but failed to explain the difference in response. Analyses of the Washington and British Columbia apples revealed little difference in fruit firmness, size, titratable acidity, and calcium. However, Washington Golden Delicious were less green and higher in soluble solids. It was not known whether these differences could account for the disparity in response to the prestorage CO_2 treatment.

Table 10.3: Fruit Quality of "Golden Delicious" Apples as Influenced by the Rates of Loading and Establishing the 2.5 per cent Oxygen Storage Atmosphere

Storage Treatment	*Duration of Loading (Days in 0°C Air)*	*Days Taken to Reduce Oxygen from 21 to 2.5%*	*Days from Harvest to 2.5% Oxygen*	*After 8-Month Storage*	
				Flesh Firmness (N)	*Juice Acidity (% Malate)*
Air storage	–	–	–	47.6	0.16
Normal CA	9	20	29	59.6	0.28
Rapid CA	9	1	10	62.7	0.31
Rapid CA	4	1	5	66.3	0.32
Rapid CA	1	1	2	69.8	0.36

Lau observed that in British Columbia the CA rooms were normally filled in 10 d, and then an additional 20 d was usually required for the desired concentration of oxygen to be established by fruit respiration. This storage management practice did not result in consistently firm CA Golden Delicious. In contrast, when the high CO_2 (and low oxygen) regime was imposed almost immediately after harvest of Golden Delicious in Washington, the flesh firmness retention was consistently high. Lau theorized that consistently high firmness retention of prestorage CO_2-treated Golden Delicious might be associated with the rapid establishment of the CA regime as well as the presence of high CO_2 in that CA regime. To test this theory, he conducted numerous laboratory- and commercial-scale experiments in which several cultivars were subjected to variable durations of CA room loading and variable durations of oxygen reduction to 2.5 per cent.

Lau determined that the shorter the interval between Golden Delicious harvest and establishment of the CA regime, the greater was the retention of flesh firmness (Table 10.3). The term "rapid CA" was applied to the storage management practice that resulted in establishment of the CA regime within 7 d of harvest. Subsequent tests indicated that soft-fleshed cultivars, such as McIntosh, Empire, and Golden Delicious, showed a greater response to rapid CA than did firm-fleshed cultivars, such as Spartan and Delicious. Although early harvest resulted in the most firm fruit after harvest, harvest dates of Golden Delicious did not significantly influence the retarding effect of rapid CA on flesh softening in storage.

Table 10.4: Effects of Harvest Date and Delays in Controlled Atmosphere (CA) on Flesh Firmness Loss of Golden Delicious Apples

Harvest Days Past Full Bloom	*Delay (Weeks) Between Harvest and 3% Oxygen in CA Storage*			
	0	*2*	*4*	*Average*
	Flesh Firmness Loss (N) During Storage			
134	3.1	14.7	13.8	10.7
141	4.0	12.0	19.1	11.6
148	0.0	10.7	19.6	10.2
Average	2.2	12.5	17.3	

The concept of rapid CA is not new. However, in many previous investigations, the significance of rapid CA may have been underestimated because the apples were frequently evaluated after the flesh firmness of rapid and slow CA apples had reached plateau values. Flesh firmness was the sole or principle measurement used to evaluate apples in those experiments. Greater attention might have been focused on rapid CA if flesh firmness measurements had been made earlier in the storage period and/or if the fruit had been evaluated by a taste panel.

Table 10.5: Poststorage Evaluations of "Empire" Apples Harvested on Six Dates and held in Controlled Atmosphere (CA) Storage until April 1985–Evaluations after 2 Weeks in Air at 0°C plus 48 h at 20°C

Orchard	*Harvest Date*					
	Sep. 26	*Sep. 30*	*Oct. 4*	*Oct. 8*	*Oct. 12*	*Oct. 16*
			Flesh Firmness (N)			
A	64.1a	64.5a	54.7b	53.8b	55.1b	54.2b
B	60.0a	60.0a	50.3b	45.8b	48.0b	48.0b
Combined Scores for Panel Evaluations of Fruit Eating Quality and Appearance–High Score Preferred						
A	1.2c	1.3c	2.7b	4.8a	5.0a	2.8b
B	1.6c	2.0c	2.6b	4.1a	4.6a	3.3b

Although rapid CA for Cox's Orange Pippin has been evaluated at East Mailing for several seasons, no beneficial effects on flesh firmness, green color, or sensory attributes were observed. The advantages of rapid CA over slow CA for other cultivars are increased retention of flesh firmness and more uniform maturity of the fruit in a CA room. Commercial experience has shown that these advantages may be accompanied by some problems.

Many of the CA rooms in eastern North America are owned and operated by small grower-shippers. CA rooms at these establishments normally have a capacity of 200 to 400 t. These growers do not normally have sufficient acreage to fill a CA room in 3 to 5 d with apples of uniform maturity and adequate red color. This problem has been solved in a few instances by reducing CA room capacity or by storing cooperatively with a neighbor.

Another problem is carbon monoxide poisoning of fossil-fueled forklift operators, who spend several hours stacking bins in a rapid CA room. In some cases, carbon monoxide-poisoned forklift operators were unaware of the early symptoms of carbon monoxide poisoning: headache, which is usually characterized by a sharp pain between the eyes; dizziness; a feeling of weakness; and finally nausea. The use of electronic forklifts in confined areas will eliminate the problem of carbon monoxide poisoning.

CA refrigeration evaporators normally having the capacity to cool apples to 0°C in 3 d, if incoming apple core temperatures are 21 to 27°C and the room is loaded in 10 d or if the room is loaded in 4 or 5 d with 10 to 55°C incoming core temperatures. If the heat load exceeds the heat removal capacity of the evaporator coil, the fruit cooling period will be extended. Since slow cooling may nullify the effect of rapid CA on flesh firmness (Table 10.6), incoming apples should be precooled in other rooms when evaporators have failed to reduce air temperature to below 3°C during the night.

Table 10.6: Effect of Cooling Rate on Firmness of Rapid Controlled Atmosphere (CA) "Empire" Apples

Days to Cool to 0°C	*Days from Harvest to 3% O_2*	*Flesh Firmness (N)*	
		Removal from CA	*After 2 Weeks at 0°C + 2 Weeks at 20°C*
1	4	62.7a	55.1a
7	4	58.3b	52.5b
114	4	51.6c	50.3b

The spread in cultivar maturity dates limits rapid CA to one or two cultivars per room because each rapid CA room must be filled and sealed in 3 to 5 d. Marketing problems are not created when CA storage is limited to one or two cultivars. Thus, rapid CA is easily incorporated into the storage management strategies in Washington State, where CA storage is limited to Red Delicious and Golden Delicious, and in some eastern North American areas where McIntosh represents a very large percentage of the total CA holdings. However, in other apple areas, where several cultivars are grown, CA storage operators have traditionally filled their CA rooms with a mixture of cultivars. Although this storage management, practice requires a long time to fill the CA rooms in the all, it makes available in the spring a mixture of cultivars from one open CA room, which is sold out in 3 to 4 weeks. If rapid, CA is used in these areas, operators would need access to several open CA rooms to obtain a mixture of cultivars for sale. Since the rooms normally require 3 to 4 weeks to be sold out, a much longer period would be required to sell out several simultaneously opened CA rooms. There would be unacceptable deterioration of fruit condition if the apples are held in air at 0°C for longer than 3 to 4 weeks. There are two solutions to this apparent conundrum: open several CA rooms, remove 10 to 25 per cent of the bins from each room, reseal the rooms, and regenerate the low-oxygen atmospheres; create smaller CA rooms by installing permanent curtain walls in existing CA rooms.

During the 1940s, CA storage demonstrated its economic worth to apple growers located in the eastern and central North American fruit growing areas. Following this period of slow initial CA growth, these was a rapid increase in the US apple crop and, not surprisingly, a parallel of very rapid increase in the total U.S. CA capacity. Although the increase in the size of the apple crop and the increase in CA capacity were most obvious per cent in Washington State, these increases were a national phenomenon.

The total US storage holdings on December 1 increased from 868,000 t in 1956 to 1960 to 1,561,000 t in 1981 to 1985. There were corresponding increases in storage holdings on February 1, April 1, and June 1. Although the increases in April 1 and June 1 storage holdings appear small in contrast with increases in December 1 storage holdings, the total quantity of apples in storage on April 1 increased from 198,000 to 538,000 t and on June 1 increased from 38,000 to 198,000 t during that period of time.

The apple industry saw that CA storage practices used during the 1950s and 1960s were, not acceptable in the 1970s and 1980s because the storage practices of the earlier period did not maintain the fruit in acceptable condition for the late spring and early summer markets, which represented a significant volume of fruit in the 1970s and 1980s. It is not surprising that several new CA strategies were developed in the 1970s and 1980s to improve the condition of CA apples that were scheduled for market in the late spring and summer months: prestorage CO_2 treatment, rapid CA, low oxygen CA, and low ethylene CA.

Now let us try to understand about the low oxygen CA.

The term low oxygen, CA refers, to the use of oxygen concentrations below 2 per cent. To avoid confusion, less precise descriptive terms, such as ultra-low and hyper-low oxygen, are not used in this chapter; where appropriate, specific percentages of oxygen are given.

The results of most tests are in agreement with the observations of Eaves *et al.* that the beneficial responses to lowering oxygen do not significantly accrue until the oxygen is less than 5 per cent. This is the reason that CA regulations in North America require the oxygen in apple CA rooms to be maintained below 5 per cent. Although there were some successful early experiments with oxygen concentrations below 2 per cent, the frequent development of low oxygen injury discouraged most researchers from further investigations into the use of oxygen concentrations below 2 per cent. Indeed, 30 per cent oxygen was recommended in central and eastern North America to give CA operators added protection against low oxygen injury.

The first breakthrough in low oxygen CA occurred at the East Malling Research Station where North and Cockbum observed that Cox's Orange Pippin apples would tolerate 1 per cent oxygen if they were picked before the respiratory climacteric. In many earlier low oxygen CA experiments at East Malling and elsewhere, low oxygen injury developed because the apples were too mature when they were subjected to a low oxygen environment. Once the variable of fruit maturity was eliminated, the remaining requirements to reduce the risk of low-oxygen injury were quite quickly determined. Subsequent experiments and a commercial, development program for Cox's Orange Pippin led to the first published recommendation for the use of oxygen concentrations below 2 per cent: 1.25 per cent O_2 < 1.0 per cent CO_2, and 3.5 to 4.0°C.

The success of the English low oxygen CA program for Cox's Orange Pippin led to research and the development of commercial low oxygen CA programs for apple cultivars grown in North America, Europe, South Africa, and Australia. Although low oxygen CA has been incorporated into many of the world CA procedures, it has not been universally accepted because low oxygen CA has sometimes resulted in impaired development of cultivar flavor and/or existing programs did not sufficiently reduce the risk of low oxygen injury.

Cox's Orange Pippin apples may develop a flesh corking disorder in CA at oxygen concentrations above 2 per cent, but lower oxygen concentrations greatly aggravate it. The occurrence of this disorder is associated with delayed harvest and flesh K/Ca ratios in excess of 30. Incidences of corky flesh browning disorder of McIntosh at 1 per cent oxygen were increased by late harvest and storage at 0.5 rather than 3.0°C.

Empire as well as McIntosh and Spartan apple are more susceptible to low oxygen injury when the average number of seeds per apple is below five. As with calcium-related storage disorders, this relationship does not exist among apples in a single storage lot, but rather among storage lots in a CA room.

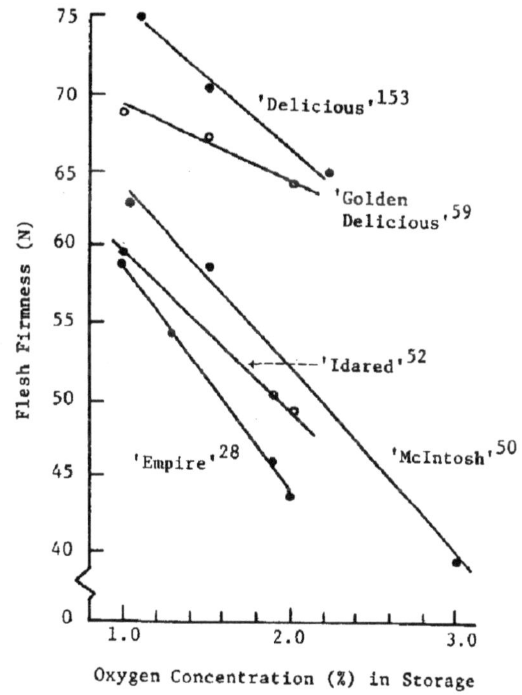

Figure 10.1: Effect of Oxygen Concentration with < 1.0% CO_2 on Flesh Firmness of Several Apple Cultivars

If harvest is delayed, the fruit in low oxygen CA may be more susceptible to internal browning disorders flesh breakdown, flesh corking, and anaerobiosis.

Postharvest drenches with diphenylamine and calcium chloride have increased the susceptibility of McIntosh and Granny Smith apples to low oxygen injury. Low oxygen injury of drenched McIntosh apples was eliminated by allowing the apples to dry before they were placed into storage.

Storage at 1.5 per cent oxygen caused low oxygen injury of Empire apples when the fruit was held in air at 0°C for 20 d before the low oxygen environment was established. When the low oxygen environment was established 4 d after harvest, comparable apples were more firm and free of low oxygen injury.

Researchers and commercial CA operators have found it was necessary to slightly raise the storage temperature when using oxygen concentrations below 2 per cent. The increase in storage temperature was required to control low temperature breakdown of Cox's Orange Pippin and of Idared. The development of low temperature breakdown of Idared was preceded by an increase in respiration rate. The increase in storage temperature is also needed to reduce the susceptibility of apples to low oxygen injury. Under some circumstances, raising the storage temperature may only delay the development of low oxygen injury.

Table 10.7: Effects of Storage Temperature and Duration on the Incidences of Low-Oxygen Injury of "McIntosh" Apples Stored at 1.0 per cent O_2 + 1.5 per cent CO_2

Storage Temperature (°C)	*Storage Duration (d)*	*Low-Oxygen Injury* (%)*
0	90	0
0	180	96
0	270	100
3	90	0
3	180	16
3	270	24

* Epidermal bluing, outer cortical browning, diffuse cortical browning, pith browning, and low-oxygen scald were included as low-oxygen injury.

The striking similarity of some forms of chilling and low oxygen injury suggest the possibility that low oxygen raises the temperature threshold for chilling injury. When ribbon-like, depressed skin browning develops in air at 0°C on chilling-sensitive cultivars such as Jonathan and Golden Delicious, it is "ribbon scald" or "soft scald". The same symptoms often appear on man apple cultivars stored at oxygen concentrations which cause anaerobiosis. Under these conditions, it is called "low oxygen injury." When the symptoms, develop at 2.0 to 2.5 per cent oxygen, but only in the coldest locations in the room, one might ask whether it is chilling injury or low oxygen injury. The injury could have been avoided by either raising the storage temperature or the oxygen concentration. The diagnosis has been given that the visible symptoms are expressions of chilling injury, which were induced by low oxygen.

Although a high, safe concentration of CO_2, 2 to 5 per cent, depending upon the cultivar, may reduce the rate of flesh softening at 2 to 3 per cent oxygen, CO_2 does not usually influence flesh softening in low oxygen CA. Flesh firmness data plotted in Figure 10.2 are quite typical.

CO_2 (1 to 2 per cent) has been shown to inhibit scald development of Delicious and Granny Smith stored at 1 to 2 per cent oxygen.

Core flush has been a problem in low oxygen CA when susceptible cultivars were held at CO_2 concentrations in excess of 1 per cent. Therefore, when oxygen is maintained at concentrations below 2 per cent, the concentration of CO_2 should be kept below 1 per cent for core flush-susceptible cultivars. CO_2 may be maintained at 1 to 2 per cent for all other cultivars except McIntosh, which will tolerate 3 per cent CO_2 under low oxygen CA conditions.

Two recent articles indicated that apples which respond to low ethylene CA at 2 to 3 per cent oxygen did not respond to removal of ethylene from the storage environment when oxygen concentrations were maintained at 1 per cent.

Causes for Low Oxygen Injury

It has been known since the days of Louis Pasteur that ethyl alcohol accumulates in plant tissue exposed to anaerobic conditions. Thomas showed that acetaldehyde will also accumulate in apple tissue if CO_2 is permitted to accumulate in the anaerobic environment.

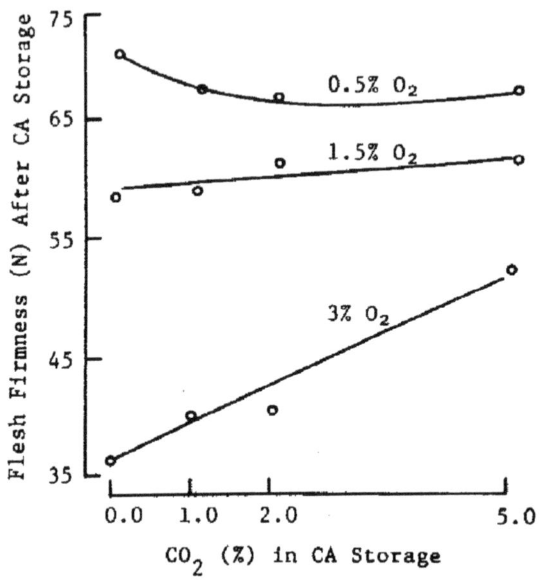

Figure 10.2: Effects of Oxygen and CO_2 Concentrations on Flesh Firmness of McIntosh Apples Stored at 0°C

Low oxygen injury has often been described as "alcoholic poisoning" and the flavor of low oxygen-injured apples has been described as "alcoholic". However, the accumulation of ethyl alcohol is not the cause of visual symptoms of low oxygen injury. Furthermore, the bad flavor, which has been described as off-flavor, alcoholic, tainted, or fermented flavor, is also not caused by ethyl alcohol accumulation. Thus, although ethyl alcohol accumulates during anaerobiosis, ethyl alcohol is not the cause of visual low oxygen injury symptoms and is not the cause of tainted flavor. Furthermore, unpublished observations of several tainted apple and grape cultivars have not verified the previous observation that tainted flavor is caused by accumulation of ethyl acetate.

Several types of injury have been associated with the storage of apples in low oxygen CA. The first visual symptom on red cultivars is frequently a change of anthocycianin pigment from red to blue or purple color. Another mild form of low oxygen injury is friction discoloration of McIntosh, which occurs as light- or dark-brown areas on the skin after the fruit has been subjected to abrasion, scraping, or brushing. Ribbon-like, depressed skin browsing develops on many apple cultivars after prolonged exposure to low oxygen concentrations. In extreme cases, the fruit may soften and split. In the flesh and core tissue, low oxygen may cause moist, chocolate-brown areas with sharply defined borders. When these lesions reach the skin, ribbon-like, depressed skin browning may develop. The development of "alcoholic" or "tainted" flavor is preceded by the loss of cultivar flavor. Loss of cutivar flavor is frequently associated with the first visual symptoms of low oxygen injury, *i.e.*, blueing or purpling of the skin.

There are not only cultivar differences in susceptibility to low oxygen injury, but there are also among-storage lot differences for the same cultivar stored in a CA room. Tests with McIntosh apples showed that among-orchard and among-harvest date differences in incidences of low oxygen injury were not correlated with differences in resistance of the skin to gas diffusion. It seems likely, therefore, that among-storage lot differences in susceptibility to low oxygen injury may be positively correlated with differences in physiological senescence. The validity of this observation has not yet been tested.

Orchard, handling, and storage factors that may weaken the fruit and advance maturity may also increase the susceptibility of the apples to low oxygen injury. These factors include, but may not be limited to; low average seed count, light crop, very high flesh K/Ca ratio, delayed harvest; slow cooling and slow establishment of the low oxygen atmospheric; repeated flushing of the room to re-establish the low oxygen atmosphere and prolonged storage under low oxygen conditions. In addition, there appears to be a negative relationship between temperature and low oxygen injury: the lower the temperature, the higher will be the oxygen threshold for low oxygen injury.

Let us try to understand about the method, how to defense against high C_2H_4 concentrations in low ethylene.

Ethylene concentrations in the exhaust of fossil-fueled forklifts are commonly 500 µl 1^{-1}. Ethylene analyses that showed catalytic mufflers did not reduce the exhaust C_2H_4 to acceptable concentrations. Therefore, fossil-fueled forklifts should not be used before low ethylene CA to handle bins of apples in enclosed areas.

Although the claim has been made that some fossil-fueled atmosphere generators do not produce C_2H_4, analyses by some workers indicated that open-flame atmosphere generators produce C_2H_4 and most catalytic atmosphere generators produce C_2H_4 when the oxygen is below 5 per cent and/or catalyst temperatures drop below 650°C. Therefore, the low oxygen atmosphere should be quickly established by flushing the atmosphere from the room with nitrogen gas. Two common sources of nitrogen gas are liquid nitrogen- and ammonia-fueled atmosphere generators. Comparative costs for atmosphere generation are available.

The fourth and final avoidance strategy is the harvest of apples at an early stage of maturity. McIntosh and Cox's Orange Pippin apples must be harvested before the climacteric rise has been initiated in the orchard. Early ripening Empire blocks may respond to low ethylene CA if harvested soon after the climacteric rise has begun. Normally, however, Empire apples should also be harvested before or at the climacteric. The effect of harvest date on Bramley's Seedling responses to low ethylene CA is not clear.

Several attempts have been made to develop a reliable method to predict C_2H_4 climacteric dates for apples in the orchard. The prediction methods were based on Liu's observation that, the closer the fruit sampling date is to the orchard climacteric date, the shorter will be the period between sampling and initiation of the C_2H_4 climacteric by the sample. These methods appear to show promise for predicting orchard C_2H_4 climacteric dates of McIntosh and Empire apples, but not for several other apple cultivars.

Two types of C_2H_4 scrubbers are currently used to remove C_2H_4 produced by apples in low ethylene CA: catalyst combustion and chemical absorption.

Ethylene is oxidized catalytically on a platinum catalyst, which is heated to 200°C to obtain maximum efficiency: $C_2H_4 + 3CO_2 \rightarrow 2CO_2 + 2H_2O$. An efficient heat exchanger is used to minimize the energy requirement for maintaining the catalyst at 200°C and to greatly reduce the temperature of the atmosphere before it is returned to the CA room. In comparison with chemisorption C_2H_4 scrubbers, heated catalyst converters (1) are much, easier to operate and (2) maintain a constant low concentration of C_2H_4 in the CA room because the conversion efficiency of the machine is normally 0.90 to 0.98. However, heated catalyst converters have a high initial capital cost and the total fixed cost plus operating cost per ton of storage fruit is constant regardless of the C_2H_4 load.

Ethylene may also be scrubbed from the storage atmosphere by chemical reaction with potassium permanganate: $4KMnO_4 + C_2H_4 \rightarrow 4MnO_2 + 4KOH + 2CO_2$. Some factors that influence the efficiency of this reaction under CA storage conditions have been described. Practically, potassium permanganate is incorporated into beads of activated alumina (Al_2O_3), which in New York State are held in a cabinet outside the CA room. The atmosphere is pumped from the CA room, through the bed of beads in the cabinet, and returned to the CA room. A sample of the beads must be removed from the cabinet periodically for visual insertion to determine when the charge of beads should be removed: fresh beads are purple and spent beads are brown in color.

The cost for C_2H_4 adsorbing beads was estimated to be three quarters of the total cost for C_2H_4 scrubbing with beads. Therefore, the cost of scrubbing C_2H_4 with beads should be highly dependent upon the C_2H_4 load, *i.e.*, the greater the C_2H_4 load, the higher the cost. Records from Empire low ethylene CA rooms in New York State indicated there was very little variation in bead usage. However, Idared, which produces C_2H_4 at a faster rate than Empire, had a higher bead usage or the storage season.

If a storage operator does not wish to make a long-term capital commitment to low ethylene CA, chemical absorption will be more attractive than catalyst combustion of C_2H_4 because the former has a lower capital investment. The disadvantages of chemisorption scrubbers are:

1. The required weekly maintenance of the scrubber and
2. The cycling of CA room C_2H_4 concentration, which is related to the reduction in scrubbing efficiency that occurs as a function of time.

A complete breakdown of costs for catalyst combustion and chemical adsorption scrubbing of C_2H_4 has been published. These costs are based on a specified, theoretical C_2H_4 removal load. The costs figures presented in Table 10.8 were calculated using the fixed operating cost parameters previously stipulated. The C_2H_4 scrubbing costs for the Idared room were higher than the C_2H_4 scrubbing costs for the three Empire rooms because the Idared produced more C_2H_4 than Empire in low ethylene CA. The C_2H_4 scrubbing costs for the three Empire rooms were higher than previous estimates, but seem to verify the observation that catalyst combustion may be slightly less expensive than chemical adsorption, if the catalyst combustion scrubber is properly sized to meet the C_2H_4 scrubbing needs.

Inhibition of Ethylene

In a low ethylene CA environment, the internal C_2H_4 concentrations of apples will be directly proportional to the rate of C_2H_4 production by the fruit. Inhibition of C_2H_4 production will, therefore, not only reduce the C_2H_4 load on the C_2H_4 scrubber, but it will also result in lower fruit internal C_2H_4 concentrations. Several strategies can be used to inhibit C_2H_4 production.

Rapid cooling and rapid CA are needed to suppress the rate of fruit ripening. Without these two strategies, the fruit ripening would quickly progress to the stage which results in the induction of rapid C_2H_4 production.

Small reductions in oxygen concentration below 3 per cent have resulted in pronounced reductions in rates of C_2H_4 production in Cox's Orange Pippin and McIntosh. At 1 per cent oxygen, the rates of C_2H_4 production are very low and C_2H_1 responses at 1 per cent oxygen are often negligible. The rapid accumulation of C_2H_4 in 1 per cent oxygen static atmosphere chambers indicated that retardation of ripening at 1 per cent oxygen was not the result of inhibited C_2H_4 production.

Oxygen at 2.0 to 2.5 per cent is used for low ethylene CA Empire in New York because higher oxygen concentrations have resulted in high average C_2H_4 concentrations and high rates of C_2H_4 production lower concentrations of oxygen increased the risk of low oxygen injury. New York low ethylene CA operators have frequently reported difficulty in maintaining the oxygen concentration at 2.0 to 2.5 per cent because CA rooms that were sufficiently tight for normal CA were found to be too leaky for low ethylene CA. In the low ethylene CA environment, the rate of fruit respiration was too low to consume the oxygen that leaked into the rooms. One or a combination of the following remedial procedures was used to maintain the oxygen at the desired 2.0 to 2.5 per cent.

Figure 10.3 indicates the scrubbing efficiency and concomitant increase of ethylene concentration.

After the study of ethylene inhibition we will try to understand about fruit internal ethylene concentration.

Estimations can be made of fruit internal C_2H_4 concentrations during low ethylene CA. Empire will be used as the model. The estimations of fruit internal C_2H_4 concentrations are calculated by using Fick's law of gas diffusion:

$$C_i = C_o + \frac{JR}{A}$$

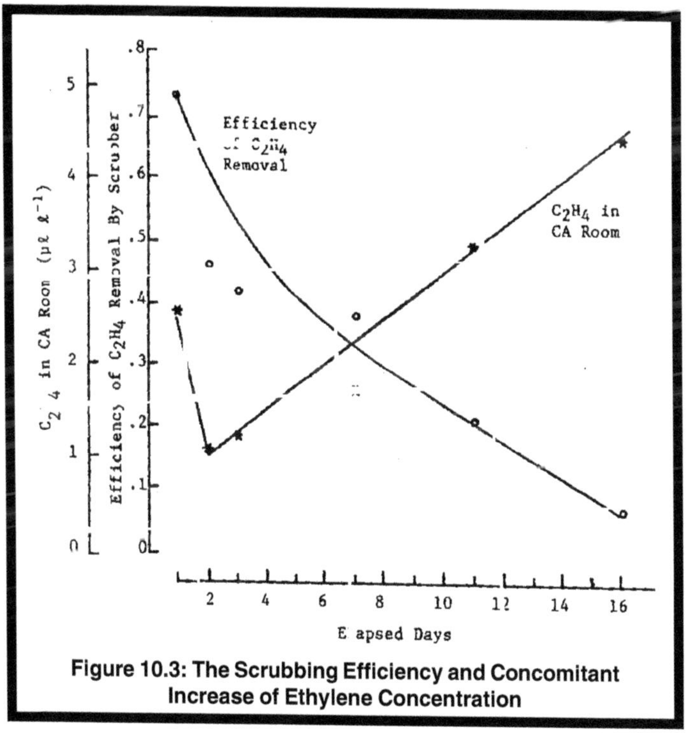

Figure 10.3: The Scrubbing Efficiency and Concomitant Increase of Ethylene Concentration

where $C_i = [C_2H_4]$ internal ($\mu l\ l^{-1}$), $C_o = [C_2H_4]$ outside ($\mu l\ l^{-1}$), J = rate of C_2H_4 efflux ($\mu l\ s^{-1}$), R = resistance coefficient ($10^4 s\ cm^{-1}$), and A = area of skin (cm^2).

C_o, the C_2H_4 concentration outside the apples, is the C_2H_4 concentration in the atmosphere of the low ethylene CA room. In the studies described below, C_o samples were transported in glass sample bottles from the commercial CA rooms to the lab, where the samples were analyzed for C_2H_4. Preliminary tests showed that C_2H_4 concentrations of samples held in plastic syringes began to decline after 1 h, but remained constant for at least 7 d when held in glass sample bottles which were fitted with a septum in the cap.

J, the rate of C_2H_4 efflux from apples, could theoretically be estimated from measurements of the rate of C_2H_4 removal by the C_2H_4 scrubber. Thus, the rate of C_2H_4 removal would be equal to the rate of C_2H_4 production if the concentration in the room remained constant. When this procedure was used, estimates of J were erratic and several fold higher than expected. An alternative method for estimating J, described below, resulted in values which were similar to expected values.

Ethylene scrubbers in three commercial Empire low ethylene CA rooms were turned off for a period of 48 h. Ethylene concentrations in the rooms were measured after 0, 24, and 48 h. The increase in C_2H_4 concentrations was linear with time. To change the concentrations of C_2H_4 to absolute quantities of C_2H_4, it was necessary to estimate the void volume of the rooms, *i.e.*, the volume of space not occupied by apples and wood bins. Two methods were used to estimate the void volumes of CA rooms.

With the first method, the void volumes of CA rooms were estimated with the formula: $V_{void} = V_{empty\ room} - (V_{apples} + V_{wood\ bins})$. $Y_{empty\ room}$ was determined by direct measurements or from architect's drawings. Empirical measurements in the laboratories showed that apples occupy about 55 per cent of the empty volume of the storage bins: $V_{apples} = 0.55\ (V_{empty\ bins})$ (number of bins). Measurements of several bins showed $V_{wood} = 0.1\ m^3$ per bin.

Several New York CA operators used liquid nitrogen to rapidly establish the low oxygen atmosphere in CA. Their liquid nitrogen usage and concomitant reductions in oxygen concentration frequently indicated close agreement with theoretical values.

Agreement of theoretical with actual values would require perfect mixing of the nitrogen gas with the atmosphere in the room. Perfect gas mixing was indicated in two subsequent tests. During the 2-h periods required to lower the oxygen from 20.8 to below 5.0 per cent, the oxygen concentrations at the CA room exhaust ports were identical to the oxygen concentrations in the centers of the rooms filled with bins of apples. This suggested that nitrogen flushing data could also be used to estimate CA room void volumes.

To use nitrogen flushing data to estimate the void volume in a room filled with bins of apples, it is first necessary to determine the number of void changes that occur when the room is flushed with nitrogen:

$$\frac{C}{C_o} = e^N$$

where C = final $[O_2]$ and C_o = original $[O_2]$ and N = number of void changes. For example, this formula can be used to determine that 1.65 void changes will be required to lower the oxygen from 20.8 to 4.0 per cent. The void volume of the room can then be calculated by dividing the volume of nitrogen gas (1 l of liquid N = 0.583 m^3 gaseous N at 0°C by the number of void changes. For example, if 590 m^3 of nitrogen gas resulted in 1.65 void changes, then the void volume of the room is 5.90/1.65 = 358 m^3.

Absolute quantities of accumulated C_2H_4 per metric ton of apples can then be estimated using C_2H_4 concentration, void volume, and the weight of the apples in the room. Determinations of J for these apples were in the expected range for Empire apples exposed to the concentrations of C_2H_4 ranging from 0.2 to 5.3 $\mu l\ l^{-1}$.

Values for the resistant coefficient of skin (R) of Empire apples were determined using the steady-state method and values were determined by Park, who used the CH_3CH_3 efflux method described by Cameron. Both methods indicated Empire values for R ranged from 1.1 to 3.7 $10^4 s\ cm^{-1}$.

Since empirical measurements of Empire apple weight and diameter were highly correlated (r = 0.997), values for the area of fruit skin, A, were determined by weighing the fruit and then calculating the area of the skin with the formula used to determine the area of a sphere (area = 12.57 r^2).

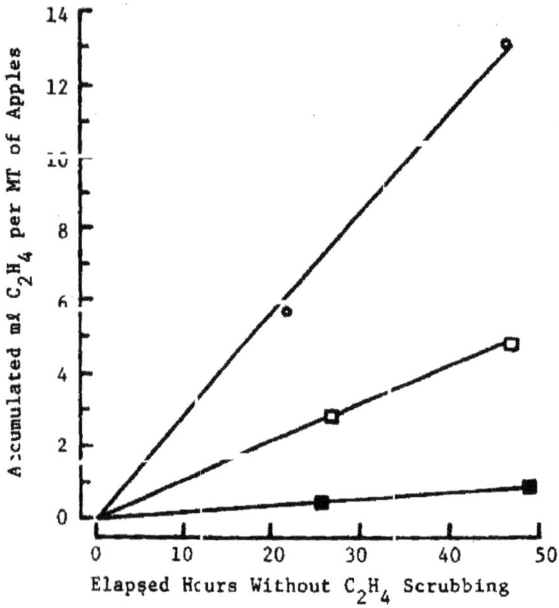

Figure 10.4: Accumulation of C_2H_4 in Three Low Ethylene CA Empire Rooms After the C_2H_4 Scrubbers were Turned Off

Estimated values of Empire apple internal C_2H_4 concentration (C_i) during low ethylene CA are show in Table 10.8. The two values for J (0.10 and 0.50 µl kg^{-1} h^{-1}) are one half and approximately twice the minimum and maximum values, respectively.

Table 10.8: Estimated Rates of Ethylene Emanation by Empire Apples During Low-Ethylene Controlled Atmosphere (CA) Storage, 1986

CA Room		*Seasonal Mean C_2H_4 Concentration in Ca (µl l^{-1})*	*C_2H_4 Emanation Rate in May (µl kg^{-1} h^{-1})*
Number	*t*		
1	252	5.3	0.29
2	357	1.2	0.10
3	260	0.2	0.02

The two values for R (1.1 and 3.7 10^4s cm^{-1}) are the extremes as previously mentioned. The two sizes of fruit (100 and 200 g) represent the normal range in Empire fruit size. The estimated fruit internal C_2H_4 concentration (C_i) shown in the table ranges from 0.12 to 6.13 $\mu l\ l^{-1}$, depending upon the values for the stipulated variables. These values for C_i are surprisingly close to the C_i values previously published and McIntosh apples held in CA storage with similar concentrations of C_2H_4.

The influence of four variables on C_i : J (rate of C_2H_4 production), R (skin resistant to gas diffusion), C_o (C_2H_4 concentration in the CA room), and A, expressed as fruit weight in the table. Tabular values indicated that fruit size has very little influence on C_i. This was verified by several observations. R has a strong influence on C_i, which may be measurably decreased by maintaining a high relative humidity (RH) in storage.

Physiology of Pears

The physiology of pears presents some interesting and challenging problems, that are distinctly different from the normal problems associated with apple postharvest physiology and technology.

In contrast with most apple cultivars, most pear cultivars are harvested while hard and green in color, have a chilling requirement for ripening, have storage lives that are significantly influenced by small differences in storage temperature near 0°C, lose water and shrivel rapidly at 0°C storage at 90 per cent relative humidity, do not ripen normally in cold storage, require a temperature of 15 to 20°C to ripen with good quality, lose the capacity to ripen when stored too long, when unrussetted are extremely sensitive to abrasion injury after removal from storage, and are at prime eating quality when the flesh firmness is 8 to 16 Newtons (N). Three of these recently studied postharvest pear attributes will be discussed in the paragraphs that follow: skin shriveling, a problem which is not unique to pears, but presents some special problems when pears are stored in CA for 6 to 9 months at – 1°C; over storage, a physiological disorder which can be delayed by CA; and brown core, a physiological disorder associated with CO_2 in storage.

Table 10.9: Estimated Empire Apple Internal Ethylene Concentration during Low- Ethylene Controlled Atmosphere (CA) Storage

J (µl kg^{-1} h^{-1})	0.1		0.50	
R (10^{4}s cm^{-1})	1.1	3.7	1.1	3.7
C_o (µl 1^{-1})	C_i (µl 1^{-1})		C_i (µl 1^{-1})	
		100-g fruit		
0.10	0.12	0.18	1.31	4.18
1.00	1.02	1.08	2.21	5.08
		200-g fruit		
0.10	0.13	0.20	1.62	5.23
1.00	1.03	1.10	2.52	6.13

During the 1950s and 1960s, the skin shrivel problem was controlled by storing pears in polyethylene film box liners. When the liners were sealed to allow the pears to generate a modified atmosphere (MA), CO_2 frequently accumulated in toxic concentrations. When the sealed liners were perforated to allow the CO_2 to escape, the MA conditions were lost because the oxygen concentrations were lose to or at the oxygen concentration in air. Shortly after this apparent conundrum was solved by inserting packages of hydrated lime into the sealed polyethylene box liners, the fruit industry switched from boxes to bins for handling pears.

Table 10.10: Internal Ethylene Concentrations or Individual McIntosh Apples Held in Two Storage Chambers at 1°C with 2.5 per cent O_2 and 2.5 per cent CO_2

Orchard Tree	Storage Chamber	C_2H_4 Chamber Atmosphere (µl l^{-1})		Apple Internal C_2H_4 (µl l^{-1}) Sep. 12	Sep. 19	Sep. 26	Oct. 3
1	A	0.2	Range*	0.7–1.1	0.7–2.4	0.6–4.1	0.8–7.0
			Average	0.8	1.0	1.5	2.7
2	B	0.9	Range*	1.6–7.3	1.2–9.4	1.2–8.8	2.8–9.3
			Average	3.3	4.6	4.5	5.5

Note: Harvest dates were in 1980.

* Range for ten apples from each harvest date.

Table 10.11: Weight Loss and Skin Shrivel Symptoms of Apples and Pears Stored in Air at 0°C, 95 per cent Relative Humidity (RH)

Fruit	*Weeks at 0°C, 95 per cent RH*		
	4	*8*	*12*
Apples*	3.8	5.4	6.7
Pears**	4.3	72	8.1
	Skin Shrivel Symptom		
Apples*	None	None	Slight
Pears**	Trace	Extremely severe	Extremely severe

* Averages for Golden Delicious. Jonathan, McIntosh, and Delicious apples.

** Averages for Magness, Moonglow, and Grant Seckel pears.

The problem of weight loss from bin-stored CA pears was investigated by fruit storage engineers at East Malling. They worked with Conference pears stored at a commercial storage facility located in Kent. They found that the three types of polyethylene film bin covers they tested increased the cooling time to 2°C by 1.5 to 4 d; increased the average storage temperature after pull down by 0.1 to 0.5°C; and had no significant effects on the poststorage ripening and incidences of flesh and core breakdown. When polyethylene film covered the tops and extended all the way down the sides of the bins, there was significantly less water loss and significantly more fruit decay in the covered bins. When loose fruit was used to hold polyethylene film sheets only at the tops of the bins, the corners of the film sheets frequently loosened, causing severe skin, shrivel of exposed pears. In the last reported test, polyethylene film covers extended 20 cm down the sides of the bins. Fruit weight loss was significantly reduced in bins located at the tops of the stacks. Pears in covered bins in the middle of the stack had fruit weight losses comparable to pears in covered bins which were located at the tops of the stacks. Pears in uncovered bins in the middle of the stacks lost 60 per cent of the weight lost by pears in uncovered bins located at the tops of the stacks. Data in Table 10.12 show the effect of polyethylene film bin covers on fruit weight loss at three locations in the bins.

Table 10.12: Effect of Polyethylene Bin Covers on Weight Loss of Conference Pears Stored in Bins at the Tops of Stacks in a Commercial Controlled Atmosphere (CA) Pear Storage Room

Location of Pears Within the Test Bins	*Polyethylene Film Bin Cover*	
	None	*Covers Stapled and Cut 20 cm Down Each Side*
Weight Loss (Percent of Original Weight)		
Top	5.4	(–) 0.1*
Middle	2.8	1.0
Bottom	2.4	1.8

* Pears gained 0.1 per cent of original weight.

Now let us try to understand about, the loss of the capacity of pears to ripen with good eating quality.

Overstorage is a term used to describe the loss of the capacity of pears to ripen with ; good eating quality when held at ripening temperatures (15 to 20°C) after prolonged storage at –1 to 0°C, Table 10.13 shows the symptoms of overstorage for four pear cultivars. The induction periods shown in Table 10.13 are not absolute, but indicate the relative differences among cultivars. When overstorage occurs, the flesh texture becomes dry and coarse rather than juicy and buttery after the pears have been ripened at 15 to 20°C. All four cultivars retain the ability to develop yellow color; flesh softening occurs in Anjou and Conference, but not in Bartlett and Bosc. Pear storage life, *i.e.*, the length of the period that pears can be stored and subsequently ripened with juicy and buttery flesh, may be influenced by a number of variables. The variables that influence overstorage must be understood before an attempt is made to develop a CA program to control or delay overstorage.

The following conclusions can be drawn from two early studies of the relationship between overstorage and total amount of CO_2 respired during storage at several non-ripening temperatures (–1 to 7°C):

1. Increasing the storage temperature resulted in proportional decreases in the storage life and proportional increases in the total amount of CO_2 respired by Bartlett and Anjou pears during their storage life and

2. Increasing the storage temperature shortened the storage life of Howell, Bosc, Packham's Triumph, and Winter Nellis pears, but the total respired CO_2 was constant for each cultivar.

Table 10.13: Induction Period and Fruit Characteristics after Overstorage Develops in Four Pear Cultivars

Cultivar	*Induction Period in Air at −1°C d*	*Fruit Characteristics After Overstorage Develops*		
		Flesh Texture Juicy and Buttery	*Flesh Softening*	*Skin Yellowing*
Bartlett	105	No	No	Yes
Bosc	110–120	No	No	Yes
Anjou	180–220	No	Yes	Yes
Conference	245	No	Yes	Yes

Recent studies at the Hood River Experiment Station in Oregon have focused on objective quantification of the symptoms of overstorage. Before these studies were completed, flesh firmness and panel evaluations of flesh texture were used to determine when overstorage occurred during a prolonged storage period at non-ripening temperatures. Flesh firmness measurements did not satisfactorily describe acceptable texture of pears, such as Anjou and Conference that soften, but have a coarse, dry texture when overstorage occurs. Professionally accepted panel evaluations of fruit texture were difficult to obtain.

In contrast with overstored Anjou pears, Anjou pears that retained the ability to ripen with buttery and juicy texture had a higher acid content, higher free amino acid content, and lower protein nitrogen and were usually more firm when they were taken from storage and placed at 20°C for ripening.

During the ripening period at 20°C, overstored and normal ripening Anjou pears frequently showed similar developments of the CO_2 and C_2H_4 climacterics, similar changes of internal C_2H_4 concentration, and similar flesh softening curves. Although the water content (percent fresh weight) of pear pulp was similar in unripened pears, more juice was extracted from ripened, overstored pears than from comparable pears that ripened with buttery, juicy texture.

The physiological characteristics of ripening Bosc and Anjou pears are similar in that the development of buttery, juicy flesh texture is associated with a reduction of extractable juice during ripening of both cultivars. In contrast with Anjou, Bosc had an increase in soluble solids when the pears were not overstored. Also in contrast with Anjou, Bosc lost less flesh firmness during the ripening of overstored fruit. With Bosc, there was little or no change in water-soluble polyurinides (WSP) during the ripening of overstored pears, but WSP increased significantly in Bosc that ripened with a buttery, juicy texture. Chen and Borgic speculated that the hygroscopic binding capacity of WSP inparts the buttery, juicy texture to ripened pear tissue.

Although the development of a procedure to measure extractable juice in ripened pears is of little interest to the commercial storage operator, it is of great importance to the researcher because it can be used to objectively quantify overstorage. The researcher must no longer rely on subjective panel evaluations to evaluate the effects of orchard, handling and storage practices on the development of overstorage.

Figures 10.5 and 10.6 illustrate the usefulness of extractable juice, measurements to quantify symptoms of overstorage.

Flesh firmness measurements are used for most cultivars to determine the proper stage of maturity for harvest. However, starch-iodine patterns are a better indicator of harvest maturity for Conference pears, which should be picked for CA when the percentage stained area is between two thirds and one third of the maximum stained percentage.

Data presented in Table 10.14 show that internal CO_2 increased dramatically in ripening Bosc pears. The later the harvest, the higher was the CO_2 in ripened pears at harvest. The longer the storage period, the higher was the internal CO_2 in ripened pears from any given, harvest date. Kidd and West also reported very high concentrations of CO_2 in ripened Conference pears. These observations suggest that orchard, handling, and storage factors that advance maturity may also increase internal CO_2 concentrations, which in turn cause brown core. However, this is probably not the case. Although factors that advanced maturity also increased the internal CO_2 concentration of ripened pears, these factors did not increase the internal CO_2 concentration in mature, unripened pears. Since brown core is present in mature, unripened pears, we may

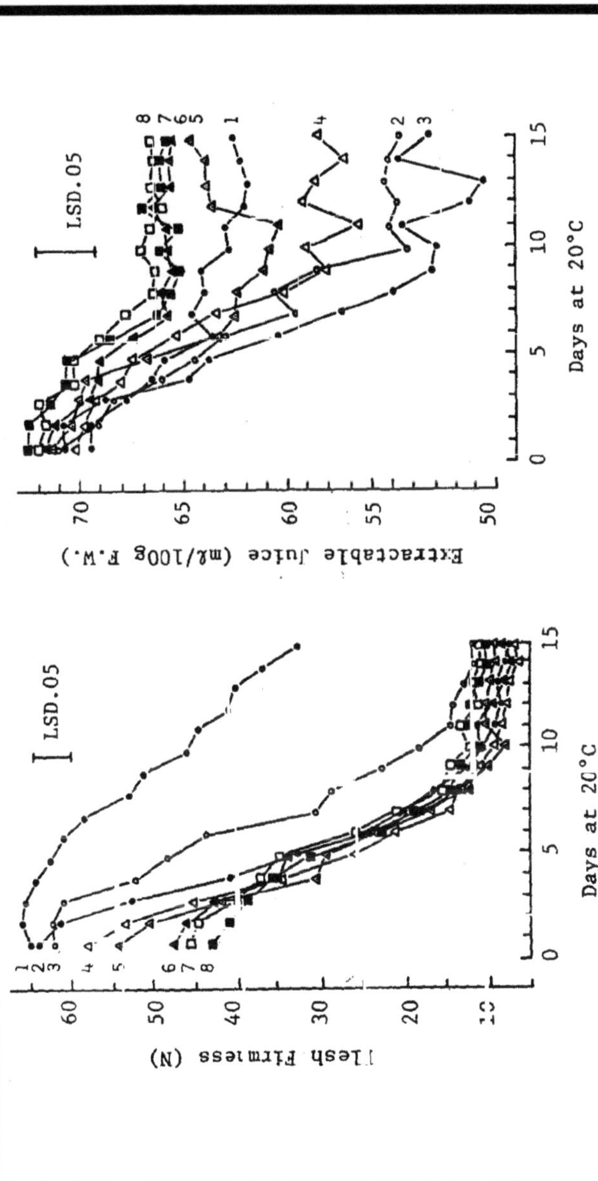

Figures 10.5 & 10.6: Usefulness of Extractable Juice, Measurements to Quantify Symptoms of Overstorage

speculate that factors which advance maturity may increase the sensitivity of pears to CO_2.

Table 10.14: Concentrations of CO_2 Frequently Recommended for Pears Stored at 2 to 3 per cent Oxygen

< 1% CO_2	*1 to 3% CO_2*	*> 3% CO_2*
Bosc	Anjou	Hardy
Clapps Favorite	Bartlett	Inverno
Conference (England)	Buona Luisa	Passe Crassane
Kaiser	Conference (Italy)	Spadona
Nejiseiki	Packham's Triumph	

Table 10.15: Etiology of Brown Core in Pears

Factor	*Influence on Brown Core*
Climate	Cool season increased injury
Tree vigor	Low tree vigor increased injury
Fruit size	Large fruit more severly injured
Harvest data	Late harvest increased injury
	Late harvest did not increase injury
Delayed storage	Increased injury
Slow cooling	Increased injury
High storage temp	Increased injury
Polyethelene box liners	Increased injury when sealed
CO_2	Increased injury
O_2	Low O_2 increased injury when CO_2 was present
Period of storage	Prolonged storage increased injury

The effects of CO_2 and oxygen concentration, harvest date, and storage removal date on sensitivity of CO_2 are illustrated by the data presented in Table 10.16. With 3 per cent oxygen, 12 per cent CO_2 caused earlier and more severe symptoms of brown core than did 0.2 per cent CO_2. Brown core incidences recorded in the last column of Table 10.16 show the increased sensitivity to CO_2 caused by the advanced maturity associated with both delayed harvest and prolonged storage. The effect of oxygen on sensitivity to CO_2 injury can be seen by comparisons of data for pears stored in air and at 3

Table 10.16: Bosc Pear Internal CO_2 and Brown Core (BC) Index when Removed from Storage

Storage Removal	*Harvest Date*	*Air*		*CA with 3% O_2*			
				2% CO_2		*0.2% CO_2*	
		Internal CO_2 (%)	*BC Index**	*Internal CO_2 (%)*	*BC Index**	*Internal CO_2 (%)*	*BC Index**
December	September 29	1.1	0	3.7	111	0.6	0
	October 6	0.9	0	4.7	223	0.5	0
	October 13	1.0	0	4.9	295	0.5	2
February	September 29	1.0	0	4.8	211	0.6	1
	October 6	0.8	0	4.8	284	0.5	15
	October 13	1.2	0	5.9	290	0.5	72
April	September 29	1.2	0	3.6	245	0.6	37
	October 6	1.1	0	3.6	277	0.5	53
	October 13	1.1	0	3.4	287	0.5	87

* BC index = (1 × % slight injury) + (3 × % moderate and severe injury).

per cent oxygen with 0.2 per cent CO_2. The latter had lower internal CO_2 concentrations, but higher incidences of brown core. That lowering the oxygen concentration increased the sensitivity of pears of CO_2 would also explain the progressive decrease of brown core from 94 to 0 per cent when oxygen was increased progressively from 1.2 to 10.8 per cent in an atmosphere containing 3.5 per cent CO_2. At 1 per cent oxygen, Bosc and Anjou pears are very sensitive to CO_2 and therefore, require that CO_2 concentrations be kept below 1 per cent in commercial CA pear storages.

Chapter 11

Postharvest Disease Development and Atmospheric Modifications

Respiratory Gases

We know well that oxygen is required for normal respiration and growth of the pathogen.

Alterations in the concentration of the respiratory gases in the storage atmosphere are primarily aimed at suppressing the respiration and other metabolic reactions of harvested fruits and vegetables to retard the ripening and senescence processes. Atmospheres are modified by elevating the carbon dioxide (CO_2) level, reducing the oxygen (O_2) tension, or both, to create a controlled atmosphere (CA) or modified atmosphere (MA) around the produce. CA storage generally implies precise control of O_2 and CO_2 in the atmosphere, whereas the term MA storage may designate any synthetic atmosphere, arising intentionally or unintentionally, but is often used when the composition of the atmosphere cannot be closely controlled. The current definition of MA refers to a self-modified atmosphere produced within sealed polymeric wrappings as a result of the respiration of the wrapped produce.

Many volatile compounds evolved by the produce accumulate in the storage atmosphere, of which ethylene is apparently the most important one. Since the accumulation of ethylene above certain levels may hasten the ripening of many fruits and enhance senescence of many tissues, its removal from the atmosphere may confer a further advantage of CA storage.

Hypobaric or low-pressure storage is another relatively new technology for postharvest preservation of perishables, involving the use of low barometric pressures and, hence, low-oxygen atmospheres. Under hypobaric storage, ethylene rapidly diffuses out of the tissues, while its production and action, like other metabolic processes, are greatly retarded. All these modifications lead to the extension of the postharvest life of commodities for which, rapid biochemical changes are an important factor in their deterioration.

However, for many fresh fruits and vegetables, the limiting factor in extending their useful life is the development of postharvest diseases. Along with the prolongation of the storage life and the advancing physiological deterioration of host tissues, the vulnerability to attack by typical postharvest organisms is greatly increased. Many investigations have proved that modifications of the storage atmosphere, along with their effects on the physiological life of the host, can also be advantageous in retarding postharvest disease development in various commodities, and a comprehensive review has been published on the subject by El-Goorani and Sommer. The effect of atmospheric modification on postharvest disease development can be direct–by suppressing various stages of the pathogen growth and its enzymatic activity–or indirect–by maintaining the resistance of the host to infection as a result of keeping it in a superior physiological condition. The present chapter will survey the effects of controlled and modified atmospheric components on pathogen development and their contribution to the control of pathogen-induced diseases in different host-pathogen interactions.

Oxygen is required for normal respiration and growth of the pathogen. Suppression of growth by low O_2 is most likely due to effects on the cytochrome system of electron transport, although other oxidative enzyme systems present in cells may also be suppressed by low oxygen. In most fungi, no growth occurs in the absence of molecular O_2. The reduction in the level of O_2 required for the inhibition of the various stages of fungal growth varies considerably among species. In general, however, lowering the level of O_2 from 21 to 5 per cent has little or no effect on spore germination or mycelium growth. In order to obtain appreciable reduction of spore germination, mycelial growth, and sporulation in many postharvest fungi, O_2 concentrations of about 1 per cent or less are required.

Exposure of 18 postharvest fungi to 2.3 per cent O_2 resulted in growth reduction of a few species only, including *Monilinia fructicola, Phytophthora parasitica,* and *Pencillium italicum,* whereas other species, even those of the same genus (*P. digitatum* and *P. expansum*), were not affected. Decreasing the O_2 atmospheric content to 0.5 per cent was found to decrease the *in vitro* rate of growth of some common postharvest fungi on solid media from one third to one half that of growth in air. Colony diameter of *Botrytis cinerea, Alternaria tenuis* (= *A. alternata*), *Cladosporium herbarum,* and *Rhizopus stolonifer* decreased significantly with each decrease in O_2 concentration from 1 to 0.25 per cent. When returned to air, all cultures resumed growth at normal rates.

Using mycelial dry weight as a criterion for fungal growth indicated that growth of these species, as well as that of *Fusarium roseum,* decreased linearly with decreasing O_2 concentration from 4 to 0.25 per cent. Mean per cent growth of *A. alternata, B. cinerea, C. herbarum,* and *F. roseum* was inhibited more than 50 per cent when the O_2 concentration was 4 per cent. For *R. stolonifer,* however, a 50 per cent inhibition was caused only at 2 per cent O_2 and some growth was recorded at 0 per cent.

In vitro effects of low O_2 on the linear phase growth of *R. stolonifer* were found to be dependent on the substrate. The O_2 requirement was less limiting to growth when glucose was the major source of carbon in the medium. Furthermore, there was evidence of growth in 0 per cent O_2 atmosphere on glucose medium, but not on pectin medium. Thus, on potato glucose medium, growth in 0.5 and 0.25

per cent O_2 averaged 21 and 15 per cent of that in air, respectively, whereas on potato pectin medium growth in 0.5 per cent O_2 averaged only 3 per cent of that in air and no growth occurred in 0.25 or 0 per cent O_2.

The existence of anaerobic pathways of metabolism in the genus *Rhizopus* has been documented. Glucose is fermented both aerobically and anaerobically to produce high yields of fumaric and lactic acid. Anaerobic metabolic pathways may, therefore, permit *R. stolonifer* to make effective use of trace or of low concentrations of O_2 for growth and for the production of extracellular enzymes.

A relationship between the composition of the medium and O_2 requirement was also recorded for *F. oxysporum* f. *cubense*. The fungus was found to grow anaerobically provided the medium contained yeast extract, MnO_2, KNO_3, Na_2SeO_3, or ferric ions.

The sensitivity to low O_2 tension of mycelial growth does not necessarily correspond with that of spore germination. Wells and Uota showed that germination of *R. stolonifer* and *C. herbarum* in 1 per cent O_2 was about 50 per cent of that in air and decreased linearly with decreasing O_2 concentration below 1 per cent. Spore germination of *A. alternata*, *B. cinerea*, and *F. roseum*, however, was not greatly affected until the concentration was 0.25 per cent or below. Completely anaerobic conditions prevented spore germination of *Gloeosporium musarum*, although it did not prevent mycelial growth on solid medium.

Bussel *et al.* showed that resting sporangiospores of *R. stolonifer* remained viable after exposure to anoxia for 72 h, but only a few germinating spores survived that period of anoxia. Electron micrographs revealed that after 72 h of anoxia, sporangiospores had only a few small mitochondria located at the periphery of the cytoplasm. The cytoplasm itself appeared mottled and had many vacuoles, vesicles, and lipid bodies. Upon returning to air, the spores recovered and mitochondria were seen in the central part of the cell.

The response of fungal sporulation to low O_2 may differ greatly with the species. *B. cinerea* produces abundant aerial mycelium when grown in 1 per cent O_2, but no spores develop in this level of O_2 or below. Sporulation of *A. alternata* and *C. herbarum*, on the other hand, takes place even in 0.25 per cent O_2, although mycelial growth is markedly reduced. Maturation of sporangia in *R. stolonifer* cultures decreases in 1 per cent O_2 and no mature sporangia are produced in 0.5 per cent O_2 or less.

Lowering of O_2 concentrations was found to reduce sclerotial formation by *Sclerotinia minor.* Sclerotial production was more sensitive to low O_2 than was radial growth and at 1 per cent O_2 no sclerotia were produced, although some growth was recorded.

Sensitivity to the concentration of O_2 in the atmosphere has been recorded for the decay-causing bacteria *Erwinia carotovora, E. atroseptica,* and *Pseudomonas fluorescence.* Significant reductions of *in vitro* growth resulted from lowering the O_2 concentration to between 1 and 3 per cent, a range tolerated by many agricultural commodities in storage. Growth decreased linearly with the logarithm of decreasing O_2 concentrations, but some growth (3 to 12.5 per cent of that in the air) was recorded at 0 per cent O_2 for these facultative anaerobic microorganisms.

Growth suppression at O_2 concentrations above 21 per cent is widespread among microorganisms, although some aerobic bacteria and fungi can grow in atmospheres of nearly 100 per cent O_2. Harrison attributed high O_2 toxicity to the accumulation of superoxide radicals which affect cell metabolism.

CO_2 is essential for the growth of many aerobic microorganisms. Although it is fixed by fungi. It cannot be used as an exclusive source of carbon for metabolism. However, elevation of the atmospheric CO_2 above a certain level may result in a direct suppression of fungal growth. At high levels of CO_2, most pathogens are suppressed by reduction in the rate of various metabolic functions, resulting in lowered respiration. The early studies of Brown and Brooks *et al.* have already demonstrated the inhibitory effect of high CO_2 atmospheres on mycelial growth or spore germination of *B. cinerea, R. stolonifer, Penicillium glaucum, Mucor* sp., and other fungi. The retarding effect of CO_2 on fungal growth was found by Brown to be greater at lower temperatures than at higher temperatures. Dilution of nutrient and density of spore suspension are other factors affecting the retardation of spore germination. The retarding effect of CO_2 was found greatest in the earliest phases of growth.

The concentration of CO_2 required to inhibit mycelial growth and spore germination varies with the species. Growth of *A. alternata, B. cinerea,* and *C. herbarum* decreases linearly with increasing CO_2

concentrations from 10 to 45 per cent and is inhibited by 50 per cent in an atmosphere of 20 per cent CO_2. Growth of *F. roseum,* however, is inhibited by 50 per cent only when CO_2 is elevated to 45 per cent. Spore germination of *R. stolonifer, C. herbarum,* and *B. cinerea* is already inhibited by over 90 per cent at 16 per cent CO_2, but levels as high as 32 per cent have no effects on *A. alternata* rate of germination.

El-Goorani and Sommer showed that growth of 18 common postharvest pathogens is slowed in air with 18 per cent CO_2 and that growth retardation is greater at 5.5 than at 12.5°C.

High concentrations of CO_2 (8 per cent and above) reduce both mycelial growth and sclerotial formation of *S. minor*-sclerotial formation being more sensitive to high levels of CO_2 than radial growth. A significant reduction in sclerotial germination occurs, however, only at about 20 per cent CO_2.

Growth of the two bacteria, *E. carotovora* and *E. atroseptica,* is inhibited both by CO_2 concentrations in excess of 10 per cent and by the absence of CO_2. Growth in 3 per cent CO_2 closely approximated that in air, but a decrease in the CO_2 concentration to 1 per cent resulted in a decrease in growth and at 0 per cent CO_2 no growth occurred at any concentration of O_2.

Littlefield *et al.* showed that increasing the concentration of CO_2 in a low O_2 atmosphere results in an increase in the suppressive effects of the atmosphere on the mycelial growth and sporulation of some postharvest pathogens. Sporulation of *Botrytis allii* was retarded about 50 per cent in an atmosphere of 2 per cent O_2 with 2 per cent CO_2, whereas *R. nigricans* (= *R. stolonifer*) was affected with respect to the height of the mycelial mat. *P. expansum,* although developing much lighter spores in this atmosphere, appeared to sporulate even better under these conditions. When 2 per cent O_2 was combined with 10.5 per cent CO_2, *B. allii* did not sporulate at all and only very slight sporulation occurred in *Rhizopus* cultures. Sporulation of *P. expansum* has already been arrested by increasing the CO_2 concentration in the CO_2–O_2 atmosphere to about 7 per cent.

Combining low O_2 with high CO_2 concentrations does not necessarily result in additive effective of the two atmospheric components. For certain fungi, growth retardation induced by low O_2 concentrations is not increased any further by elevating the CO_2 content of the atmosphere. Thus, retardation of growth rates recorded for *Phytophthora parasitica, Botryodiplodia theobromae,* or *Whetzelinia sclerotiorum* (= *Sclerotinia sclerotiorum*) in an atmosphere of 2.3 per cent O_2 plus 5 per cent CO_2 was not greater than that caused by 2.3 per cent O_2 alone. Certain fungi, however, such as *A. alternata* and *Penicillium expansum* have been found to grow more slowly in CA composed of 2.3 per cent O_2 and 5 per cent CO_2 than in air, although neither O_2 nor CO_2 alone caused any significant growth retardation.

Stimulations by CA Combination

The growth stimulation by CA combination is rather the exception, one cannot exclude such a possibility when evaluating the potential utility of atmospheric modifications.

Studying O_2 effects on *Phytophthora parasitica* var. *nicotianae* growth on a solid medium under static conditions, Dukes and Apple found that maximum growth occurs at about 5 per cent O_2. However, a decrease in *P. parasitica* rate of growth in low O_2 was reported by Klotz *et al.* on a liquid medium using a flowing gas system. Further studies with *Phytophthora* sp. showed that the growth of *P. cactorum* and *P. citrophthora* was stimulated by low O_2 concentrations on solid media, whereas no stimulation was recorded for these species in liquid media. The slow diffusion through the agar of toxic compounds, which may be produced at low O_2 levels, was suggested as a possible explanation for the stimulatory effects of O_2 on a solid medium.

Atmospheres low in O_2, however, were found to be stimulatory to *Geotrichum candidum* in liquid as well as on solid media. Submerged growth of *G. candidum* in a liquid medium was greater in 1 or 3 per cent O_2 atmospheres, with or without CO_2, than growth in air. In 21 per cent O_2, approximately 50 per cent growth reduction was found compared to that in the low O_2 levels. Fungal growth at all concentrations of O_2 was stimulated by the presence of 3 per cent CO_2 and was about 33 per cent greater than that in the corresponding O_2 atmosphere alone. At 0 per cent O_2, no growth occurred without CO_2, whereas in the presence of 3 per cent CO_2 limited anaerobic growth was detected.

At certain concentrations higher than 0.03 per cent, CO_2 was found to stimulate the growth of several pathogenic fungi. Mycelial growth of *F. roseum* was stimulated at 10 per cent CO_2, whereas, at 20 per cent CO_2 growth was similar so that in air. High CO_2

concentrations were also stimulatory to spore germination, and germination rates similar to that in air were recorded only when CO_2 was elevated up to 32 per cent. CO_2 was stimulatory to *Gloeosporium album* growth when applied at 5 or 10 per cent with 7.5 to 15 per cent O_2 although low concentrations of O_2 in the absence of CO_2 reduced mycelial growth. Similarly, growth of *F. oxysporum* was reduced by decreasing O_2 concentration, but was stimulated when the reduced O_2 atmospheres were supplemented with 2.5 to 15 per cent CO_2.

Stimulation of growth by CO_2 was recorded for several fungi when the O_2 level was less than normal. When the O_2 concentration is 2 per cent, and thereby limiting to mycelial growth, CO_2 at 4 to 16 per cent stimulates growth of *A. alternara. B. cinerea,* and *C. herbarum,* as well as *F. roseum.* When the accompanying O_2 level is 1 per cent, CO_2 induces stimulation of spore germination of all fungi for which this concentration functions as a limiting factor.

CO_2 stimulation of growth has been attributed to CO_2 fixation within fungal cells, with a number of enzymes being involved. Energy utilization derived from CO_2 fixation into acids of the Krebs cycle may be of significance to some species of fungi when O_2 is limiting to growth and when CO_2 concentrations are not in themselves inhibitory.

The effect of gas mixtures used in the storage of apples on the pectolytic activity *in vitro* of *Gloeosporium perenans* and *G. album* was studied by Edney to determine whether the differences obtained in rates of rotting in CA storage were a reflection of the direct atmospheric effect on the invading fungi. CA composed of 5 per cent CO_2 + 3 per cent O_2 was the most inhibitory atmosphere tested and markedly reduced mycelial growth as well as enzymatic activity of polygalacturonase (PG) and polymethylgalacturonase (PMG). The inhibitory effects on *G. album* were greater than those on *G. perenans,* but neither species exhibited any tissue maceration when grown in the gas mixture.

The effect of low O_2 atmospheres on the *in vitro* production of extracellular enzymes by *R. stolonifer* was related to the extent of the growth response and to the influence of glucose-induced repression of enzyme activities. *In vitro* effects of low O_2 on linear-phase growth of *R. stolonifer* depended on the substrate. PG, pectin methyl-esterase (PME), and cellulase (Cx) activities in potato-pectin culture filtrates were directly related to the growth response, with highest activities in the normally aerated cultures. In the potato-glucose medium, however, PG activity was higher in cultures held in low O_2 atmospheres following 48 h of growth in normal atmosphere than in cultures continually exposed to normal atmosphere. This was attributed to the repression of PG activity in late linear-phase growth in the glucose medium. Low O_2 atmospheres in restricting growth also interfered with PG repression. Enzyme repression does not occur in the pectin medium, but can be induced if the pectin is amended with glucose.

Various volatile emanations which may be accumulated in the storage atmosphere may affect disease development. Ethylene, which is the most important of these compounds, is produced by most plant organs as well as by a large number of microorganisms. Since initiation of natural ripening in climacteric fruits is accompanied by an increase in ethylene evolution, the fruit itself may be the source of ethylene in the storage atmosphere. However, enhancement of ethylene evolution also generally occurs in wounded tissues and in decaying plant organs. In the latter case, the ethylene evolution can be either infection induced, but of host origin, or of both host and pathogen origin.

The direct effects of various concentrations of ethylene on postharvest fruit pathogens has recently been studied by El-Kazzaz *et al.* Ethylene was found to stimulate spore germination of *Penicillium digitatum, P. italicum,* and *Thielaviopsis paradoxa* as was indicated by the increased rate of germination at 20°C. However, little or no effect on spore germination was recorded for the other seven fungi tested (*A. alternata, B. cinerea, Botryodiplodia theobromae. Colletotrichum gloeosporioides, Monilinia fructicola, P. expansum,* and *R. stolonifer*). On the other hand, exposure to ethylene stimulated germ tube elongation of most of the fungi tested, although it was not consistently reflected in the rate of increase in colony diameter. However, using fungal chitin determination as an index of fungal growth, the glucosamine content of *Botrytis cinerea* and *P. italicum* significantly increased in response to ethylene treatments. For *P. italicum,* a relationship between glucosamine value and ethylene concentration was observed.

Now let us try to understand about hypobaric pressure, which is employed to dealy the ripening process.

Hypobaric pressure, designed to delay ripening and senescence processes and to extend the storage life of fruits and vegetables, directly inhibits spore germination and mycelial growth of postharvest fungi on solid media. Inhibition of growth was exhibited by induction of a lag period in the growth curve, reduction in growth rate, and subsequent delay in sporulation. Apelbaum and Barkai-Golan have shown that the degree of inhibition increased with the decrease in pressure below 150 mm Hg. Reducing the pressure from 760 to 100 mm Hg inhibited spore germination of *P. digitatum* only.

Further reduction of the pressure to 25 mm Hg resulted in 100 per cent inhibition of germination of *P. digitatum, B. cinerea,* and *A. alternata,* but only in a 5 per cent inhibition in the population of *Geotrichum candidum* var. *citri-aurantii* (Figure 11.1). Colony growth was markedly reduced at 50 mm Hg, while at 25 mm Hg *Geotrichum* was the only fungus exhibiting some development (Figure 11.2). Delay in fungal sporulation was recorded under 50 and 25 mmHg. When cultures were transferred from hypobaric to atmospheric pressure, a normal rate of growth was resumed, suggesting that no irreversible damage to the fungi was caused.

The retardation of fungal growth under hypobaric pressure can be attributed to the reduction in the partial O_2 tension. By reducing the pressure to 100 or 50 mm Hg, the amount of available O_2 has been lowered from 21 per cent at atmosphere pressure to about 2.8.and 1.4 per cent, respectively. A comparison of *B. cinerea* growth at 100 and 50 mm Hg to growth at normal barometric pressure with low O_2, concentrations suggested that the suppressive effect at low pressures is similar to CA at equivalent O_2 tension. However, Wu and Salunkhe found that the growth and sporulation of several postharvest fungi including *P. expansum, R. stolonifer. Aspergillus niger, B. allii,* and *Alternaria* sp. was less at an atmosphere of 102 mm Hg, which contained the equivalent of 2.7 per cent O_2, than at normal atmospheric pressure (646 mm Hg) with 2.7 per cent O_2. Apelbaum and Barkai-Golan have shown that admission of O_2 to fungal cultures held at 50 mm Hg only partially reversed the growth inhibition effect. Furthermore, when cultures were held at atmospheric pressure (760 mm Hg), but with O_2 partial pressure similar to that prevailing at 50 mm Hg with air ventilation, mycelial growth was greater than at the same O_2 tension at 50 mm Hg. This led to the suggestion that growth inhibition under low pressure conditions may be due, at least in part, to effects other than the

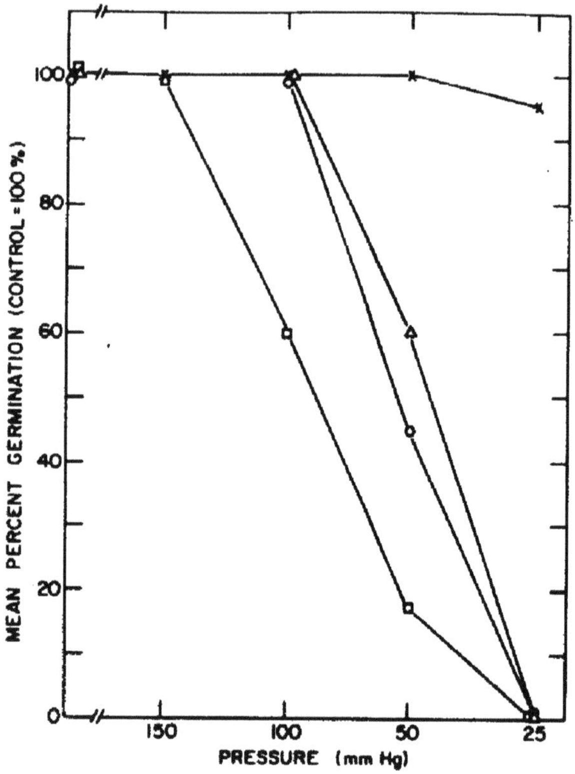

Figure 11.1: Effect of Hypobaric Pressure on Spore Germination of *Penicillium digitatum* (□), *Botrytis cinerea* (O), *Alternaria alternata* (Δ), and *Geotrichum canadium* var. *Citri-aurantii* (x)

reduction of O_2 tension. A reduction in atmospheric pressure causes a reduction in the CO_2 partial pressure as well. The additive effect of simultaneous reduction of O_2 and CO_2 tension and, perhaps, the acceleration of the escape of volatile compounds from within the tissue may account for the pronounced inhibition of fungal growth under subatmospheric pressures.

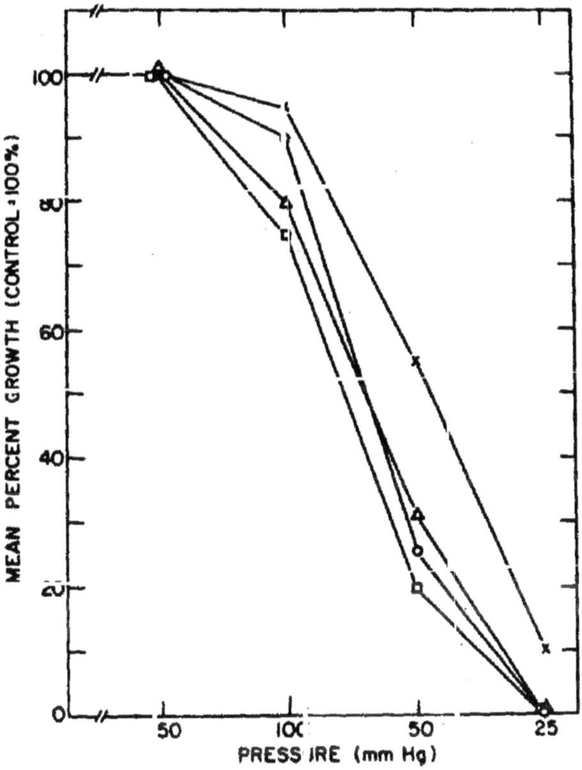

Figure 11.2: Effect of Hypobaric Pressure on Colony Growth of *Penicillium digitatum* (□), *Botrytis cinerea* (O), *Alternaria alternata* (Δ), and *Geotrichum canadium* var. *Citri-aurantii* (x)

Carbon monoxide is used effectively to suppress postharvest pathogens. This gas functions physiologically as an inhibitor of cytochrome oxidase and as a competitor of O_2. When added to a low O_2 (4 per cent) atmosphere, CO (5 to 10 per cent) retarded the *in vitro* growth of *B. cinerea* as was evidenced by the significant reduction in germ tube elongation and the radial growth of the colony. The suppression was less pronounced if CO was added to air rather than to low O_2 CA. It was clearly shown that the 4 per cent O_2 atmosphere alone was not adequate to influence *Botrytis* growth rate.

EI-Goorani and Sommer showed that growth of most common postharvest fungi was either not affected or was only slightly slowed by the addition of 9 per cent CO to air. However, certain fungi, including *P. digitatum, P. italicum, W. sclerotiorum,* and *Verticillium theobromae,* were found to be sensitive to CO and their growth was reduced by 50 to 70 per cent of that in air alone. The effect of CO is generally much greater when added to an atmosphere low in O_2 (2.3 per cent). In most cases growth is further suppressed when CO_2 (5 or 18 per cent) is added to the atmosphere, because of the additive effects of CO and CO_2.

CO (9 to 11 per cent) was found to strikingly suppress growth of *B. cinerea, M. fructicola,* and *P. expansum* only if the atmosphere contained less than 5 or 6 per cent O_2. The reduction in fungal growth in low O_2 atmosphere increased with the increase in CO concentration. Growth of *B. cinerea* and *M. fructicola,* which is almost the same at 2.2 per cent O_2 as in air, was found to be reduced to two thirds and one third, respectively, with the addition of 10 per cent CO. The response to CO at about 1 per cent O_2 was even more striking. At about 1 per cent O_2 *Monilinia* growth was 80 per cent of that in air, but with CO added, the rate was only 4 per cent of that in air. Similarly, CO in 2.5 per cent O_2 atmosphere markedly reduced growth of *P. expansum,* although at this O_2 level without CO the fungus grew as well as in air.

Control of Microorganisms

The effects of decreased O_2 concentration, increased CO_2 level, or both on the *in vitro* growth of microorganisms may sometimes serve as a basis for evaluating the effect of CA composition on infected plant tissues. However, the growth of a pathogen within a host may differ greatly from its growth under *in vitro* conditions, not only because of the different medium posed by each kind of fruit or vegetable, but also as a result of the interrelationship between the host and the pathogen. Under *in vivo* conditions, the same atmospheric gases affect (1) the host tissues, (2) the pathogen propagules, and (3) host-pathogen relationships, of putting it schematically:

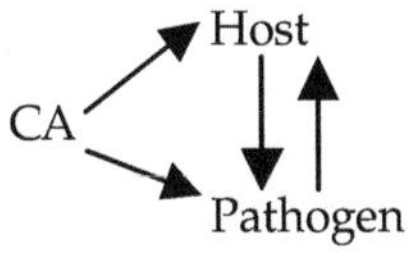

The storage atmosphere can therefore, influence disease development both by directly suppressing the growth of the pathogen and by altering the resistance of the host to parasitism.

Reduced O_2 and/or elevated CO_2 retard ripening and senescence of the host by reducing rates of respiration and other metabolic reactions. Low O_2 and high CO_2 might also exert their effect on ripening via slowing or inhibiting the production and action of ethylene which is considered to be the triggering agent for ripening.

It is well recognized that the susceptibility of fruits to postharvest decay organisms generally increases with ripeness. Theories proposed to explain the increase in the susceptibility of the host with ripening and aging involve:

1. An increase in the membrane permeability resulting in the release of nutrients and water into the intercellular spaces
2. A decrease in the ability of the host tissue to synthesize antifungal compounds with age of the produce, and
3. An increase in the susceptibility of the plant cell wall to attack by macerating enzymes of the pathogen.

As fruits ripen, the pectic substances of the middle lamella, which are initially laid down by the cell in a relatively insoluble form, protopectin, become more soluble and the tissue begins to soften. The increased solubility of the pectic substances in the soft tissue makes it more vulnerable to macceration by the pectolytic enzymes of the pathogen. CAs retard changes in pectic constituents and keep the texture firmer for a longer period. Retardation of changes of protopectin to soluble pectin by CA was shown in several cultivars of apples. Decay suppression under CA conditions that delay ripening can therefore be related, fully or partially, to the retention of a high level of the natural host resistance. When softening is delayed by CA storage, the fruit might also be less prone to injury and thereby be less exposed to wound inoculations.

An exception to the correlation between ripeness and disease development is the bacterial soft rot of tomatoes, caused by *Erwinia carotovora,* which develops even more rapid in mature-green fruits than in mature fruits. The susceptibility of the green tomatoes to bacterial soft rot may explain why CA conditions, which delayed ripening and reduced the development of *Rhizopus* and *Alternaria* rots in mature-green tomatoes, do not result in bacterial soft rot control.

Atmospheres containing about 2.5 per cent O_2, a level which is commonly maintained in CA storages, are most likely to act upon host resistance rather than directly inhibit growth of the pathogen, since most pathogenic fungi grow under these conditions. It was thus found that decay of tomatoes by *Rhizopus* sp. and *Alternaria* sp. was markedly reduced when stored in a 3 per cent O_2 atmosphere, although both fungi grow well in this atmosphere or in even lower O_2 levels.

As a matter of fact, growth of common postharvest pathogens is reduced greatly only when O_2 is lowered to 1 per cent or less. However, excessively low O_2 concentrations result in anaerobic or fermentative respiration and the accumulation of alcohols and aldehydes in the tissues. These lead to the development of off-flavors and to irreparable damage of the tissues.

The development of persistent off-flavors was found in strawberry fruits at low O_2 concentrations which reduced decay caused by *B. cinerea* and *R. stolonifer.* For Shamouti oranges, atmospheres containing 2.5 or 5 per cent O_2 were mostly effective in reducing postharvest decay caused mainly by *Alternaria citri, Diplodia natalensis, P. digitatum,* and *P. italicum.* However, under these conditions, rind injury appeared on most of the young immature fruits.

Elevating the CO_2 level of the atmosphere above 5 per cent suppresses fruit respiration, but has almost no direct effect on fungal growth. In order to inhibit growth significantly, the level of CO_2 has to be increased to about 10 per cent or more. The rate of decay incited by either *M. fructicola* or *R. stolonifer* in inoculated peach fruits was markedly reduced only at CO_2 levels above 15 per cent which have been shown to harm fruit quality. Fruits and vegetables may tolerate high CO_2 levels for short periods, but only few fruits can tolerate very high levels, during extended storage or transit. Under excessive concentrations of CO_2, off-flavors may develop and the fruit is injured.

Strawberries and sweet cherries are quite tolerant to high CO_2 levels and elevating the CO_2 content of the atmosphere to 20 to 30 per cent is a well-known procedure for suppressing decay in these fruits during transit. Brooks *et al.* have shown that 10 to 13 per cent CO_2 had little effect on the development of *B. cinera* and *R. stolonifer* in inoculated strawberries, whereas 23 per cent CO_2 almost completely inhibited their growth. A level of 20 per cent CO_2 was found not only to reduce decay in stored strawberries, but also to retard softening without injuring the berries or impairing their delicate flavor. At this concentration, CO_2 can reduce decay both by retarding the physiological deterioration of the berries and by directly inhibiting fungal growth.

A CO_2-enriched atmosphere, averaging 12 to 15 per cent, was found to present an alternative to rapid pre-cooling for blueberries stored at 2°C. This atmosphere compensates for the lack of pre-cooling by suppressing postharvest decay (mainly *B. cinera, Alternaria* sp., and *Colletotrichum* sp.) during the time required to cool the berries to the required cold temperature. For best results, however, the CO_2 should be a supplement to rapid pre-cooling.

Reduction of Decaying of Food Grains

Tolerance to low O_2 or high CO_2 levels varies considerably not only among different commodities but also among cultivars of the same species. Other factors such as the physiological state at harvest, growing conditions, and duration of storage may also affect the reaction to CA storage. Because of the widely varying and generally unpredictable responses of different commodities to atmosphere modifications, ideal combinations need to be determined experimentally for each case.

The use of CA is frequently designed primarily to retard ripening or aging processes. In some cases, however, reduction of decay is the main reason for using CA.

Apples are an example of the group of crops for which CA is aimed at retardation of ripening. The low O_2 and high CO_2 in combination with low temperature suppress and delay the climacteric of the fruit more than does low temperature alone. Striking benefits were found in apple cultivars that suffer from chilling injury at temperatures near 0°C and those which have a short storage life. Ethylene reduction or removal in CA storage frequently resulted in better firmness retention in certain cultivars of apples, mainly if the fruit was picked in its pre-climacteric state. In addition to fruit cultivar and harvest date, the use of growth regulators in the orchard was also found to influence the effectiveness of low-ethylene CA storage.

Along with the prolongation of the physiological life of apples, CA was also found to suppress decay such as apple scab lesions caused by *Venturia inaequalis* and the blue mold rot caused by *P. expansum*. When *P. expansum*–inoculated Delicious apples were held in either 3 per cent O_2 with 2 per cent CO_2 or 1 per cent O_2 CA storage at 0°C, the fruit exhibited about 45 per cent less area of decay than the air-stored fruit.

Edney showed that the efficiency of CA in controlling *Gloeosporium* rot in apples depended upon inoculum concentration. When the fruit were heavily infected, either naturally or following inoculation, rot development in CA was similar to that found in air. Furthermore, as a consequence of the differences between *G. album* and *G. perenans* response to CA, the former tended to predominate in air, while the latter predominated in CA storage. Differences in response to CA were also found between *Pezicula alba* and *P. malicorticis,* which are responsible for extensive damage during cold storage. These differences were attributed to the ability of *P. malicorticis,* but not of *P. alba,* to develop in the unripe fruit.

Recent technologies involving short-term pre-treatment with high levels of CO_2, or the reduction of O_2 levels immediately after harvest to 1 to 1.5 per cent, have greatly improved the store-ability of apples. However, in order to control postharvest decay of apples, satisfactory fungicidal treatments should precede CA storage.

Pre-treatment with high CO_2 prior to CA storage was also found to improve the keeping quality of pears. Couey and Wright showed that prestorage CO_2 treatment did not reduce firmness loss in storage, but resulted in reduced stem decay and improved flavor retention in d'Anjou pears.

Avocados benefit greatly by CA storage. Early studies have shown that the respiration climacteric peak of avocado fruit is delayed by decreasing the O_2 concentration from 21 to 2.5 per cent or elevating the CO_2 level to 10 per cent. Storing Florida avocados for 45 to 60 d at 10°C in an atmosphere of 2 per cent and 10 per cent CO_2

resulted in a remarkable reduction in anthracnose (*C. gloeosporioides*), the major cause of spoilage during storage and marketing, compared to fruit stored at the same temperature in air. This atmosphere also prevented chilling injury, which is generally accompanied by a greater susceptibility to fungal attacks, and permitted storage of the cold-sensitive fruit at 7.2°C. Atmospheres of low O_2 without CO_2 or high CO_2 with normal O_2 were more effective than air alone, but not as effective as the combination of the two for controlling decay and chilling injury. Since atmospheres of 2 per cent O_2 to 10 per cent CO_2 had only a slight inhibitory effect on the growth of *C. gloeosporioides*, decay control in this atmosphere was attributed to maintenance of the fruit in the firm mature-green stage at which it is resistant to the latent fungus.

Reduction of anthracnose in avocados stored at 12°C was recently obtained by applying CO_2 pulses every 5 d for up to 4 weeks, to give an atmosphere of 20 per cent CO_2, before removal to normal atmosphere.

In contrast to avocados, CA storage had no to slight benefit for papayas, for which decay is, similarly, the chief limiting factor in storage extension.

CA delayed fruit ripening of several cultivars of mango even at the relatively high temperature (13 to 14°C) required for this cold-sensitive fruit and extended their storage life to 6 to 10 weeks. However, postharvest fungicidal treatments were needed for suppression of decay caused mainly by *A. alternata.*

Decay control is the main reason for using CA in strawberries, which are characterized by a very limited postharvest life. Although the development of *B. cinerea,* the main postharvest pathogen, is greatly reduced at about 2°C, a rapid growth is gained upon removal of the fruit to shelf-life conditions. Decay caused by *B. cinerea* and *R. stolonifer* in California-grown strawberries is reduced markedly by lowered O_2 concentrations or raised CO_2 concentrations. However, exposure of the berries to reduced O_2 (>0.5 per cent) resulted in the development of off-flavors, whereas CO_2 atmospheres of 10 to 20 per cent reduced decay without causing damage or objectionable off-flavors. High CO_2 have been found to suppress decay not only during cold storage but also after removal of the fruit to shelf-life conditions in air. An atmosphere of 15 per cent CO_2 was much more effective in suppressing *B. cinerea* lesions in inoculated berries than was CA composed of 2.3 per cent O_2 with 5 per cent CO_2 and almost completely prevented infection by contact with mycelium-bearing fruit.

In experiments to simulate conditions in commercially treated polyethylene-covered pallets of strawberries, Browne *et al.* found that the effectiveness of 10 per cent CO_2 in reducing fungal decay due to *B. cinerea* and *Mucor piriformis* varied according to season and time of harvest. Furthermore, 10 per cent CO_2 also caused persistent off-flavors in the U.K.-grown cultivar, indicating a difference in varietal response or an effect of growing conditions.

Beneficial effects of CA for tomatoes had already been observed in 1933 by Kidd and West, who reported that an atmosphere of 5 per cent O_2 with 5 per cent CO_2 at 12°C was effective in retarding both ripening and fungal growth. Different CA compositions, however, were suggested by Eaves and Lockhart as optimal for delaying ripening and for reducing decay. Tomato decay caused by *B. cinerea*, *Rhizopus* sp., *Penicillium* sp., *S. sclerotiorum*, and *Colletotrichum coccodes* was reduced by CA storage at 3 per cent O_2 and 3 per cent CO_2. *A. tenuis* (= *A. alternata*), although being retarded by CA, developed rapidly when the fruit was removed to the ripening room.

Decay by *Rhizopus* sp. and *Alternaria* sp. was reduced considerably when mature-green tomatoes were stored in an atmosphere of 3 per cent O_2 without an increased level of CO_2. Combining the low O_2 atmosphere with high CO_2 did not further affect decay incidence and sometimes resulted in CO_2 injury.

Studying the potential for long-term storage of tomatoes, Mansfield *et al.* found that CA of 4 per cent O_2 without CO_2 or with 2 per cent CO_2 reduced decay significantly in both uninoculated and *B. cinerea* inoculated tomatoes after 7 weeks at 12.5°C. A reduction in decay incidence was also reported for the decay-sensitive late fall fruits that had been subjected to chilling in the field before harvest. On the other hand, there are data indicating that *Geotrichum* rot and *Fusarium* rot of tomato were stimulated by atmospheres of 3 per cent O_2 with 3 per cent CO_2 and of 3 per cent O_2 with or without 5 per cent CO_2, respectively.

CA has considerable benefits over air for the long-term refrigerated storage of white cabbage. These are exhibited by lower trimming losses, retardation of general senescence, such as yellowing, toughening, and loss of flavor, inhibition of physiological disorders, and repression of decay development. Lower incidence and severily of *B. cinerea* rots were found in cabbages stored at 0 to 2°C in CA containing 5 to 6 per cent CO_2 and 3 per cent O_2 as compared to those found in air at the same temperature. Adair showed, however, that while the development of *B. cinerea* and *S. sclerotiorum*, the two important postharvest pathogens of cabbage, were inhibited in CA with 4.7 per cent CO_2 and 1.4 per cent O_2, *F. roseum*, which has not been considered an important pathogen of cabbage stored in air, was consistently isolated under CA conditions. A study of the direct effects of CA on inoculated cabbage leaf disks has shown that the virulence of *B. cinerea* in 4.7 per cent CO_2 and 0.8 per cent O_2 was much reduced from that in 4.7 per cent CO_2 and 21 per cent O_2, whereas *F. roseum* was equally virulent at either O_2 concentration. The incidence of several other pathogenic fungi *A. alternata, A. brassicicola,* and *Mycosphaerella brassicicola* was not consistently controlled by CA storage and, in some cases, these pathogens were more evident in CA storage where they were not masked by *B. cinerea* development.

Accumulation of Ethylene and Hypobaric Pressure

The accumulation of ethylene at certain levels in the storage atmosphere may result in the initiation of ripening in fruits or in the enhancement of tissue senescence and has generally been regarded as detrimental to prolongation of the storage life of fruits and vegetables. Since susceptibility to disease generally increases with ripeness, exposure of the fruit to ethylene may also be conducive to increasing its susceptibility to fungal attack.

Applying ethylene to various varieties of oranges significantly increased stem-end rot caused by *D. natalensis* during storage by accelerating senescence and abscission of the fruit button and by stimulating the development of the quiescent infections of the fungus in this organ. The increase in stem-end rots was directly related to both the concentration, of ethylene used in degreening and the duration of exposure to a given ethylene concentration. Brown and Barmore showed that the development of anthracnose in tangerines inoculated with *C. gloeosporioides* occurred only when the fruits were exposed to ethylene immediately after inoculation. Ethylene was suggested to induce disease by stimulating appressoria present on the fruit to form infection hyphae that are capable of penetrating the epidermal cells. An increase in *Alternaria* rot was observed on tomato fruits following treatment with ethylene. A stimulatory effect of ethylene was also exhibited for *B. cinerea* in strawberry fruit and *Penicillium italicum* in Valencia oranges when the glucosamine content of the infected tissues was used as criterion for fungal growth.

The addition of 20 ppm ethylene to either air or CA (2.3 O_2 + 5 per cent CO_2) resulted in enhanced disease development in *B. cinerea*-inoculated strawberries of three cultivars.

In a recent study, Aharoni *et al.* found that ethylene plays a primary role in the initiation of broccoli influorescence senescence and that the senescence-retarding effects of a CO_2-enriched atmosphere are related to its ability to block ethylene action. Furthermore, the marked decrease in rot development observed in broccoli florets inoculated with *B. cinerea* spores and stored in a CO_2-enriched atmosphere was completely nullified by the addition of 10 ppm ethylene to the atmosphere.

Several studies, however, have indicated that ethylene might also induce greater plant resistance to fungal development. A retarding effect of ethylene on disease development has been reported on sweet potato roots infected by *Ceratocystis fimbriata* and on McIntosh apples infected by *Gloeosporium album* and held in 0 per cent CO_2 and 3 per cent O_2. A resistance in green tangerines to anthracnose was induced by ethylene when the fruit was exposed to the gas for 3 d before inoculation with *Colletotrichum gloeosporioides* spores.

Table 11.1: Effect of CO_2 and Ethylene on Decay Development on Broccoli Florets Inoculated with *Botrytis cinerea* Spores

Treatment	*Index of Decay* (After 120 h of Incubation)*
Air	4.0
CO_2 (10%)	2.6
C_2H_4 (10µl/l)	4.2
CO_2 + C_2H_4	4.2

* 1: none; 5: maximum development.

El-Kazzaz *et al.* observed that ethylene did not affect lesion diameters on naval oranges inoculated with *P. italicum* before ethylene exposure was initiated. However, when the fruit was treated with ethylene for 2 to 6 d at 20°C prior to inoculation, induced resistance to the blue mold was noted and smaller lesions were developed. The increased resistance to disease was suggested to be due to the induced formation of phytoalexins or other antifungal compounds in the ethylene-treated tissues.

Now let us try to understand about low pressure storage.

Hypobaric or low-pressure (LP) storage consists of maintaining the commodity in a vacuum-right and refrigerated container continually ventilated with humidified air, relative humidity (RH) = 80 to 100 per cent, at subatmospheric pressure. The oxygen level in the atmosphere varies in direct response to the absolute pressure. In the continuously ventilated partial vacuum, CO_2, ethylene, and various volatile by-products of metabolism rapidly diffuse out of the commodity and are flushed from the storage chamber. As a consequence of the low partial pressure of O_2 and the hyponormal levels of ethylene, ripening and senescence of fresh fruits and vegetables are delayed and storage life is extended.

Following the disclosure of the hypobaric storage technology by Burg and Burg in 1966, extensive investigations have been conducted with a wide range of fruits and vegetables. Although the high RH conditions under LP storage appear ideal for fungal growth and resultant decay, several studies report on the beneficial effects of the subatmospheric pressure on decay control. Since hypobaric pressure exhibits a direct fungistatic effect, it can retard disease development both directly, by inhibiting fungal colonization, and indirectly, by delaying ripening and senescence processes.

Beneficial results reported for apples under LP storage include, along with delayed softening, higher acids and sugars and control of physiological disorders, as well as reduced decay and the consequent increased shelf life after removal from storage. On the other hand, the incidence of decay was quite severe in mature-green tomatoes which were stored at 76 mm Hg for 56 d. Bangerth has pointed out that a postharvest application of fungicides was necessary to suppress fungal decay in cucumbers and peppers under LP conditions.

Latent Infections

Latent infections, which originate from infection of the fruit in the field but remain dormant until fruit ripening, have been observed in many host-pathogen interactions. Two types of fungal growth inhibitors are believed to be involved in the formation of latent infections on the unripe fruit–preformed compounds and induced compounds which are synthesized by the host in response to attempted infections (phytoalexins).

C. gloeosporioides, the causal agent of avocado anthracnose, forms a typical latent infection in avocado fruits. Exposure of the fruits to CA of 2 per cent O_2 and 10 per cent CO_2 at 10°C resulted in a remarkable reduction in anthracnose, although it had almost no direct effect on colony growth of *C. gloeosporioides.* The control of decay on avocados by CA storage was attributed to the induced retardation in the rate of softening, which would serve to keep the fungus dormant. The removal of ethylene from the atmosphere, which may lead to a further retardation of the fruit ripening, results in a further reduction in decay development. Prusky *et al.* showed that hypobaric pressure delayed softening and prolonged the latency of *C. gloeosporioides* on avocado fruit disks. Since *Colletotrichum* development on fruit disks was completely prevented at 150 mm Hg, while fungal growth *in vitro* was affected only slightly, inhibition of disease development was related to the delayed ripening process.

Recently, a preformed antifungal compound–*cis,cis*–1–acetoxy–2–hydroxy–4–oxo–heneicosa–12,15–diene–isolated from the peel of unripe avocado fruits, was suggested to account for the latency of *Colletotrichum* infection. This compound, which inhibits, vegetative growth and spore germination of *C. gloeosporioides* at concentrations even lower.than those found in the unripe fruit peel, decreases during ripening to subtoxic levels. The degradation of the antifungal diene during ripening was suggested to result from lipoxygenase activity which increases with fruit ripening owing to the decline in the level of its endogenous inhibitor epicatechin. Storing avocado fruits in a CO_2-enriched atmosphere (9 per cent CO_2 + 13 per cent O_2), produced within sealed plastic bags, delayed fruit softening, eliminated the reduction of the antifungal diene, and maintained latency.

Latency of *A. alternata* infection in mango fruits has recently been related to the presence of preformed antifungal resorcinols in the peel of unripe fruit. The concentration of the antifungal compounds was found to decrease during ripening, coinciding with the renewed development of previously latent *Alternaria* infections. Ethylene treatments, which enhanced ripening, induced an earlier reduction in the concentration of the antifungal resorcinols and concomitantly enhanced disease expression in inoculated fruits. On the other hand, delayed ripening caused by hypobaric pressure delayed the decrease in the concentration of the antifungal resorcinols and suppressed disease development due to the maintenance of latency.

An involvement of induced antifungal compound in disease latency was reported for *Nectria galligena* in Bramley's Seedling apples. Inoculation of apples with *N. galligena* results in areas of necrotic tissue in which the antifungal benzoic acid can be readily detected. Synthesis of benzoic acid, which is elicited by an extracellular protease produced by *N. galligena,* is toxic only at the low pH values found in immature apple where it prevents extensive development of the fungus. As the fruit ripens and becomes less acidic, the benzoic acid dissociates to the anionic form and the fungus resumes growth. Swinburne observed that the incidence of natural decay of apples by *N. galligena* was lower in 4 per cent CO_2 than in air or in CO_2 concentrations above 4 per cent. The accumulation of benzoic acid was found to be stimulated by concentrations of CO_2 up to 2.5 per cent but was inhibited in apples stored at CO_2 concentrations above 4 per cent.

The suitability of atmospheres containing CO for the suppression of disease during storage or transit has been evaluated for several commodities. The effectiveness of CO as a fungistatic gas is pathogen dependent and is greatly enhanced when combined with low O_2 atmospheres.

Applying varying mixtures of O_2 and CO to *B. cinerea* inoculated apples showed that suppression of lesion development in atmospheres of reduced O_2 with 10.5 per cent CO was closely related to the decrease in the O_2 level from 8 to 2 per cent. When O_2 remained constant at 2.1 per cent, and CO varied between 2 and 10.4 per cent, the lesions appeared to be suppressed linearly as the CO concentration increased.

Kader *et al.* showed that the addition of 5 or 10 per cent to CO to a low O_2 atmosphere (4 per cent) retarded growth of *B. cinera in vitro* and reduced decay incidence and severity of inoculated tomatoes harvested at the mature-green or pink stage and held at 12.5°C. This reduction was still evident after removing the fruit to 20°C in air. Suppression was less pronounced when CO was added to air than to a 4 per cent O_2 atmosphere. When added to air, CO increased CO_2 and ethylene production rates and hastened ripening of mature-green fruits, while CO with a low O_2 atmosphere had no or a very small effect on these physiological responses. In all treatments, fruit ripened normally after transfer to air at 20°C and attained full color.

Under low O_2 conditions, CO greatly reduced the rate of lesion development in *B. cinerea*-inoculated kiwi fruit. The presence of CO_2 further slowed lesion development. The most striking benefit of CO with low O_2 was the prevention of the formation of aerial mycelia, thus eliminating fruit-to-fruit infection.

The addition of 10 per cent to CA composed of 2.3 per cent O_2 + 5 per cent CO_2 markedly suppressed the development of *B. cinerea* lesions in strawberry fruits and reduced rates of mycelial infection during storage. This suppression was comparable to that recorded in air with 15 per cent CO_2. However, off-flavors which were found in fruit held in the latter atmosphere were not detected in the CO-treated fruit.

Lesions of *Botrytis* rot were also markedly suppressed in inoculated sweet cherries and actinidia fruits when stored in CA with 10 per cent CO. Actinidia fruits held under this atmosphere kept well for 158 d as compared with only 60 d of storage in air.

El-Goorani and Sommer showed that an atmosphere of CO in CA resulted in an 80 to 90 per cent reduction in the rate of rot development caused in strawberries by *B. cinerea,* in apples by *Penicillium expansum,* in lemons by *W. sclerotiorum,* and in oranges by *P. italicum* and *P. digitatum.* Slight off-flavors which developed in oranges and strawberries were attributed to low O_2 and elevated CO_2 and not to added CO.

The development of the CO-sensitive *Monilinia fructicola* in inoculated peaches was completely inhibited at 5°C by the addition of CO (11 per cent) to a reduced O_2 atmosphere (4 per cent), with or without CO_2 but the normal rate of rot development resumed once fruits were transferred to air at 20°C. Kader *et al.* found that CO in air mimicked ethylene effects on stored peaches by stimulating CO_2 and ethylene production in fruits. However, when added to low O_2 (4 per cent), high CO_2 (5 per cent), or their combination, CO effects on ethylene production were insignificant. CO added to a reduced O_2 atmosphere did not affect the advantageous delay of softening caused by low O_2 alone and was considered the optimum fungistatic treatment.

The effects of CO on decay developments as well as on the physiological responses of certain commodities showed that this fungistatic gas can be a useful supplement to low-O_2 atmospheres or to CA, which combines low O_2 with high CO_2, to reduce decay during storage and transit without causing undesirable effects on ripening or quality. However, the use of CO is limited primarily by its high toxicity to humans and by the fear of accidental inhalation of this poisonous gas by workers. CO is additionally limited by its tendency to mimic ethylene effects as was found in tomatoes, strawberries, sweet cherries, peaches, and probably kiwi fruits. This may limit the use of CO for long-term storage of ethylene-sensitive commodities.

Modified Packaging

Sealing certain commodities in polymeric film packages enables the extension of their storage life as a result of the formation of a MA within the package.

A produce package is a dynamic system in which respiration and permeation are occurring simultaneously. There is uptake of O_2 by the produce and evolution of CO_2, C_2H_4, H_2O, and other volatiles and, at the same time, specific, restricted permeation of these gases through the packaging film. The level of the gases surrounding the sealed produce is determined by the respiration rate of the commodity, the storage temperature, the permeability of the film to the atmospheric gases, the ratios of the produce to the weight area of semi-permeable film, and other factors.

A recent extension of this approach is aimed at the application of semipermeable films to create a MA around individual fruits (microatmospheric packaging or "Unipack"). Since different commodities require different films, the success of the development will again depend upon the availability of suitable polymeric films.

For certain commodities, the atmospheres produced within semipermeable wrappings are beneficial both in delaying the senescence of the produce and in retarding decay development. We will look at the suppressive effect of MA on postharvest diseases in a number of crops representing different plant organs.

Looking for a suitable method to export strawberries by sea, Barkai-Golan *et al.* found that sealing freshly harvested fruit in baskets or fiber board trays in relatively thick polyethylene or polyvinyl chloride (PVC) film enables fruit storage for 10 d at 2°C followed by 2 d at 20°C. The prolongation of the storage life was evidenced by better maintenance of fruit firmness and the fresh appearance of the fruit and the calyx, along with a reduction by half in the incidence of decay, caused mainly by *B. cinerea.* The incidence of decay was correlated with the level of the built-in CO_2 which accumulated in the atmosphere around the fruit. During storage at 2°C, the level of CO_2 ranged from 1.5 to 2.5 per cent, but decay was controlled by the low temperature. Following transfer to 20°C, CO_2 levels rose up to 7.5 to 12 per cent. The influence of CO_2 on decay development was attributed mainly to its primary effect on delaying the senescence processes of the fruit and maintaining host resistance. Combining preharvest fungicidal sprays–applied during the flowering season–with suitable PVC wraps almost eliminated pathological deterioration during prolonged storage.

An initial charge of CO_2 plus that produced by respiration served to maintain the CO_2- enriched atmosphere during transit of film-covered pallets of California strawberries. Transit tests conducted under commercial conditions indicated that MAs were beneficial if an adequate concentration of CO_2 and optimum temperatures were maintained.

Table 11.2: Keeping Quality of Broccoli Heads Packed in Either Nonsealed PVC Wraps or Polyethylene (PE)-Sealed Bags During Simulated Transit by Air or Sea

Simulated Transit	*Treatment*	*Yellowing Index**	*Decay (%)*
Air			
3d at 15°C + 2d at 20°C	PVC wrap	4.8	8.5
	PE-sealed bag	1.0	0
Sea			
14d at 0.5°C + 2d at 20°C	PVC wrap	4.1	32.0
	PE-sealed bag	1.1	0

* 1: None; 5: maximum yellowing.

The keeping quality of broccoli flower heads is very low, especially at temperatures above 5°C. Respiration rates of the florets increase sharply, they rapidly turn yellow, and, in an advanced stage of senescence, decay caused mainly by *Alternaria alternata* and *B. cinerea* starts to develop. Packaging of broccoli heads in sealed polyethylene bags inhibited both senescence and decay in the flower buds and kept the produce green and fresh during prolonged cold storage followed by shelf-life conditions.

This prolongation was attributed to the high respiration rate of this vegetable, which resulted in the formation of MA within the sealed bags. Respiratory CO_2 increased from 2 to 3 per cent in cold storage (0.5°C) to 6 to 8 per cent after transfer of the broccoli to shelf-life conditions (20°C). Under these conditions, O_2 concentrations decreased at shelf-life to about 6 per cent. Neither offensive odors nor off-flavors were detected in the PE-sealed bags. The accumulation of respiratory CO_2 rather than O_2 depletion was concluded to be the main factor controlling the keeping quality of the broccoli.

Experiments *in vitro* have shown that 12 per cent CO_2 caused a decrease of about 25 and 15 per cent in the mycelial growth of *B. cinerea* and A. *alternata*, respectively. Since less than 10 per cent CO_2 was sufficient for complete inhibition of fungal development on the broccoli, the suppression of rot development was related mainly to the effect of CO_2 in retarding the senescence processes in the host tissue, thereby maintaining its natural resistance to decay.

The senescence-retarding effect of CO_2 was related to its ability to block the action of endogenous ethylene, which has been shown to play a primary role in the regulation of senescence processes of broccoli influorescence. The inhibitory effect of the CO_2-enriched atmosphere on decay development in *B. cinerea*-inoculated broccoli florets was completely nullified by addition of ethylene to the CO_2-treated florets.

Storage of lettuce for about 2 weeks at low temperature and high humidity is limited mainly by the development of decay organisms. Packaging of romaine lettuce in polyethylene-lined cartons markedly reduced yellowing and decay of leaves. The incidence of rot caused by *B. cinerea, Stemphylima botryosum* f.*lactucae,* and *Sclerotinia sclerotiorum* in inoculated leaves was reduced by 21, 37, and 52 per cent, respectively, after 14 d at 10°C and 2 additional days at 10°C. Under shelf-life conditions, polyethylene liners not only reduced the incidence of *Stemphylium* rot, but also inhibited further growth in incipient infections, resulting in the appearance of restricted lesions.

The reduced incidence of rot was related to the alteration of the atmosphere within the polyethylene-lined cartons. Concentrations of CO_2 and O_2 in the atmosphere were stabilized during 14d of cold storage (1°C) at 4 and 16 per cent, respectively. On transforming the lettuce to 20°C, O_2 concentration was reduced to about 4 per cent, whereas CO_2 was increased up to 12 per cent. This level of CO_2, at which no injury was induced in romantic lettuce, seems to reduce decay by retarding the physiological deterioration of the lettuce, although some direct effect on fungal growth may be involved.

Packaging of green herbs in nonperforated polyethylene-lined cartons resulted in the retardation of yellowing and decay in coriander, dill, parsley, chives, chervil X2 watercress, and sorrel. Retardation of senescence processes in these yellowing-susceptible herbs was attributed to the accumulation of respiratory CO_2 within the package rather than to the depletion of O_2. This implies that the senescence-retarding effect of CO_2 in green herbs, as has been shown for other green tissues, can be attributed mainly to its ability to antagonize ethylene action.

Alteration of Pathogens

Alteration in the concentration of the atmospheric gases can be of benefit only when it delays deterioration of a fruit or vegetable without injuring it. Since even closely related fruits or vegetables, or even different varieties of the same crop, may differ greatly in their tolerance of atmosphere modification, a combination of O_2–CO_2 beneficial for one given produce may be hazardous for another. Furthermore, the reaction to CA combinations is temperature dependent, and a concentration of O_2 or CO_2 that is beneficial for a given crop at a given temperature may be harmful at another temperature. Tissue injury induced by unsuitable CA treatment may affect host-pathogen interactions and lead to enhanced disease development by increasing the susceptibility of the host to microbial infection.

The tendency of CA to increase fungal decay in citrus fruit was shown by Harding. CA of 3 to 3.5 per cent O_2 and 1.5 to 2 per cent

CO_2 was deleterious to lemons, grapefruit, and oranges, resulting in rind breakdown and the formation of a fermented odor and flavor. This was accompanied by an increase in the incidence of *P. digitatum* and *P. italicum* decay. Similarly, low O_2 atmospheres were harmful to limes, resulting in severe injury and excessive decay.

With tomato fruits, excessively low O_2 or high CO_2, apart from preventing proper ripening, enhances the danger of chilling injury which is well known to render the fruit more susceptible to *Alternaria* rot and other postharvest diseases.

Holding freshly harvested potatoes in high CO_2 (10 per cent or more) and particularly low O_2 (less than 5 per cent) greatly increases bacterial soft rot of tubers. These atmospheres were found to inhibit effective healing of superficial wounds by preventing rapid periderm formation even at high humidities and temperatures favorable to its formation. This alteration of texture leads to the easy penetration of soft rot organisms. This increase in soft rot under CA, treatment was also related to the enhanced susceptibility of the tuber cortical tissue to invasion by *E. carotovora.* Increasing the level of CO_2 during storage at 0 and 5°C was found to increase both the sucrose content of the tubers and membrane permeability. A correlation between sucrose content, membrane permeability, and rate of decay caused by *E. carotovora* var. *atroseptica* was drawn. In various table and industrial cultivars of potatoes, a marked increase in *Fusarium* rot was recorded when the CO_2 concentration was elevated from 1 to 12 per cent in a low O_2 atmosphere at 8°C.

Some of the deleterious effects of CA may result from an accumulation of volatile compounds, including ethylene, in the storage atmosphere. Application of exogenic ethylene has been observed to increase *Alternaria* rot on tomato fruits through ripening acceleration and to increase *Diplodia* stem-end rot on citrus fruits by accelerating senescence and abscission of the fruit button.

The development of decay, particularly stem-end decay incited by *Fusarium* sp., in Valencia oranges stored in CA (3 per cent CO_2 and 5 per cent CO_2) appeared to be primarily associated with the presence of ethylene in the storage atmosphere. The involvement of ethylene is probably by hastening senescence of the fruit and by rendering it more susceptible to attack by a weak pathogen. Decreasing the ethylene content resulted in a significant reduction in *Fusarium* stem-end decay under CA conditions. Mold incidence in lemons was found to be high in CA storage (3 to 5 per cent O_2 and 0.1 to 0.2 per cent CO_2) where ethylene accumulated. The use of an ethylene absorbent during CA storage greatly reduced rot development presumably by maintaining the fruit in an immature, more resistant condition. Removal of evolved ethylene from the CA storage of avocados resulted in a further reduction in anthracnose and in improvement in the keeping quality of the fruit over that found in CA prior to ethylene removal.

In various crops, CA storage results in the alteration of the natural population of the postharvest pathogens.

In Shamouti oranges, CO_2-enriched atmosphere changed the pathogenic flora by dramatically increasing the incidence of stem-end rot caused by *D. natalensis.*

In white cabbage, CA storage (4.7 per cent CO_2 and 1.4 per cent O_2) markedly reduced the two important postharvest organisms *B. cinerea* and *S. sclerotiorum.* However, *F. roseum,* which has not been considered an important pathogen of cabbage stored in normal atmosphere, was consistently isolated from cut stalks and leaves of CA-stored cabbage. Similarly, CA conditions (3 per cent O_2 and 5 to 6 per cent CO_2) which consistently reduced the incidence and severity of cabbage rots caused by *B. cinerea* did not suppress development of other cabbage pathogens–*Alternaria* sp. and *Mycosphaerella brassicicola,* which in some cases predominated on the CA-stored cabbage.

Lipton and Harris observed that, whereas species of *Alternaria* and *Cladosporium,* were the main postharvest pathogens in broccoli stored in air, *Mucor* sp. predominated in a low O_2 atmosphere. Experiments carried out with inoculated potato tubers revealed that fungal species which were weakly virulent in normal atmosphere, such as various species of *Fusarium* and *Phoma exigua* var. *exigua,* were clearly more aggressive under CA conditions. Species not pathogenic in normal atmosphere (*C. coccodes, Cylindrocarpon* sp., and *Phoma eupyrena*) caused rotting in CA, while species virulent in normal atmosphere (*Fursarium sulphureum* and *P. exigua* var. *foveata*) were less virulent in CA storage. Storing tomatoes in an atmosphere of 3 per cent O_2 without CO_2 or with 3 or 5 per cent CO_2 resulted in enhanced development of *G. candidum,* but this seems to be attributed to the direct stimulatory effects of CA on fungal growth.

There are promising indications that various fruits and vegetables could benefit from the MA produced within suitable sealed polymeric films by delaying both their physiological and pathological deterioration during transit and marketing.

CA storage and MA formed within sealed plastic films are generally combined with low temperatures which may similarly delay disease development both by maintaining the resistance of the host to infection and by directly retarding microbial growth. In various crops, however, satisfactory disease suppression may only be achieved as a result of the combination of fungicidal treatments with MAs and low temperatures.

In contrast to conventional postharvest fungicides, which are generally of low penetration ability into the infected host, modified storage atmospheres that inhibit the onset of ripening or senescence can suppress the progress of deep-seated infections. A renewed growth of latent infections is successfully delayed in various host-pathogen interactions under CA storage. Antifungal compounds which are active in the immature fruit and decrease with ripening can, in several cases, be related to changes in fruit resistance during prolonged CA storage. In many host-parasite interactions, postharvest pathogen development is associated with the secretion of pectolytic enzymes that cause tissue maceration and lead to cell death. However, only little information is available on the effect of atmospheric modifications on the production and activity of pectolytic enzymes during pathogenesis or on factors affecting enzymatic activity.

The resistance of fruits and vegetables to postharvest diseases is highly related to their state of maturity and to the physiological changes associated with tissue senescence. CAs or MAs can indirectly reduce postharvest diseases to a large extent, by maintaining host resistance via delaying the climacteric rise in fruit respiration and by inhibiting the onset of ripening and senescence. Removal of ethylene from the storage atmosphere may further reduce disease by leading to an increased retardation in the ripening process. A reduced O_2 or elevated CO_2 level also has a direct inhibitory effect on pathogen growth. However, under common CA combination, the direct atmospheric effects are very restricted or negligible. For a few crops only, such as some varieties of strawberries and sweet cherries, that tolerate high CO_2 concentrations, level of 20 per cent

CO_2 applied during transit can suppress disease development both directly and by retarding physiological aging.

Ethylene as a Plant Hormone

Although first recognized as a product of ripening fruit, ethylene is now seen to be one of the plant hormones regulating the growth, development, or function of various organs in many plant species. In the postharvest context, its roles in promoting abscission, senescence, and fruit ripening are important. Because of its gaseous nature, ethylene can readily escape from plant tissues; thus, in the open environment, it must be continuously produced close to its site of action to be effective. However, in the postharvest environment in general and in CA storage in particular, the aeration of plant produce is restricted so that ethylene will tend to accumulate and can pass from one fruit or vegetable to another. This may mean that particular individuals within a harvested crop can influence other individuals or that one crop can influence another in mixed commodity storage. The seriousness of the effects will depend upon the rates of ethylene production relative to rates of ventilation and the responsiveness of a particular crop to ethylene under the storage conditions.

Data for rates of production and response of plant crops to ethylene under typical commercial storage conditions are very limited. Most experiments have been conducted at ambient temperatures (around 20°C). Among the highest ethylene-producing crops are the climacteric fruits, such as apple, banana, avocado, tomato, and melon. As these fruits ripen, their respiration rises and, concurrently, their ethylene production increases from 0.1 to 0.2 to 10 to 100 $\mu l\ kg^{-1}\ h^{-1}$. Although this high rate of ethylene synthesis probably coordinates and promotes ripening, it is difficult, if not impossible, to influence the rate of ripening at this stage by adding or removing ethylene in the atmosphere around the fruits. Because of the diffusion barrier at the fruit surface, the atmosphere inside the fruit contains excess ethylene even at zero external concentration. However, these fruits will respond to ethylene before the onset of natural production and it is in this phase that ethylene removal can be effective in delaying ripening.

The fruits which do not show a climacteric pattern of respiration produce up to 0.1 $\mu l\ kg^{-1}\ h^{-1}$ of ethylene and the rate does not change with ripening. These fruits may be further subdivided into those which respond to exogenous ethylene and those which do not. The responsive fruits include the citrus group and the "vegetable fruits" of the cucurbit family, such as cucumber and courgette. Unresponsive fruits include cherry, grape, and bell peppers (*Capsicum* spp.). Some other fruits are generally thought to be unresponsive, although there are some contrary indications. *e.g.*, strawberry and pineapple. Vegetables, excluding those which are botanically fruits, generally produce less than 0.1 µl kg^{-1} h^{-1} of ethylene, although broccoli and cauliflower may produce more. Leafy vegetables show the most obvious responses to ethylene, generally yellowing and abscission. However, the gas can induce sprout initiation in potato tubers, toughening of asparagus stems, and off-flavor development in carrots.

Fresh cut flowers produce low levels of ethylene, but petal senescence is associated with a rise in ethylene synthesis reminiscent of ripening fruit. Fading, wilting, and abscission of petals and the swelling of the ovary can be accelerated by ethylene treatment. Ethylene accumulation in the storage of bulb crops is usually caused by decaying bulbs and can cause death of the flower bud initials.

Ethylene synthesis is accelerated during the invasion of plants by many bacterial and fungal pathogens. Ethylene from higher plant or microbial origin can stimulate the metabolism and development of the pathogen and the host species. The full significance of ethylene in host parasite interactions is not yet clear.

It is now clear that ethylene in higher plants is synthesized from methionine via the intermediates, *S*-adenosyl methionine (SAM) and 1-aminocyclopropane-1-carboxylic acid (ACC). The first enzyme unique to the ethylene pathway is ACC synthase which converts SAM to ACC. ACC synthesis seems to be possible in the absence of oxygen, whereas the fragmentation of ACC to yield ethylene has a strict oxygen requirement. The enzyme catalyzing this step has not been isolated, but studies with tissue slices suggest that it is a relatively low-affinity oxidase with half-maximal activity at about 2 per cent oxygen. Carbon dioxide can increase ethylene production by leaf tissue, perhaps by activation at the ACC to ethylene step; however, this effect seems to be saturated below 1 per cent CO_2, and most bulky plant organs such as fruits and tubers would contain higher levels of CO_2. A short exposure to 20 per cent CO_2 depresses ethylene synthesis and increases ACC levels in apples, suggesting inhibition of the ethylene forming enzyme at very high CO_2 levels.

Response to ethylene must be dependent upon binding to a receptor which presumably undergoes a conformational change leading indirectly to changes in tissue metabolism and development, probably involving transcription of an altered population of messenger RNA molecules. Detail of these processes is almost entirely unknown; various workers are investigating binding and metabolism of ethylene by plant tissues and these may prove to be processes related to ethylene action. It is noteworthy that ethylene is metabolized by oxidative processes with comparatively low affinities for oxygen and that metabolism is inhibited by CO_2. Much has been deduced about the properties of ethylene receptors from the relation of response to ethylene concentration. Burg and Burg concluded that the receptor in a number of plant tissues was half saturated at about 0.1 $\mu l\ 1^{-1}$ ethylene and that CO_2 was a competitive inhibitor and oxygen was a cooperative effector of ethylene action. The relative rates of response to a range of olefins were similar to the dissociation constants of the equivalent silver-olefin complexes and this led to the concept that the receptor is a metalloprotein. Further analyses of ethylene effects on growth of a number of plant species revealed a diversity of apparent affinities from 0.1 to 1.35 $\mu l\ 1^{-1}$. The true affinity of the, receptor may be less variable as the endogenous ethylene, oxygen, and carbon dioxide concentrations in a plant organ can all affect the shape of the dose-response curve. The rates of various processes in ripening fruits also respond differentially in relation to ethylene concentration; respiratory rises in citrus and immature melons appear to be half-maximal at around 5 $\mu l\ 1^{-1}$, whereas ripening of a number of climacteric fruits is half-maximal between 0.05 and 0.5 $\mu l\ 1^{-1}$ ethylene. Since chlorophyll loss in citrus fruits appears to be a high-affinity process, modulation by internal gases cannot be invoked to explain the low affinity of the respiratory response. It is possible that there is more than one receptor present and the concept of the universality of a high-affinity receptor is brought into question.

Ethylene binding is significant in practice only if it leads to physiological or morphological changes in a plant commodity. A measurable response depends upon reactions consequential to the primary event of ethylene binding. Commonly, discussion of the "sensitivity" of plant material conflates all aspects of the ethylene response, possession and properties of receptors, and rate and nature of response processes. It is arguably preferable to retain the term "sensitivity" to describe the concentration dependency of an ethylene response and "responsiveness" to describe the rate or extent of response in a given time. Thus, adopting the analogy of enzyme kinetics, a plant tissue is *sensitive* if it shows a response which is half-maximal at low-ethylene concentrations; but the *response* to the saturating ethylene level could be high or low depending upon the circumstances.

In principle it is possible that environmental conditions experienced by a plant commodity could affect ethylene synthesis, sensitivity, or responsiveness independently, and it will be useful to consider these aspects in relation to CA storage. In practice it may prove difficult to distinguish between receptor affinity and response reactions. The validity of the analysis in terms of saturation kinetics requires that the binding of ethylene to its receptor should be the rate-limiting step. If, however, the rate of response reaction is limiting, the process may be saturated at relatively low ethylene concentrations; the system will appear to be sensitive, although not very responsive. In view of this, the approach may have empirical value in that it will describe an ethylene response of a commodity in terms of concentration dependency and rate, but this will not be directly interpretable in terms of affinity of ethylene receptors. In an extreme case, it may not be possible to show an effect of ethylene treatment because of the slowness of a response. This may lead to a conclusion that a process does not involve ethylene, when in fact endogenous levels are sufficient. Such problems can partly account for earlier contention that ethylene had no role in the ripening of fruit at storage temperatures.

Laboratory experiments on CA storage commonly use premixed gases to purge small containers of plant produce. Under these conditions, ethylene produced is continuously diluted and vented to waste. The concentrations measured for vegetables and preclimacteric fruits may well be below the limit of detection. High CO_2 and low O_2 concentrations delay the onset of lipid ethylene production in climacteric fruits under flow-through conditions. If it is accepted that preclimacteric ethylene levels influence the rate of progress to the climacteric and the associated rise in ethylene synthesis, the delay by CA could be explained by effects on ethylene synthesis or action. However, effects on other aspects of metabolism cannot be excluded. Once initiated, the maximal rate of ethylene production is lower in high CO_2 or low O_2 than in air for apples and in bananas low O_2 is inhibitory, but CO_2 has little effect. This appears to be the basis of the suggestion that inhibition of synthesis could be a significant factor in the preclimacteric phase. Ripening climacteric fruits contain ethylene concentrations considerably in excess of that required to saturate the high-affinity sites believed to be involved in most ethylene responses. Thus, the inhibition of maximum ethylene synthesis by CA should not be physiologically significant unless low-affinity sites were involved.

The concentrations of CO_2 and O_2 commonly employed in CA storage of plant produce inhibit the effect of ethylene on the growth of etiolated pea stems. Thus, the effects of CA in flow-through conditions could to some extent result from inhibition of ethylene action. However, at least for fruit, the high levels of ethylene in a sealed store would be expected to overcome the inhibitory effects of CA.

To understand the role of ethylene in CA conditions, it is desirable to determine the extent to which effects of CA in flow-through systems can be reversed by ethylene and this has not often been directly examined. Low O_2 and high CO_2 delay the onset of ethylene production and, consequentially, ripening in bananas; these effects can be reversed by ethylene treatment. Low oxygen also slows ripening in this fruit after initiation and, again, the effect is largely reversed by ethylene; there is a residual effect of levels below 5 per cent O_2, but this may be because the fruit is approaching anoxia internally; CO_2 has little effect on ripening in the postinitiation phase. In apples ethylene does not reverse the effect of low oxygen on softening or chlorophyll loss. Ethylene is not controlled in standard commercial CA storage of apples and since this is demonstrably successful there must be CA effects apart from those on ethylene synthesis and action. The relation between CA and ethylene varies even for apples. Chlorophyll loss in apples and other fruits is accelerated by ethylene in air at high temperatures, but it is unaffected by addition or removal of ethylene in CA storage. In contrast, fruit softening can be accelerated by ethylene treatment or delayed by ethylene removal in CA storage, although rates of softening are slower in CA with ethylene than in air storage. Synthesis of the cuticular hydrocarbon, farnesene, is particularly affected by ethylene levels in CA storage of apples and it may be that the CA effect on this process is largely mediated via ethylene.

The effectiveness of CA in delaying ethylene-promoted processes irrespective of ethylene levels implies that CA can affect metabolic steps after synthesis and binding of ethylene to its receptor. In other words, low oxygen and high CO_2 can inhibit reactions involved in ripening processes such as chlorophyll destruction and softening. The identity of these oxygen- or CO_2-sensitive metabolic steps is not clear. Although mass action effects on respiratory oxidases and decarboxylases provide an immediately obvious hypothesis, the affinities of the isolated enzymes do not match the concentrations known to be effective in CA storage. The overall slowing of metabolism in CA may be an adaptive response to minimize demand for respiratory energy in an adverse gaseous environment. In addition, oxygen may be directly involved in processes such as chlorophyll destruction and aspects of lipid metabolism.

Ethylene removal during storage of horticultural crops is desirable where quality criteria cannot be satisfied with simpler technology and where the benefit of removal exceeds the extra cost. The technology is now available to remove ethylene from store atmospheres by chemical oxidation with potassium permanganate or catalytic oxidation. Permanganate is probably more suited to small-scale uses, whereas the catalytic systems are more appropriate for quantities above a tonne. Most of the experimental success with ethylene removal has involved climacteric fruits in CO_2-rich or O_2-depleted atmospheres. These atmospheres probably aided ethylene removal both by suppressing synthesis and by decreasing sensitivity to ethylene. When ethylene removal is ineffective, this may be because changes in the produce are so retarded by CA that ethylene has no accelerating effect or that the removal system cannot prevent a rise in synthesis. This second type of failure may be a result of lack of control of developmental processes leading to ethylene synthesis rather than a result of limited ethylene removal capacity. The developmental commitment to ethylene production varies with the type and variety of fruit. Some apple varieties can be held in CA in the preclimacteric state for long periods provided ethylene is continuously removed; others show an early onset of rapid ethylene synthesis under the same conditions. If it is not possible to control ethylene accumulation, chemical treatments to delay rapid ethylene synthesis may be helpful. An existing agrochemical, daminozide, has this effect on apples, so that a combination of daminozide spray in the orchard, low-oxygen CA storage, and ethylene removal has been advocated for McIntosh apples. The inhibitor of ACC synthesis, aminoethoxyvinylglycine, has also been used experimentally to the same purpose, but its approval for food use seems unlikely.

Most workers have found it necessary to prevent ethylene levels rising above about 1 $\mu l\ l^{-1}$ in order to achieve beneficial effects by ethylene removal. A retardation of fruit softening has been the most general result, although more specialized effects, such as delay of the development of physiological disorders, have also been reported. There have been some reports of effects resulting from reduction of ethylene to levels in excess of 10 or even 100 $\mu l\ l^{-1}$ from higher concentrations. However, this seems unlikely in relation to the physiological activity of ethylene and the more general impression is that, once fruit has begun to produce ethylene rapidly, removal is no longer effective in delaying ripening. The tendency to produce large amounts of ethylene makes fruit an obvious but difficult subject for ethylene removal. Much less is known about the role of ethylene in CA storage of vegetables and nonclimacteric fruits, although the task of controlling ethylene levels should be considerably easier for these.

The application of CA is usually supplementary to refrigeration and it is appropriate to consider first the effects of temperature on the ethylene physiology of crops. If the long-held view that ethylene is ineffective in low-temperature storage were substantiated, then there would be little point in considering how CA could influence crops via ethylene.

First, plant produce at low temperature continues to synthesize ethylene. Even pears which are stored at –1°C produce up to 1 $\mu l\ kg^{-1}\ h^{-1}$. The rate of synthesis generally declines with temperature, although Arrhenius plots may show more rapid decline below a threshold temperature. This is arguable evidence for the location of at least one step in synthesis on a membrane which is subject to phase transition at the threshold temperature; the step commonly held to be membrane associated is the fragmentation of ACC to yield ethylene.

In chilling-sensitive commodities, rise in synthesis of ethylene can accompany chilling injury, *e.g.*, in avocados and cucumbers. Where a rise in synthesis is associated with a developmental stage in the life of a plant organ, this may be variously affected by temperature. In carnation flowers, the rise was found to be delayed progressively by lowering the temperature to 7°C below which a rise was not observed. In some varieties of apple, the rise in ethylene production is similarly delayed; in others temperature has little effect on the timing, while in at least one apple variety and many pears the rise occurs earlier at low temperatures. The basis of this response does not appear to be an induction of ACC synthesis prior to conversion to ethylene as in cucumbers. The implication is that there is a cold-stimulated step in the pathway leading to induction of rapid ethylene synthesis and that this is rate limiting in the fruits that show the response. Whether it is not cold stimulated or is not rate limiting in other fruits is not clear. A superficially similar enhanced rate of ethylene synthesis occurs on completion of the chilling requirement for floral bud development in tulip bulbs.

In principle, ethylene can be active at low temperature, although the effects of its action may not be commercially significant. The apparent insensitivity of apple fruit to ethylene in air under refrigerated conditions can be explained by the presence of endogenous ethylene and the slowness of ripening processes under these conditions. In other words, ethylene may not be the limiting factor in ripening and there might be little commercial risk in allowing ethylene to accumulate; but this does not mean that ripening becomes independent of ethylene under refrigeration. Slight effects of ethylene on fruit softening can be demonstrated in apples stored in air at 3°C and softening of kiwi fruit at 0°C can be accelerated by ethylene. Similarly, leaf abscission can be induced in cabbages at 1°C and isocoumarin synthesis can be induced in carrots at 2°C. The lack of effect of ethylene on carnations at 0°C implies that the maximal rates of developmental processes at this temperature are limiting in this commodity.

Storage at subatmospheric pressures has similarities to CA methods in that the partial pressure of oxygen is lowered, unless the intake gas is deliberately enriched with oxygen. Thus, the usual effects of low-oxygen storage are to be expected, but, in addition, diffusion of ethylene out of plant produce is accelerated and the treatment can have some spectacular effects on delay of senescence.

The relative significance of effects of low oxygen and ethylene removal has not always been estimated. When ethylene has been added back under hypobaric conditions, the effects have been completely abolished in bananas, partially so in apples, and not at all in tomatoes. Hypobaric storage should also facilitate CO_2 diffusion from produce which might make it more sensitive to ethylene but could conceivably reduce activity of ethylene forming enzyme. Hypobaric treatment is certainly a useful experimental method for investigation of ethylene involvement in plant developmental processes. The cost of large-scale installations discourages application other than to very high-value crops.

Since the first convincing demonstration that ethylene removal during CA storage of apples had beneficial effects, there have been numerous studies on a range of apple varieties. Control of ethylene accumulation has only been effective if apples were harvested in advance of the natural onset of rapid ethylene synthesis. The most common storage atmospheres have been both CO_2 enriched and depleted of O_2, *i.e.*, 2 to 5 per cent CO_2/2 to 3 per cent O_2, although simple high CO_2 or low O_2 atmospheres have also been employed according to varietal tolerance. Apples have often been sprayed with daminozide in the orchard prior to low-ethylene storage and the interaction of spray concentration, harvest date, and ethylene removal has been studied. The usual benefit reported and the motive for much of the research has been a slower rate of flesh softening. This effect has usually been demonstrated with the penetrometer traditionally used in fruit research and a correlation between high penetrometer readings and consumer acceptability has been assumed or occasionally demonstrated. Other effects of low-ethylene storage include inhibition of the development of certain storage disorders. Thus, flesh breakdown was inhibited in McIntosh apples and the skin browning, known as superficial scald, was delayed in Bramley apples; this last effect appeared to be related to the role of ethylene to induce synthesis of α-farnesene, already mentioned. A less desirable effect is the inhibition of synthesis of volatile alkyl carboxylate esters, which contributes to flavor, observed in Cox and McIntosh apples. The use of ethylene removal in conjunction with CA storage has developed to the level of commercial application for McIntosh and other apple varieties in the eastern US. Ethylene removal has been applied on a commercial scale for Bramley apples in England.

Chapter 12

Preservation of Tree Nuts and Dried Fruits

Dried Fruits and Tree Nuts

The tree nuts and dried fruits are generally mature and become physiologically senescent prior to harvest.

Many insects that infest these commodities initiate their-attack on the ripening of mature fruit and nuts in the field and remain in them when they enter storage. Other insects are more adapted to storage conditions and initiate their infestations when the fruit is in storage. To minimize losses, control techniques must be applied while the commodity is in the field, when it is received at the storage, while it is in storage, while it is in shipment, and when it arrives at destination stores and warehouses.

World wide annual production of dried fruit and tree nuts is over 1.7 million tonnes, the production of each being nearly equal. The high nutritive value of these commodities makes them important for sustaining and improving man's diet, and they are easily stored for use when food production is reduced, as in times of drought. Insecticidal protectants and fumigants are currently recommended to reduce losses in these foods.

The use of modified atmospheres (MAs) is an age-old method of food preservation in which natural respiration of the stored product or microbial respiration alters composition of the storage atmosphere. Drawbacks to this method are the long time needed for controlling the pests and the loss of some of the commodity if infested before storage. In modern times, controlled atmospheres (CAs), or artificially altered storage atmospheres, have been further developed. Generators quickly alter the storage atmosphere to the desired levels of oxygen and carbon dioxide for a given commodity and therefore provide more protection than a non-controlled MA.

Since dried fruits and tree nuts are stored at a low moisture content, often at ambient conditions, they are ideally suited to the application of CAs. The commodity (with the exception of the embryo) is non-living and therefore, tolerant of the atmospheric conditions that are needed to kill insects. As a general rule, the CA should be either low oxygen (O_2) ($\leq$ 0.5 per cent) or high carbon dioxide (CO_2) (approximately 60 per cent in air) or a combination of low O_2 with CO_2, other atmospheres, such as slightly higher O_2 levels, also can kill insects, but the treatment time must be extended.

While controlling insects we must remember the following factors:

1. No toxic pesticide residues remain in the commodity (a factor useful in marketing the commodity).
2. The gases are chemically inert and do not corrode handling equipment.
3. Fire hazard in storage is reduced.
12. CAs are generally less hazardous to workers than toxic fumigants; however, they are asphyxiating at effective concentrations.
5. Rodents and birds, in addition to storage insects, are controlled.

Application of CAs to dried fruit or tree nut storages may be by single pass or recirculation. Single pass utilizes a purge phase in the displacement of the atmosphere inside the storage.

Dried fruits are normally stored in boxes or bins of various sizes, but usually are less than 0.6 m in depth within the bins to avoid excessive compression of fruit. Although some compression does occur over a period of time, CAs have been found to readily penetrate the boxes/bins of raisins and figs. Raisins are normally stored with the stems intact in bins that are placed into stacks, either in buildings or tightly covered with waterproof tarpaper.

Tree nuts may be stored in bags or bins, but are mainly stored in bulk. Bulk storage may be in silos or flat storages. In bulk storages, dust may accumulate in some areas, which will slow down but not prevent the penetration of the CAs.

Permanent storages currently used for dried fruit and tree nuts may be wood buildings, corrugated steel buildings, welded steel silos; concrete buildings, or concrete silos. Storage size may vary from 85 to over 28,400 m^3. Most of these storages require considerable sealing to make them acceptable for CA storage.

Temporary storages for bins can be made by covering the stacked bins with polyethylene or more frequently with a weatherproof paper laminate. These storages generally are used for less than 1 year and are commonly used by the raisin industry. In this type of storage, the bins are stacked ten wide and eight high and the stacks vary in length. A wood framework with a gable 1.1 m high (at the midridge) is nailed to the outside of the stack. Two layers of No. 222.03 Fibreen paper (Fortafiber Corporation, Tracy, CA) are then placed over the framework. The paper is held securely by nailing slats vertically at about 1.2 m intervals. The Fibreen cover is sealed to the concrete base with oiled dirt. Fibreen is made of two layers of Kraft paper (12.7 kg/100 m^2) laminated with 8.8 kg of tar per 100 m^2 and fiberglass reinforcement. The laminate is fused on one side to a 0.025 mm. thick white, polyethylene film. The Fibreen is placed with the polyethylene layer toward the outside of the stack.

Quality and Suitability of Storage

To determine the suitability of storages for CA use, the storage walls, floor, and ceiling should be inspected visually for obvious defects, such as holes, cracks, wall, separations, and damaged or poorly designed hatch and door closures. Special attention should be given to joints where expansion and contraction may break seals. Elevator entrances should be designed to allow sealing. A quick check of possible leakage areas may be determined by using the soap bubble test. In this test, the storage is closed and then pressurized with air using an air source such as the outlet of a vacuum cleaner.

Care must be taken not to over pressurize the structure beyond its stress limits. Our experience has shown that less than 2.5 cm of water column is sufficient. Next, a soap or detergent solution (less than 1 per cent in water) is sprayed on the outer surfaces of the storage and is followed by a visual observation for soap bubbles.

A second method for determining seal quality of a storage is by pressurization of the storage as above and measuring the pressure loss vs. time. This method is currently used for quarantine fumigation chambers by the state of California. They recommend a time lapse of 22 s or more for a 50- to 5-mm pressure drop as measured in a kerosene-filled manometer. However, this degree of gas tightness is probably not sufficient for holding CAs. Banks and Annis have recommended decay times of 9 min or longer for empty storages of various sizes with a pressure drop from 25 to 15 cm of water column. A third method is to determine the volume of air required to maintain a constant pressure in a storage at various pressures. This method may be used to measure the actual leakage (volume) by employing instruments to record air flow at two or more pressures.

Sealing of storages for grain has been extensively investigated in Australia, but relatively little has been done with this type of application in the US. The cost of constructing new storage facilities, suitable for CAs, would be only slightly greater than that of unsealed facilities. Retrofitting existing storages is much more costly.

CAs should be introduced into the bottom of the storage and vented or recirculated from the top. However, atmosphere introduction may be made from either the top or the bottom. To simplify the change in flow rate from the purge to maintenance phase, a dual manifold with two orifice sizes is desirable. The flow rate can be measured with a calibrated flowmeter inserted into the inflow duct. AU-tube manometer filled with kerosene is usually connected to a small tube that penetrates through the storage wall to measure the internal pressure of the storage. The manometer should be sheltered from wind, which may affect readings.

Care must be taken, especially with combustion-generator atmospheres, to prevent the condensation of moisture on the stored commodity at the point of atmosphere introduction. Condensation will occur when the relatively warm, introduced atmosphere is cooled by contact with the cool commodity in the storage room. To prevent condensation, the incoming atmosphere can be dried below the expected dew point by refrigeration or other methods.

After proper sealing of the interior of the storage, an atmosphere distribution manifold may be installed. The purpose of a manifold is to distribute a uniform layer of the incoming atmosphere, which displaces the existing storage atmosphere without excessive mixing. Channeling through the commodity creates pockets of the existing atmosphere that can be eliminated only by diffusion, which is a slow process. If the flow rate is too high, eddy currents are formed that mix the incoming atmosphere with the existing atmosphere and therefore require a longer purge time. Ideally, a slow upward displacement with little mixing of the incoming atmosphere with the existing atmosphere is desired. With this method, only a narrow bond of mixed atmosphere will exist, and if the purge is from the bottom of the storage, a layer effect will occur and rise toward the top of the storage. One must consider differential heating within a storage, such as may occur when a stack of raisins has its longitudinal axis parallel to the sun. Heating of one side of the stack causes convection currents, which mix the introduced and existing atmospheres in the stack, thus showing the purge time.

The storage atmosphere may be monitored with sample tubes placed at various locations. A small air pump is needed to withdraw an air samples for analysis. Samples may be analysed with a paramagnetic O_2 analyser, a gas chromatograph, or other CO_2 or O_2 monitoring equipment.

Although a recirculation system is one method of applying CAs, a single-pass (or purge-maintenance) system will allow a more rapid method of obtaining the conditions necessary to kill insects. With this method, the CA is introduced at a high rate initially and then is reduced to a maintenance rate when the desired conditions are achieved.

The purge volume required is determined by the type of storage configuration. Bulk storages have less free air space and therefore, require a smaller volume of purging atmosphere. Bin storages have larger air spaces and therefore, require larger volumes of the purging atmosphere. Our experience indicates that a bulk silo containing almonds would need 2 to 2.5 storage volumes to reach a desired O_2 concentration of 0.5 per cent when purged with an atmosphere containing 0.3 per cent O_2. To achieve the same O_2 level in a

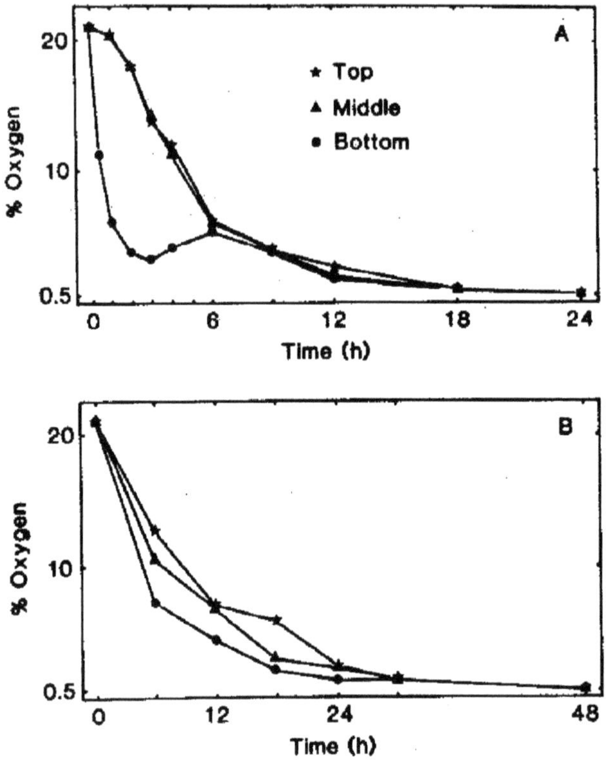

Figure 12.1: Average O_2 Concentration at Three Levels in an 1869 m[3], Fibreen-covered Raisin Storage

rectangular bin storage of raisins, 3.5 to 12.0 storage volumes were required. The volume of the introduced atmosphere required will be determined by its O_2 or CO_2 concentration, the final concentration desired, the interstitial space in relation to total volume, the storage shape, and the leak rate. During the purge phase, the storage is vented to prevent excessive pressure buildup and subsequent structural damage.

Once the desired atmosphere concentration is reached in the storage, the vent is sealed and the maintenance phase is initiated.

The purpose of this phase is to maintain the storage with the desired composition of gases with the least amount of replacement atmosphere. With a tightly sealed structure, a very small maintenance rate will maintain a small positive internal pressure and prevent inflow of air.

Insects and Environmental Factors

Insect species that infest dried fruits and tree nuts are present in all commodities, a few insects may be restricted to a single or limited number of crops. Some insects may be present only by inadvertent contamination and do not feed or reproduce in the stored product. Insect species important in dried fruit storage have been reviewed by Simmons and Nelsons. The same insects generally infect tree nuts. The most important insects are the Indian meal moth, *Plodia interpunctella* (Hiibner), almond moth, *Cadra cautella* (Walker), navel orangeworm, *Amyelois transitella* (Walker), carob moth, *Spectrobates ceratoniae* (Zeller), merchant grain beetle, *Oryzaephilus mercator* (Fauvel), dried fruit beetle, *Carpophilus hemipterus* (Linnaeus), and several other species of nitidulidae. Another important species is the codling moth, *Cydia pomonella* (Linnaeus), that may be found in English walnuts.

Insect responses to CAs are affected by many factors, including insect species, life stages present, relative humidity (RH), temperature, and composition of the displacement atmosphere. Within the nonlethal temperature range, higher temperatures reduce insect kill time with CAs. Lower RH in the storage environment reduces the insect kill time compared to higher humidity.

Insect species vary in the exposure time required for complete mortality. Those insects that breed in areas that are high in CO, or low in O_2 require the longest exposure periods in CAs; an example is *Drosophila* spp. which develop in fermenting fruit.

The morality curves for *plodia interpunctella* pupae which is treated with low O_2 atmosphere can be shown in Figure 12.2.

A particular life stage of insect is dependent on its respiration rate. Very young eggs, mature larvae, and pupae are usually the least susceptible to CAs and therefore require the longest treatment times. Since insects are adversely affected by atmospheres of low RH, storages containing commodities with low-moisture contents will require shorter treatment times than those with high-moisture

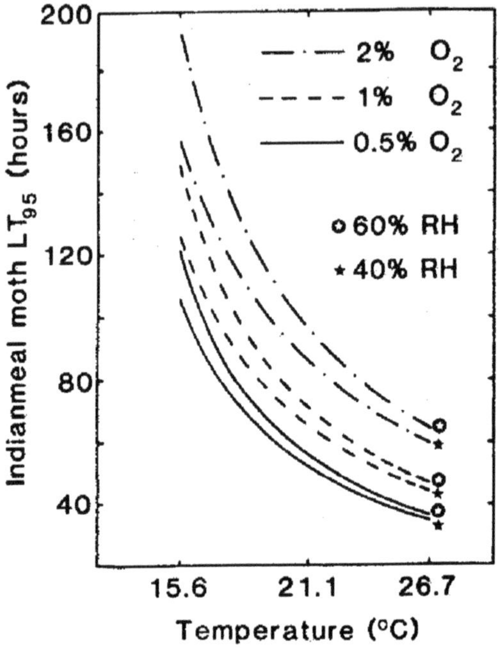

Figure 12.2: The Morality Curves for *Plodia interpunctella* Pupae which is Treated with Low O_2 Atmosphere

content. The moisture content of the incoming atmosphere should be maintained below the dew point of the atmosphere in the storage facility. In general, the CA will quickly attain a RH that is in equilibrium with the stored commodity.

An insect's rate of respiration is directly related to temperature and therefore to susceptibility to CAs. A plot of the effect of temperature on the susceptibility of insects to a low O_2 atmosphere is in the form of a hyperbola, in which the higher temperatures produce the fastest kill times. The atmosphere composition of the storage and the rate of its change from a normal atmosphere to a CA will affect the insect kill time. Ideally, the atmosphere should be altered as rapidly as possible, but the economics of doing so may outweigh the treatment time saved.

Timing of treatment should start at the end of the purge phase, when the desired atmospheric content is attained. The treatment time is also governed by the lowest temperature in storage and the time required to kill the least susceptible insect present. Times need for insect mortality with low O_2[9] and with 60 per cent CO_2[10] treatments at 27°C are shown in Table 12.1.

Table 12.1

Insect Species	*Life Stage*	*Time to Kill (d)*	
		Low O_2	*High CO_2*
Cadra figulilella	Pupae	3	–
Plodia inrerpunctella	Pupae	2	2.5
Drosophila melanogaster	Pupae	5	–
Carpophilus hemipterus	Mixed stages	2.5	–
Oryzaephilus surinamensis	Adults	2	–
	Larvae	2	–
Tribolium castaneum	Adults	2	–
	Larvae	2.5	–
Ameylois transitella	Pupae	2	3

Low-temperature thresholds for effective low-O_2 atmospheres have been calculated by Soderstrom *et al.* to be 9 to 11°C for *Plodia interpunctella* (Hubner) and 7 to 9°C for *Amyelois transltella* (Walker). Jay showed that CO_2 was not effective below 15.6°C for grain disinfestation.

CAs may be utilized for quality preservation of dried fruits and tree nuts since oxidative processes are reduced and rancidity of the product is prevented or reduced. For this purpose, CAs are generally applied in sealed packages, in van containers, or in well-sealed storages.

Almonds and raisins treated with a generated atmosphere consisting of 0.5 per cent O_2, 10 per cent CO_2, and 89.5 per cent N_2 for up to 1 year showed no adverse effects on flavor. English walnuts and Nonpareil and Mission almonds treated with a CO_2 enriched atmosphere for 4 d did not change in flavor. Walnut flavor was normal after 150 d of storage in N_2 or CO_2, according to Verma and Wadhi.

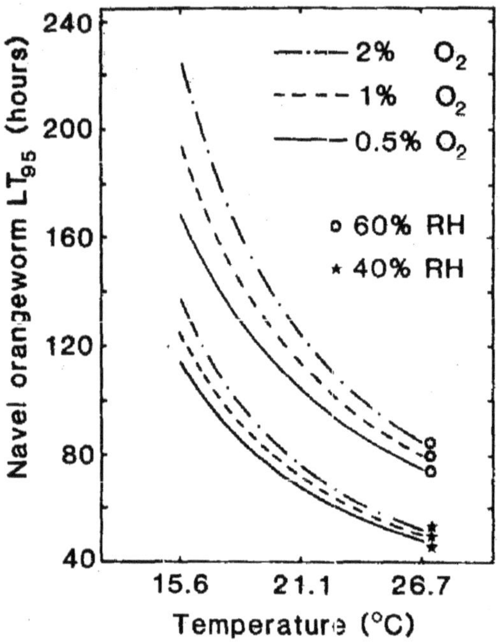

Figure 12.3: Calculated Mortality Curves for *Amyelois transitella* pupae Treated with Low-O_2 Atmospheres (Each Containing 10 per cent CO_2) at Three Temperatures and Two RHs. The curves estimate the lethal time necessary to kill 95 per cent (LT_{95}) of the treated insects.

Although CAs have been thoroughly researched for controlling insects in dried fruits and tree nuts, only the finished, packaged product is being treated with this technology. Bulk storages require extensive sealing before CA can be applied and the industry considers the process too slow and costly in comparison to other control methods. Effort is now being made to improve the equipment for producing CAs more efficiently and to improve the methods of sealing storages. As efforts are made to more efficiently use our energy resources, the energy needed to produce CAs may be obtained from cogeneration plants, biological fermentators and wineries, space and water heaters, and other combustion sources. There is a high probability that CA storage will be used, especially if insect resistance to phosphine increases and if other conventional fumigants are prohibited from use for insect control. The fact that CA leaves no toxic residues in food and public pressure to obtain pesticide-free food will undoubtedly lead to use of CAs in the future.

The U.S. Environmental Protection Agency has granted an exemption from the requirement of tolerance for use of O_2-deficient and CO_2-enriched atmospheres for use on raw and processed agricultural products. Atmosphere-generating equipment and compressed gases for providing O_2-deficient and CO_2-enriched atmospheres are commercially available from several sources. Airtight containers such as cans and jars are readily available and are currently being used for marketing dried fruits and tree nuts.

CAs may be dangerous to persons applying them due to the high pressures or low temperatures encountered with compressed or liquified gases. Consideration must be given to use of proper safety equipment for application of high-pressure gases and for prevention of exposure to ultracold temperatures associated therewith.

A potential for asphyxiation by entering treated areas or adjoining areas where the atmospheres may have leaked must be recognised. After CA treatment, the storage should be thoroughly aerated and analysed for gases that may desorb from the product, storage walls, etc. Also, adjoining rooms or closed areas, elevator pits, or low-lying areas must be analysed to determine that sufficient O_2 is present to sustain the life of personnel entering these areas. Low-lying areas are to be more suspect for asphyxiation when CO_2 is used because CO_2, being heavier than air will flow into low areas and displace O_2. All treated areas must be secured and, preferably locked during the treatment and aeration periods. For safety, when using an exothermic generator (one using a flame for generation), the unit should be placed away from the storage and the atmosphere should be piped to the storage.

Part IV
Emerging Trends in Food Processing Technologies

Chapter 13

Drying of Foods

Drying is a method of food preservation that is simple, safe and easy to learn. This is one of the oldest, simplest, and most effective ways of preserving foods. Drying is the removal of majority of water from food material and provides maximum concentration and microbial safety, lowers water activity for preservation, minimises rate of biochemical reactions, stabilizes nutritional quality and increases shelf life of the food. Drying is effective because both micro-organisms and enzymes need water to be active. It removes enough moisture from the food so bacteria, yeasts and molds can't grow. Also, many enzymes that cause chemical changes in food and other biological materials cannot function without water. When the water content is reduced below about 10 per cent, the micro-organisms are not active. However, it is usually necessary to lower the moisture content below 5 per cent in foods to preserve flavour and nutrition. Dried foods can be stored for extended periods of time.

Drying will never replace canning and freezing because these methods do a better job of retaining the taste, appearance and nutritive value of fresh food. But drying is an excellent way to preserve foods that can add variety to meals and provide delicious and nutritious snacks. One of the biggest advantages of dried foods is that they take much less storage space than canned or frozen foods.

With the renewed interest in gardening and natural foods and because of the high cost of commercially dried products, drying foods at home is becoming popular again. Drying is not difficult, but it does take time and a lot of attention.

Drying of fruits and vegetables under controlled conditions has assumed importance in recent years. The first patent for dehydration process was applied by Grayler in 1780. Artificial drying (Dehydration) of fruit is now an important method of preservation and production of a wide variety of products.

Principle and Purpose of Drying of Foods

Drying usually means removal of relatively small amounts of water from material. The purpose of drying food products is to allow longer periods of storage with minimised packaging requirements and reduced shipping weights. As a result of drying, the food shrinks and it becomes lighter in weight. In addition, drying concentrates the soluble ingredients in foods and this high concentration prevents the growth of bacteria, yeasts, and moulds. Dried food will deteriorate rapidly if allowed to become moist, but provided they are suitably packaged, products will have a long shelf life.

Evaporation refers to removal of relatively large amounts of water from material. In evaporation the water is removed as vapour at its boiling point. In drying the water is usually removed as a vapour by air. It is a process of moisture removal due to simultaneous heat and mass transfer. Surface moisture is evaporated due to heat transfer from the surrounding environment. The moisture can be either transported to the surface of the product and then evaporated, or evaporated internally at a liquid vapour interface and then transported as vapour to the surface.

The transfer of energy (heat) depends on:

1. Air temperature
2. Air humidity
3. Airflow rate
4. Pressure
5. Exposed area of food material

The rate of moisture transfer is governed by the physical nature of the food, including temperature, composition, and in particular moisture content. The dehydration equipment generally utilises conduction, convection, or radiation to transfer energy from a heat source to the food material. The heat is transferred directly from a hot gas or indirectly through a metal surface.

Generally the typical drying cycle consists of the following three stages:

1. Heating the food to the drying temperature
2. Evaporation of the moisture from the product surface. It occurs at a rate proportional to the moisture content.
3. Once the critical moisture point is reached, the falling of the drying rate occurs.

The critical moisture point depends greatly on the drying rate since high drying rates will raise the critical point and low drying rates will decrease them.

Goals of Drying

What so ever is the method chosen for drying food but one should keep the following goals into consideration:

Product Quality

1. Minimal chemical and biochemical degradation reactions
2. Selective removal of water over other salts and volatile flavour and aroma substances
3. Maintenance of product structure (for a structured food)
4. Control of density
5. Rapid and simple rehydration or re-dispersion
6. Storage stability: less refrigeration and packaging requirements
7. Desired colour
8. Lack of contamination or adulteration

Process Economics

1. Minimal product loss
2. Rapid rate of water removal (high capacity per unit amount of drying equipment)

3. Inexpensive energy source (if phase change is involved)
4. Inexpensive regeneration of mass separating agents
5. Minimal solids handling problems
6. Facility of continuous operation
7. Noncomplex apparatus (reliable and minimal labour requirement).

Other: Minimal Environmental Impact

Guidelines for Drying Foods

Although there are different drying methods, the guidelines remain the same which are discussed as under:

Speed

For a good-quality product, vegetables and fruits must be prepared for drying as soon as possible after harvesting. They should be blanched, cooled, and laid out to dry without delay. Foods should be dried rapidly, but not so fast that the outside surface becomes hard before the moisture inside has a chance to evaporate.

Drying must not be interrupted. Once drying the food is started, don't let it cool down in order to start drying again later. Mold and other spoilage organisms can grow on partly dried food.

Temperature

During the first part of the drying process, the air temperature can be relatively high, that is, 150-160°F (65-70°C.), so that moisture can evaporate quickly from the food. Because food loses heat during rapid evaporation, the air temperature can be high without increasing the temperature of the food. But as soon as surface moisture is lost and the rate of evaporation slows down, the food warms up. The air temperature must then be reduced to about 140 °F (60 °C).

Toward the end of the drying process the food can scorch easily, so one must watch it carefully. Each fruit and vegetable has a critical temperature above which a scorched taste develops. The temperature should be high enough to evaporate moisture from the food, but not high enough to cook the food. Carefully follow directions for regulating temperatures.

Humidity and Ventilation

Rapid dehydration is desirable. Higher the temperature and lower the humidity, the more rapid the rate of dehydration will be. Humid air slows down evaporation. Keep this in mind if one plans to dry food on hot, muggy summer days. If drying takes place too fast, however, "case hardening" will occur. This means that the cells on the outside of the pieces of food give up moisture faster than the cells on the inside. The surface becomes hard, preventing the escape of moisture from the inside.

Moisture in the food escapes by evaporating into the air. Trapped air soon takes on as much moisture as it can hold, and then drying can no longer take place. For this reason, be sure the ventilation around oven or in food dryer is adequate.

Uniform Drying

Drying the food evenly takes a little extra effort and attention. Stirring the pieces of food frequently and shifting the racks in the oven or dryer are essential because heat is not the same in all parts of the dryer. For the best results, spread thin layers of uniformly-sized pieces of food on the drying racks.

Types of Food to Dry

Many kinds of fresh fruits, vegetables, herbs, meat, and fish can be dried. Fruits are easier to dry than vegetables because moisture evaporates more easily, and not as much moisture must be removed for the product to keep. Ripe apples, berries, cherries, peaches, apricots, and pears are practical to dry. Fruits may be dried as whole (*e.g.* grape, various berries, apricot, alum, etc.), in sliced form (*e.g.*, banana, mango, papaya, kiwi etc.), in puree form (*e.g.*, mango, apricot etc.), as leather or as a powder by spray or drum drying.

Vegetables that are also practical to dry include peas, corn, peppers, okra, onions, and green beans. It is a waste of time and energy to dry vegetables such as carrots that can be kept for several months in a cool, dry basement or cellar.

Fresh herbs of all types are suitable for drying. The parts of the plant to dry vary, but leaves, seeds, or blossoms usually give the best results.

Preparing Foods to Dry

Preparation of raw materials for dehydration, sun–drying or other evaporation techniques is similar. Some common steps to be applied are as follows.

1. Select fresh, good quality fruits and vegetables. For dried fruit that is naturally sweet and flavorful, be sure to start with good-quality fruit. Select fruit that is fresh, fully ripe and sound–the same quality one would choose for table use.
2. Sort and wash the fruits and vegetables thoroughly. Peel and trim away seeds, core or damaged portions.
3. Discard any bruised or overripe pieces. Decay on one piece may give a bad flavor to the whole batch. Sanitation during the handling and drying process is very important.
4. Cut into halves, strips, or slices about 1/8-inch to 1/4-inch thick that dry readily.
5. Spread thinly on trays initially so there is a third to a half of the screen area clear for the passage of air. Can be more compact after the first moisture is gone.
6. Label all foods on the trays and carry the label along through processing to storage.
7. Thick vegetables need to be blanched or lightly pre-cooked. Leafy vegetables may be wilted slightly with steam, or dried directly from the garden. Fruit is pretreated.
8. Pretreatment of fruits and vegetables is essential before drying which has been mentioned below in detail.
9. To test if food is sufficiently dried, remove a piece and let it cool. Vegetables should be brittle. Fruits because of their sugar content may never get beyond a firm bend or leather quality.

Pretreating Fruits for Drying

Pre-treatments prevent fruits from darkening. Many light-coloured fruits, such as apples, darken rapidly when cut and exposed to air. If not pre-treated, these fruits will continue to darken after they're dried. For long-term storage of dried fruit, sulphuring or using a sulphite dip are the best pre-treatment. However, sulphites found in the food after either of these treatments have been found to cause asthmatic reactions in a small portion of the asthmatic population. Thus, some people may want to use the alternative shorter-term pre-treatment. Pretreatment as given to fruits are discussed below.

Sulfuring

Sulphuring is an old method of pretreating fruits. Sulfur is a mineral that occurs naturally in foods and is necessary for life. Sulfur forms sulfurous acid when it combines with the water in the fruit, but the acid evaporates during drying. The residue is a harmless compound that the body easily excretes.

To keep fruit from discoloring, one should expose it to sulfur immediately after preparing it. There are two methods of sulfuring, each with its own advantages and disadvantages: using (1) sulfur fumes or (2) a sulfite solution.

Sulfur fumes are more effective than sulfur solutions, but this method takes more time and equipment. Several of the recommended references describe methods for making a sulfur box. They also explain how to load and operate the box.

Sublimed sulphur is ignited and burned in an enclosed box with the fruit. The sulphur fumes penetrate the fruit and act as a pre-treatment by retarding spoilage and darkening of the fruit. The sulphur fumes also reduce the loss of vitamins A and C. Sulfuring is the best antioxidant treatment for preserving color. Without a permanent antioxidant treatment, apples and other fruits with light-colored flesh turn dark during drying and storage. Sulfur also helps to prevent the loss of vitamins A and C. Sulfur is not a preservative in itself, but it discourages insects and microbes, which can cause spoilage.

The sulfuring time for each type of fruit is different. Fruits sulfured by this method should not be dried indoors because the odor of the fumes is unpleasant. And take care as sulfur fumes are irritating to the eyes and nose.

Sulfite Dip

Sulphite dips can achieve the same long-term anti-darkening effect as sulfuring, but more quickly and easily. Either sodium bisulfite, sodium sulfite or sodium meta-bisulfite that are USP (food grade) or Reagent grade (pure) can be used.

Table 13.1: Schedule for Sulphuring of Some Fruits

Fruits	*Schedule for Sulphuring of Fruits*		
	Preparation	*Pretreatment (Dipping in Lye etc)*	*Time of Sulphuring*
Apple	Peel, core, trim and cut into 3 to 5 mm thick slices	NIL	30 min.
Apricot	Wash, halve, destone	NIL	30 min.
Peach	Halve, peel and pit	NIL	30 min.
Grape	Wash, cut into small bunches	10-20 sec. 0.3% lye and wash	1-2 hrs
Prune	Wash, use whole fruit	3-5 seconds and wash	–
Litchi	Wash, remove skin	Dip in 1% KMS sol. for 16 hrs	–
Banana	Wash, peel, cut into 12 mm thick slices	NIL	2 hrs
Mango	Wash, peel, cut into 12 mm thick slices	NIL	2 hrs
Pineapple	Wash, punch, core, cut into 6 mm thick slices	NIL	2 hrs
Papaya	Wash, peel, cut into 6 mm thick slices	NIL	2 hrs
Fig	Wash	NIL	1 hr
Aonla	Wash, grate/blanch for 10 min.	Dip in 70°B syrup for 10 hrs	1 hr

Soaking fruit in a sulfite solution is easy. The pieces of fruit are, however, less thoroughly sulfured than they are by fumes. Because of the soaking involved, the fruit absorbs some water, so the drying time is lengthened. Fruit that is sulfured by this method may be dried indoors or out.

To make a sulfite solution, add 1 to 2 tablespoons of sodium bisulfite to 1 gallon of water. Mix thoroughly. Soak the prepared fruit in the solution for 5 to 10 minutes. Soak lighter fruit longer.

Keep the fruit submerged in the solution. Drain the pieces of fruit and then blot them dry on absorbent towel. Do not rinse the fruit in water. Start the drying procedure immediately.

Ascorbic Acid

Ascorbic acid (vitamin C) mixed with water is a safe way to prevent fruit browning. Ascorbic acid reacts easily with oxygen and diminishes oxygen that will be used by phenolase. Ascorbic acid also reduces the o-quinones formed by phenolase to the original o-dihydroxyphenolic compounds. Protection against browning lasts as long as any ascorbic acid remains. Ascorbic acid is not satisfactory for apples since internal atmosphere of the slices contains oxygen. When apples, peaches, pears, apricots, or other light-colored fruits are cut and exposed to the air, the flesh turns brown rapidly. This darkening is caused by a chemical change called oxidation. If oxidation is not stopped, it will injure the texture, flavor, aroma, and appearance of the fruit.

While working with light-colored fruits, treat them with an antioxidant to keep them from turning brown. Mix a small amount of ascorbic acid (2 teaspoons for apples and 1 teaspoon for other light-colored fruit) in one cup of water. Sprinkle the solution over the cut fruit. Stir lightly to coat all pieces. This amount is enough for about 5 quarts of fruit. Commercial antioxidants contain ascorbic acid plus other ingredients, but they are more expensive than pure ascorbic acid.

It's not a good idea to soak fruit in salt or vinegar water because this adds water to the fruit and lengthens the drying time. Soaking also dissolves out some of the water-soluble vitamins. The vinegar-salt solutions also tend to dull the color of the fruit.

The ascorbic acid coating is only a temporary treatment. For permanent, antidarkening action the fruit still needs to be specially treated before drying.

Fruit Juice Dip

A fruit juice that is high in vitamin C can also be used as a pretreatment, though it is not as effective as pure ascorbic acid. Juices high in vitamin C include orange, lemon, pineapple, grape and cranberry. Each juice adds it own colour and flavour to the fruit.

Ethyl Oleate Dip

Ethyl oleate increases the evaporation rate of water in the initial stages of drying, acts as surfactant by increasing the spreading of free water within the sample and removes the skin wax / cell wall of certain products such as grapes, maize grains, starch pastes etc.

Honey Dip

Many store-bought dried fruits are dipped in a honey solution. A similar dip can be made at home. Honey dipped fruit is much higher in calories.

Syrup Blanching

Blanching fruit in syrup helps it to retain colour fairly well during drying and storage. The resulting product is similar to candied fruit. Fruits that can be syrup blanched include apples, apricots, figs, nectarines, peaches, pears, plums, and prunes.

Steam Blanching

Steam blanching also helps to retain colour and slow oxidation. However, the flavour and texture of the fruit is changed. Steam blanching of fruit is an alternative to sulfuring, but it is not as effective. More vitamins are lost and drying takes longer. For these reasons steam blanching is not recommended.

Cracking Skins

Blueberries, cherries, grapes, plums, and a few other fruits have relatively tough skins with a waxlike coating. The skin must be "cracked" or "checked" in many places to remove the waxy coating and to let the inside moisture come to the surface to evaporate. To crack the skin, put the fruit in-to boiling water for 30 to 60 seconds. Then dip in very cold water. Drain thoroughly on absorbent towel.

Cracking the skins in water is recommended rather than dipping the fruit in lye solutions because handling lye can be dangerous.

Pretreatment applications for some fruits are given in Table 13.2.

Preatment for Drying Vegetables

For vegetables, drying time is crucial to tenderness. The longer the drying time is, the less flavourful and poorer the product. Drying time can be hastened by drying small, uniformly cut pieces. Because

Table 13.2: Pretreatment Applications for Some Fruits

		Pretreatment (Choose One)				
Fruits	*Preparation*	*Sulfur (hours)*	*Blanch* *Steam (min)*	*Blanch* *Syrup (min)*	*Other*	*Drying Times,* Dehydrator (hours)*
Apples	Peel and core, cut into slices or rings about 1/8-inch thick.	¾	3 to 5 (depending on texture)	10	Ascorbic acid solution, ascorbic acid mixture, fruit juice dip, sulfite dip	6 to 12
Apricots	Pit and halve. May slice if desired.	2	3 to 4	10	Ascorbic acid solution, ascorbic acid mixture, fruit juice dip, sulfite dip	24 to 36 **
Bananas	Use solid yellow or slightly brown-flecked bananas. Avoid bruised or overripe bananas. Peel and slice 1/4-inch to 3/8-inch thick, crosswise or lengthwise.				Honey dip, ascorbic acid solution, ascorbic acid mixture, fruit juice dip, sulfite dip	8 to 10
Berries, Firm:	Wash and drain berries. With waxy coating-blueberries, cranberries, currants, gooseberries, huckleberries				Plunge into boiling water 15 to 30 seconds to "check" skins. Place fruit in ice water. Drain on paper towels.	24 to 36
Berries, Soft:	Wash and drain berries. Boysenberries and strawberries				No treatment necessary	24 to 36

Contd...

Table 13.2–Contd...

		Pretreatment (Choose One)				
Fruits	*Preparation*	*Sulfur (hours)*	*Blanch*		*Other*	*Drying Times,* Dehydrator (hours)*
			Steam (min)	*Syrup (min)*		
Cherries	Stem, wash, drain and pit fully ripe cherries. Cut in half, chop or leave whole.			10 (for sour cherries)	Whole: Dip in boiling water 30 seconds or more to check skins. Cut and pitted: no treatment necessary.	8 to 12
Figs	Select fully ripe fruit. Immature fruit may sour before drying. Wash or clean whole fruit with damp cloth. Leave small fruit whole, otherwise cut in half.	Whole: 1			Whole: Dip in boiling water 30 seconds or more to check skins. Plunge in ice water to stop further cooking. Drain on paper towels.	6 to 12**
Grapes, Seedless	Leave whole				Whole: Dip in boiling water 30 seconds or more to check skins. Plunge in ice water to stop further cooking. Drain on paper towels.	12 to 20
Grapes, With Seeds	Cut in half and remove seeds				Halves: No treatment necessary	12 to 20

Contd...

Table 13.2–Contd...

		Pretreatment (Choose One)				
Fruits	*Preparation*	*Sulfur (hours)*	*Blanch Steam (min)*	*Blanch Syrup (min)*	*Other*	*Drying Times,* Dehydrator (hours)*
Peaches	When sulfuring, pit and halve; if desired, remove skins. For steam and syrup blanching, leave whole, then pit and halve. May also be slice and quartered.	Halves: 2 to 3; Slices: 1	8	10	Ascorbic acid solution, ascorbic acid mixture, fruit juice dip, sulfiting	36 to 48 **
Pears	Cut in half and core. Peeling preferred. may also slice or quarter.	Halves: 5 Slices: 2	Halves: 6	10	Ascorbic acid solution, ascorbic acid mixture, fruit juice dip, sulfiting	24 to 36 **
Pineapple	Use fully ripe, fresh pineapple. Wash, peel and remove thorny eyes. Slice lengthwise and remove core. Cut in ½-inch slices, crosswise.				No treatment necessary	24 to 26 **
Plums and Prunes	Leave whole or, if sulfuring, halve the fruit.	1			Sun drying (whole): Dip in boiling water 30 seconds or more to check skins. Oven or dehydrator drying: Rinse in hot water.	24 to 36 **

* Because of variations in air circulation, drying in conventional ovens could be up to twice as long. Drying times for sundrying could range from 2 to 6 days, depending on temperatures and humidity.

** Drying times are shorter for slices and other cuts of fruit.

Source: Reynolds, Susan and Paulette Williams, *So Easy To Preserve*. Cooperative Extension Service, The University of Georgia. Revised by Judy Harrison, 1993.

they contain less acid than fruits, vegetables are dried until they are brittle. At this stage, only 10 percent moisture remains and no micro-organisms can grow.

Vegetables should be dried immediately after harvesting. To prepare them, wash in cool water to remove soil and chemical residues. Trim, peel, cut, slice or shred vegetables according to the directions for each vegetable (Table 13.3). Remove any fibrous or woody portions and core when necessary removing all decayed and bruised areas. Keep pieces uniform in size so they will dry at the same rate.

Pretreating Vegetables

Blanching

Blanching is a necessary step in preparing vegetables for drying (Table 13.3). Almost all vegetables need to be blanched. Blanching stops the enzyme action, which drying cannot stop. If vegetables are not blanched, enzymes will destroy the color and flavor during drying and storage. A few vegetables such as mushrooms, okra, and onions do not need to be blanched before drying.

By definition, blanching is the process of heating vegetables to a temperature high enough to destroy enzymes (catalase and peroxidase) present in the tissue. It stops the enzyme action, which causes loss of colour and flavour during drying and storage. It also sets the colour and shortens the drying and rehydration time by relaxing the tissue walls so moisture can escape or re-enter more rapidly in water blanching, the vegetables are submerged in boiling water. In steam blanching, the vegetables are suspended above the boiling water and heated only by the steam. Water blanching usually results in a greater loss of nutrients, but it takes less time than steam blanching. Steam blanching may prevent leakage of nutritive compounds such as water-soluble vitamins. Steam blanching takes more time, but fewer water-soluble nutrients are lost. To minimize the loss of nutrients, blanch only for the required length of time. But don't underblanch; the enzymes will not be inactivated, and the quality of the dried vegetables will be inferior.

After blanching, dip the vegetables briefly in cold water, only long enough to stop the cooking action. However, they should not be cooled to room temperature. When the blanched vegetables are only slightly hot to the touch, they should be cooled to about 120°F.

Table 13.3: Pretreatment of Vegetables

Vegetables	*Preparation*	*Blanching Time (minutes)*		*Drying Time Dehydrator* (hours)*
		Steam	*Water*	
Asparagus	Wash thoroughly. Cut large tips in half.	4 to 5	3½ to 4½	4 to 6
Beans, green	Wash thoroughly. Cut in short pieces or lengthwise. (May freeze for 30 to 40 minutes after blanching for better texture.)	2 to 2½	2	8 to 14
Beets	Cook as usual. Cool; peel. Cut into shoestring strips 1/8-inch thick.	No further blanching required.		10 to 12
Broccoli	Trim and cut as for serving. Wash thoroughly. Quarter stalks lengthwise.	3 to 3½	2	12 to 15
Brussels sprouts	Cut in half lengthwise through stem.	6 to 7	4½ to 5½	12 to 18
Cabbage	Remove outer leaves; quarter and core. Cut into strips 1/8-inch thick.	2½ to 3 until wilted	1½ to 2	10 to 12
Carrots	Use only crisp, tender carrots. Wash thoroughly. Cut off roots and tops; preferably peel, cut in slices or strips 1/8-inch thick.	3 to 3½	3½	10 to 12
Cauliflower	Prepare as for serving.	4 to 5	3 to 4	12 to 15
Celery	Trim stalks. Wash stalks and leaves thoroughly. Slice stalks.	2	2	10 to 16
Corn, cut	Select tender, mature sweet corn. Husk and trim. Cut the kernels from the cob after blanching	5 to 6	4 to 5	6 to 10
Eggplant	Use the directions for summer squash.	3½	3	12 to 14
Garlic	Peel and finely chop garlic bulbs. No other pretreatment is needed. Odor is pungent.	No blanching needed.		6 to 8

Contd...

Table 13.3–Contd...

Vegetables	*Preparation*	*Blanching Time (minutes)*		*Drying Time Dehydrator* (hours)*
		Steam	*Water*	
Greens (chard, kale, turnips, spinach)	Use only young tender leaves. Wash and trim very thoroughly.	2 to 2½	1½	8 to 10
Mushrooms	Scrub thoroughly. Discard any tough, woody stalks. Cut tender stalks into short sections. Do not peel small mushrooms. Peel large mush- rooms, slice.	None		8 to 10
Okra	Wash, trim, slice crosswise in 1/8- to ¼-inch disks	None		8 to 10
Onions	Wash, remove outer "paper shells." Remove tops and root ends, slice 1/8- to ¼-inch thick.	None		3 to 9
Parsley	Wash thoroughly. Separate clusters. Discard long or tough stems.	None		1 to 2
Peas, green	Shell.	3	2	8 to 10
Peppers	Wash, stem, core. Remove "partitions.") Cut into disks about. 3/8 by 3/8 inch.	None		8 to 12
Potatoes	Wash, peel. Cut into shoestring strips ¼-inch thick, or cut in slices 1/8-inch thick.	6 to 8	5 to 6	8 to 12
Pumpkin	Cut or break into pieces. Remove seeds and cavity pulp. Cut into 1-inch strips. Peel rind. Cut strips crosswise into pieces about 1/8-inch thick.	2½ to 3	1	10 to 16
Tomatoes	Steam or dip in boiling water to loosen skins. Chill in cold water. Peel rind. Cut into sections about ¾-inch wide, or slice. Cut small pear or plum tomatoes in half.	3	1	10 to 18

* Drying times in a conventional oven could be up to twice as long, depending on air circulation.

The heat left in the vegetables from blanching causes the drying process to begin more quickly.

Types of Dryers

Over 200 types of dryers are available commercially to process the diverse physical forms of foods and required product characteristics Menon and Mujumdar (1987) have given detailed schemes for classification of dryers based on:

1. Mode of processing (batch, continuous, semi-batch)
2. Operating pressure (vacuum, atmospheric, high pressure)
3. Mode of heat transfer (Conduction, convection, radiation, directive and its combination)
4. Physical state of product being dried (*e.g.* stationary, moving, conveyed, fluidized, dispersed, vibrated, atomized, stirred etc.)

For drying of fruits and vegetables conventional types of dryers are commonly used because of their simplicity of construction and operation as well as low cost. Most such products are seasonal in nature, which means that dryers are operated only over a fraction of the year. Thus, the unit must have low capital costs for economic reasons.

Drying Methods and Processes

Drying methods are not as precise as those recommended for canning and freezing. Food can be dried several ways, for example, by the sun if the air is hot and dry enough, or in an oven or dryer if the climate is humid. Traditionally, foods were dried in the sun and wind, but commercially today, products such as dried milk and instant coffee are made by spraying the liquid into a rising column of dry heated air; solid foods, such as fruit, are spread in layers on a heated surface.

Drying methods and processes can be classified in several different ways. Drying processes can be classified as *batch,* where the material is inserted into the drying equipment and drying proceeds for a given period of time, or as *continuous,* where the material is continuously added to the dryer and dried material is continuously removed.

Depending upon the physical conditions used to add heat and remove water vapour, drying processes can be categorized as:

1. In the first category, heat is added by direct contact with heated air at atmospheric pressure, and the water vapour formed is removed by the air. It can be oven drying or solar drying.
2. In vacuum drying, the evaporation of water proceeds more rapidly at low pressures, and the heat is added indirectly by contact with a metal wall or by radiation (low temperatures can also be used under vacuum for certain materials that may discolour or decompose at higher temperatures); and
3. In freeze drying, water is sublimed from the frozen material.

Oven Drying

Oven drying is the simplest way to dry food because there is almost no need for special equipment. It is also faster than sun drying or using a food dryer. In this type of drying, the heat needed for drying is provided by convection with hot air in direct contact with the products. But oven drying can be used only on a small scale. An ordinary kitchen oven can hold only 2 to 3 kg of food at one time.

Set the oven on the lowest possible setting and preheat to 60°C. Do not use the broiler unit of an electric oven because the food on the top tray will dry too quickly. Remove the unit if it has no separate control. Some gas ovens have a pilot right, which may keep the oven warm enough to dry the food.

It is important to keep the oven temperature at 140 to 160° F (60 to 70°C). Check the temperature about every half hour. Arrange the prepared food (500g) in a single layer on each tray. Put one tray on each oven rack. Allow 1-1/2 inches of space on the sides, front and back of the trays so that air can circulate all around them in the oven. A lighter load dries faster than a full load.

Keep the oven door open slightly during drying. A rolled newspaper, a block of wood, or a hot pad will keep the door ajar so that moist air can escape while the heat stays in the oven. Four to six inches for electric ovens or 1 to 2 inches for gas ovens is usually enough space for ventilation, but use a thermometer to check the oven temperature to make sure it stays at 140° F. An electric fan placed in front of the oven door helps to keep the air circulating.

Shifting the trays often is important for even drying because the temperature is not the same everywhere in the oven. Rotate the trays from top to bottom and from front to back every half hour. It helps to number the trays. Stirring fruit or vegetables every half hour or so also helps the food to dry evenly.

Food Dryer

A commercial or homemade food dryer or convection oven provides automatically controlled heat and ventilation. At a commercial scale, the most common of convection heat dryers used for fruits and vegetables are kiln, cabinet, tunnel and continuous belt dryers.

The kiln dryer is basically an oven heated by gas burners in which the product is placed in slotted trays. The equipment is operated in a batch mode and is still widely used to dry slices of apples and several other fruits and vegetables. It takes about 6-8 hrs to dry a batch of slices of apples.

Cabinet dryer is also batch equipment, but there is a forced flow of air along or through the trays in which the material is placed. Heated air is entered in the cabinet. The operation can be controlled and drying is much more uniform. The unit is useful for small scale drying of cut fruits.

The tunnel dryer is the most used equipment to dry fruit. It is similar to the cabinet dryer, but the trays are made to move along a tunnel in which hot air flows in parallel or counter flow to the product movement. It is most suitable for industrial applications. This method is often used to dry apricots, peaches, pears, apples, figs, dates and so on.

A food dryer takes less electricity than drying the same amount of food in an electric oven. However, the temperature is usually lower (about 120°F or 50°C), so drying takes a little longer than in an oven. In a dehydrator, it takes six or more hours to dry foods. In the oven, it takes eight or more hours. Drying time depends on the type of food, the thickness of the cut, the moisture content of the food and the drying method. Don't speed up the drying time by turning up the oven. Always dry foods in the oven or a food dehydrator by using the right combination of warmth, low humidity and air currents. In drying, a warm temperature allows the moisture to evaporate. Air currents speed up drying. Low humidity allows moisture to move from the food to the air.

When using a dryer, preheat it to 125°F (52°C). Place the food on trays and stack the trays in the dryer. Gradually increase the temperature to 140°F (60°C). It takes 4 to 12 hours to dry fruits or vegetables in a dryer. Do not use space heaters to dry food. These stir up dust and dirt, which contaminate the food. For the same reason, do not try to dry food on a furnace vent or clothes dryer.

Vegetables are dry when they are brittle, fruits when they feel like leather. After foods are dried, allow 30 minutes to one hour cooling time. Too long a cooling period allows moisture from the air to re-enter the food.

Sun Drying

Solar drying is a very simple, ancient skill. The drying of fruits and vegetables using solar energy is both direct and indirect. In the direct solar drying commonly called open sun drying or just 'sun drying' the product is placed directly on trays in open are where only the incident solar radiations is used as the energy source for drying. It requires a safe place to spread the food where dry air in large quantities can pass over and beside thin pieces. Dry, clean air including dry cold air from any source will dehydrate food. But bright sun, low humidity, and temperatures around 100°F are necessary. This process is slow and requires a good deal of care. Other disadvantages of direct sun drying are:

1. Large area requirement
2. Inability to control
3. Degradation of product by biochemical and microbiological reactions
4. Insect infestation
5. Uneven drying
6. Long drying time

The indirect solar dryer uses solar energy to heat drying air in special heat exchangers (Solar collectors) and then this air flows through over the product by natural or forced convection. The air conditions at the inlet of the collector (air humidity and temperature) vary with time depending upon the hour of the day and other factors such as the weather conditions the main advantage of these dryers is that the energy source is renewable, free and nonpolluting.

Natural-draft dryer can be used to sun dry the foods. The advantage of this kind of dryer is that it hastens drying by trapping heat from the sun. It also protects the food from insects and birds. Place pieces of food on drying trays and then cover them with a layer of cheesecloth or netting to keep off dust and insects. Place the dryer in direct sunlight on a roof or high surface away from animals, traffic exhaust, and dust. After the food is almost dry, put it in an airy, shady place to prevent scorching during the final stage of drying.

Be sure to bring the dryer indoors at night if the temperature drops more than 20°F. Dew and sudden temperature changes put moisture back into the food and lengthen the drying time. Fruits and vegetables take 3 to 7 days to dry in the sun. The length of time depends on the type of food and the atmospheric conditions. Natural heat is slower and less dependable than controlled drying in an oven or food dryer.

A single layer of trays outdoors may be covered with sheets of cotton, glass or plastic through which the sun falls on the food. Sunlight heats the food driving out moisture. The moisture-laden air falls down from the bottom of the screened trays. In this generic design, the food is usually exposed to direct sunlight. Direct sunlight destroys some of the more fragile vitamins and enzymes and the food loses color. The better quality food is produced by flat screen designs having a dark sheet of cloth or metal that shades the food. This metal shield slows the drying but these designs are still very productive. Their disadvantages are they spread out over a larger area of ground than the cabinets and they tend to blow over in gusty wind.

Updraft solar dryer designs are the most frequently seen cabinet form. In this design, the hot air flows upward through a solar heat collection trough and enters the bottom of a cabinet underneath the food. The dry air rises through the trays and around the food, exiting through a vent at the top or near the top of the shadowed side. The theoretical basis for this design is that hot air rises and therefore when heated, the air flows naturally upward through the trays of food. This much is true.

There is a basic problem with this theory. It was identified by Dennis Kline working with a research group at the Rodale Institute in the early 1980s. As hot air leaves the collector, it rises through the food gaining moisture and dropping in temperature. Under alternating sun and cloud cover in an updraft solar food dryer the hot air rises into the cabinet bathing the food with hot air. Yet the heated air, cooling and becoming moist, may fail to rise sufficiently to move all the way through the trays and out the exit at the top. Food can be heated repeatedly without sufficient air movement to remove the moisture. It progressively loses momentum and has a natural tendency to fall back rather than to continue to rise and exit through the top vent. Inadequate ventilation is a primary cause of loss of food to mildew in solar food dryers. This is made worse by intermittent heating.

Dennis Kline designed a different dryer to utilize both the tendency for hot air to rise and for cool moist air to fall. His downdraft solar dryer design (also known as a Solar Chimney Dryer) uses three passive natural forces in sequence. First, hot air in the solar collector rises to the top of the cabinet, and then cooling air falls downward over the food. The stagnant cool, moist air at the bottom of the dryer is exhausted by passive exhaust chimney. While this arrangement works well all the time, this design is particularly useful in areas with alternately sunny and cloudy days. Significant amounts of heat do not reach the food unless there is sufficient heat to establish an updraft in the chimney. As soon as it functions, the chimney draws from the stagnant, moist air pocketed at the bottom, which is precisely what is needed.

In addition to this advantage, in the downdraft design the cabinet drops below the collector which places trays at a much more convenient height than in an updraft design where the trays start at the top of the collector and rise up from there. Some updraft solar dryers require climbing a ladder to place, tend or remove loaded trays. In the downdraft configuration, trays can start shortly above ground level. This generic design can be sized for a home or enlarged to adapt to small business needs.

Solar food dryers can be hybridized so they continue to dry during cloudy weather. Removing a tray or two from the bottom, a very small flame from a 16 oz. or picnic size propane burner can be placed on the bottom. Alternately a small electric heating unit may be used. Heat rises and triggers the same effect as the solar heat. It is important to avoid overheating the food, maintaining temperatures below 120°F.

Nutritionally, dried food is ranked as better than canning, just under freezing. The tastes are related to the food, but there is some uniqueness in their flavor and texture. This is similar to the differences between fresh, frozen, and canned foods-another variation in taste.

Dried food should be used within the next months as the nutritional value slowly drops over time. It is customarily eaten the following winter and then any remainder composed as soon as new foods become available.

Drum Drying

Fruit and vegetable juices and purees may be dried with drum dryer. These products are allowed to flow onto the surface of two heated stainless steel drums rotating in opposite directions with little clearance between them. The product dries on the drum and the dried product is scrapped off by blades. This method has been replaced by other newer improved methods such as spray drying.

Spray Drying

This is preferred method for liquids, paste, and slurries. These products are sprayed into the top of a chamber simultaneously with hot air, which is also blown at the top. The cool moist air exists near the bottom and is collected by gravity flow or by the aid of scrappers.

Explosion Puffing

The incorporation of explosion puffing into hot air dehydration process leads to a final product with faster dehydration, rapid rehydration, and a lightly porous structure. Explosion puffing has been successful with apples, blueberries, and banana slices. The process starts with conventional hot air drying of fruit pieces until a certain water content is attained. Then the pieces are placed in a closed chamber. The 'gun' where the pressure is increased by means of super heated steam and after some time, the pressure is suddenly released by opening the lid of the 'gun'. Due to the high temperature and the sudden decompression, water within the particles flashes explosively. Then the resulting solids are again placed in the air drying equipment to be dried to the desired final water content.

Microwave Drying

Microwave heating increases the temperature of the interior, water parts of the solid. Microwave heating has three main advantages.

1. A penetrating quality that leads to uniform drying
2. Selective absorption by liquid water, which leads to uniform moisture profile within the particle
3. Case of control due to rapid response of such heating.

Apple, mango, pineapple and grapes more investigated recently for microwave drying under vacuum. Use of microwave applied in a vacuum for dehydration heats the product to cause water to vapourize, without causing changes in composition. Microwave energy penetrates deeply into food products and can reduce typical process time by 90 per cent. Microwave vacuum dehydration was first used for concentration of citrus juice in France.

Osmotic Drying

Hot air drying reduces notably the quality of processed food stuffs. Osmotic dehydration is an alternative technology to reduce the water content, as well as to improve the quality of final product. This process is being used in industry to dehydrate fruits, vegetables etc.

Osmo-air dried fruits are the dehydrated fruit products based on the novel approach towards dehydration. Osmo-air dehydration involves dehydration of the fruit slices like pineapple, jackfruit and mango in two stages. The first phase is the removal of water using sugar syrup as an osmotic agent. Osmotic dehydration happens in the immersion of cut foods in concentrated solutions of sugars or salts. A flux of water out of the food and another of the solute into food stuff developed due to the difference in the osmotic pressure. The product thus loses some water to the external solution. Osmotic dehydration reduces the weight of slices upto 50 per cent prior to vacuum drying. Osmotic dewatering rate can be enhanced by increasing the concentration of the osmotic solution or temperature.

The second phase of the dehydration is the air-drying where the moisture content is further reduced to about 15 per cent. The product of the osmo-air-dehydrated product is found to be near to the fresh fruit in terms of colour, flavour and texture. It can be consumed as a snack. It can also find use in military rations in suitable packing. The osmo-air dried fruit products can be used in ready-to-eat type of foods, ice cream, fruit salad, kheer, cakes and bakery products. Such osmo-air-dried fruit product based units, can be set up in rural areas nearby fruit gardens, to the benefit of rural people.

The process is simple and involves operations like selection of fruits, cleaning, washing, peeling, curing and slicing/dicing. The prepared fruit slices are soaked in sugar solution to remove water by osmotic pressure. Then the slices are drained and dried in hot air drier. The dried fruit is packed in flexible pouches.

Freeze Drying (Vacuum Drying)

Freeze drying, also called lyophilization is characterized by heating *i.e.* conduction, radiation, radio frequency, microwaves etc. Accelerated Freeze-Drying (AFD) is carried out under vacuum. This is the most modern method of commercial drying. It produces a dried product of excellent quality. It is less damaging to food than straight dehydration in the sense that foods reconstitute better, and it is used for quality instant coffee and dried vegetables. The food is frozen and then the temperature is increased, to vaporize the ice that turns to steam as it dries out the food. The colour, texture, and most of the flavour are kept, the nutritional profile is retained, the food does not shrink as much as with other methods, and the food is preserved for longer than with other drying methods. The foods lose much of their weight, but retain the original size and shape. They have a sponge like texture and rapidly reabsorb liquid when reconstituted. Refrigeration is unnecessary during storage; the shelf-life is similar to dried foods, provided the product is not allowed to become moist. The success of the method is dependent on a fast rate of freezing, and rapid conversion of the ice to vapour. Therefore, the most acceptable results are obtained with thin pieces of food, and the method is not recommended for pieces thicker than 3 cm/1 in. Fruit, vegetables, meat, and fish have proved satisfactory. This method of preservation is commercially used but the products are most often used as constituents of composite dishes, such as packet meals. The process is, however, a more costly method of drying, and AFD foods need to be handled with care as they crumble easily.

The main advantage of this method is the high quality product compared to other drying methods. These advantages are:

1. High flavour and aroma retention
2. High retention of nutritional value
3. Minimal shrinkage
4. Minimal change in shape, colour appearance
5. No damage in structure and texture
6. Porous final structure
7. Easily rehydratable

The industrial application of this method has been limited by the high capital and operating coots. The commercial application has been limited to exotic fruit and orange juices.

Storage of Dried Foods

Dried foods keep four months to one year depending on storage conditions. Store in cool, dry and dark areas. For best quality, store under 60°F. Although it is not necessary to store dried food in a refrigerator or freezer, except for meat jerky, low temperatures extend the shelf life. Proper storage prevents insects and rodents from eating the food. It also keeps moisture out and saves nutrients. Check containers within seven to 10 days to see if moisture is present. If moisture is there, remove food and re-dry at 140 to 160°F.

Nutritive Value

Dried fruits are a good source of energy because they contain concentrated fruit sugars. Fruits also contain a rather large amount of vitamins and minerals. The drying process, however, destroys some of the vitamins, especially A and C. Exposing fruit to sulfur before drying helps retain vitamins A and C. Sulfur destroys thiamine, one of the B vitamins, but fruit is not an important source of thiamine anyway. Many dried fruits are rich in riboflavin and iron.

Vegetables are a good source of minerals and the B vitamins including thiamine, riboflavin and niacin. Both fruits and vegetables provide useful amounts of the fiber (bulk) we need. Save the water used for soaking or cooking dried foods because this nutrient-rich water can be used in recipes to make soups, sauces, and gravy.

Dried fruits can be eaten as nutritious snacks, or can be soaked in water or juice for one to two hours and used in favorite recipes. Sliced vegetables can be eaten dried as tasty vegetable chips for dips. Most vegetables are refreshed with water before use. The water can be added back to vegetables by soaking in tap water for one to two hours. Use two cups of water for each cup of dried food. Using boiling water speeds up the soaking period. Dried vegetables can be added directly to soups or stews without pre-soaking, drawing on the liquid in the soup or stew for rehydration during cooking.

Chapter 14

Water Activity (a_w) and IMF Concept: Role in Food Preservation

The concept of a_w has been very useful in food preservation and on that basis many processes could be successfully adapted and new products designed. Water has been called the universal solvent as it is a requirement for growth, metabolism, and support of many chemical reactions occurring in food products. Free water in fruit or vegetables is the water available for chemical reactions, to support microbial growth, and to act as a transporting medium for compounds. In the bound state, water is not available to participate in these reactions as it is bound by water soluble compounds such as sugar, salt, gums, etc. (osmotic binding), and by the surface effect of the substrate (matrix binding). These water-binding effects reduce the vapour pressure of the food substrate according to Raoult's Law. Comparing this vapour pressure with that of pure water (at the same temperature) results in a ratio called water activity (a_w). Pure water has a_w of 1, one molal solution of sugar–0.98, and one molal solution of sodium chloride–0.9669. A saturated solution of sodium chloride has a water activity of 0.755. This same NaCl solution in a closed container will develop an equilibrium relative humidity (ERH) in a head space of 75.5 per cent. A relationship therefore exists between ERH and a_w since both are based on vapour pressure.

$$a_w = \frac{\text{ERH}}{100}$$

The ERH of a food product is defined as the relative humidity of the air surrounding the food at which the product neither gains nor loses its natural moisture and is in equilibrium with the environment.

Microorganisms vs. a_w value

The definition of moisture conditions in which pathogenic or spoilage microorganisms cannot grow is of paramount importance to food preservation. It is well known that each microorganism has a critical a_w below which growth cannot occur. For instance, pathogenic microorganisms cannot grow at a_w <0.86; yeasts and moulds are more tolerant and usually no growth occurs at a_w <0.62. The so-called intermediate moisture foods (IMF) have a_w values in the range of 0.65-0.90 (Figure 14.1).

Enzymatic and Chemical Changes Related to a_w Values

The relationship between enzymatic and chemical changes in foods as a function of water activity is illustrated in Figure 1. With a_w at 0.3, the product is most stable with respect to lipid oxidation, non-enzymatic browning, enzyme activity, and of course, the various microbial parameters. As a_w increases toward the right, the probability of deterioration of the food product increases.

According to Rahman and Labuza (1999), enzyme-catalyzed reactions can occur in foods with relatively low water contents. Their results are summarized as under:

2. The rate of hydrolysis increases with increased water activity but is extremely slow with very low activity.
2. For each instance of water activity there appears to be a maximum amount of hydrolysis, which also increases with water content.

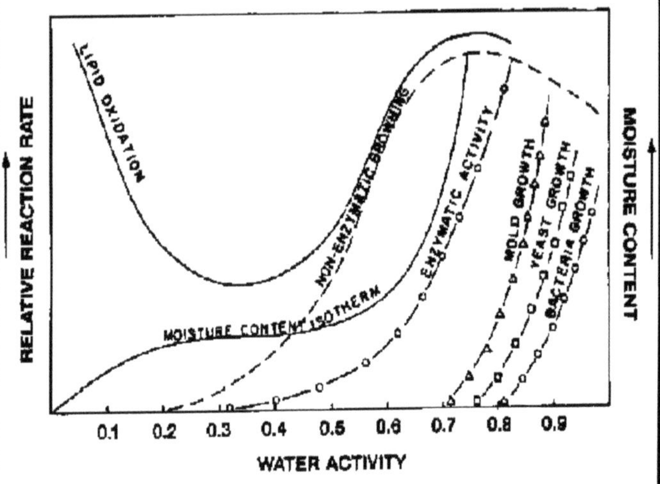

Figure 14.1: Relationship of Food Deterioration Rate as a Function of Water Activity

The apparent cessation of the reaction at low moisture cannot be due to the irreversible inactivation of the enzyme, because upon humidification to a higher water activity, hydrolysis resumes at a rate characteristic of the newly attained water activity. The relationship between water content and water activity is complex. An increase in a_w is usually accompanied by an increase in water content, but in a non-linear fashion. This relationship between water activity and moisture content at a given temperature is called the moisture sorption isotherm. These curves are determined experimentally and constitute the fingerprint of a food system.

Recommended Equipment for Measuring a_w

Many methods and instruments are available for laboratory measurement of water activity in foods. Methods are based on the colligative properties of solutions. Water activity can be estimated by measuring the following:

1. Vapour pressure
2. Osmotic pressure

3. Freezing point depression of a liquid
4. Equilibrium relative humidity of a liquid or solid
5. Boiling point elevation
6. Dew point and wet bulb depression
7. Suction potential, or by using the isopiestic method
8. Bithermal equilibrium
9. Electric hygrometers
10. Hair hygrometers

Water Activity and Intermediate Moisture Foods (IMF) Concept

Intermediate moisture foods (IMF) also referred to as shelf-stable foods generally have moisture content higher than their dry counterparts and are stable without refrigeration. IMF foods are those with a_w in the range of 0.65 to 0.90 and moisture content between 15 per cent and 40 per cent. Adjustment of water activity is necessary to prevent food spoilage. This can be done by drying or incorporating, additions or a combination of drying and additions *e.g.* glycerol, sugar and salt. Fruits contain large amount of soluble solids, water activity is adjusted by drying. Addition of sugar has more impact on flavour than glycerol. Glycerol is very good additive.

Intermediate moisture foods (IMF) may be defined as the food which can be eaten as such, without rehydration and is shelf-stable without refrigeration or thermal processing. These foods are soft and create a moist mouth feel. Food products formulated under this concept are stable at room temperature without thermal processing and can be generally eaten without rehydration.

Some processed fruits and vegetables are considered IMF foods. These include cabbage, carrots, horseradish, potatoes, strawberries, etc.; their water activities at 30°C are as under:

Foods	a_w	Foods	a_w
Cabbage	0.64	Potatoes	0.75
	0.75		0.64
Carrots	0.64	Strawberries	0.65
	0.75		0.75
Horseradish	0.75		

Under these conditions, bacterial growth is inhibited but some moulds and yeast may grow at a_w greater than 0.70. In addition, chemical preservatives are generally used to inhibit the growth of moulds and yeasts in fruits and vegetables.

The application of IMF technology has been very successful in preserving fruits and vegetables without refrigeration in most Latin American countries. For instance, the addition of high amounts of sugar to fruits during processing will create a protective layer against microbial contamination after the heat process. The sugar acts as a water activity depressor limiting the capability of bacteria to grow in food.

The first solid food ever to be consumed by man while walking on the moon was an intermediate moisture food, which was a gelatin-fruit-sugar bar fitted inside the pressure suits. The most IMF space foods have been jellied fruit candy, pecans, peaches, pears, apricots, fruit cakes and nutrient defined foods. Such products prove very useful in events of natural disaster, military persons, mountaineers, astronauts etc. These products can be stored at room temperature conveniently. The IMF foods can be moulded to desired shape, packed and transported easily. Such foods do not give sensation of harsh dryness and has a mixture similar to normal food. After concerted research efforts during last three decades, newer IMF or fabricated IMF have been developed using humectants (Salt, sugar, glycerol etc.) and fungistats (Potassium sorbate, sodium benzoate, calcium propionate) to improve their colour, flavour and storage stability.

Technology of Intermediate Moisture Foods

In an effort to apply this technology to develop ready to eat products a variety of intermediate moisture fruit slices have been prepared by soak-infusion using solutions containing glycerol and sugar with potassium sorbate as preservative at appropriate concentrations (Figure 14.2).

Replacement of glycerol totally with the much cheaper sucrose was also found feasible (Figure 14.3). This process is contrast to the well known osmo-air/vacuum dehydration eliminates the prolonged hot air vacuum drying step, subsequent to osmosis with sugar and also the prolonged rehydration required prior to consumption. The technique requires minimal processing, is low in energy and capital costs and is easy to carry out by any cottage industry.

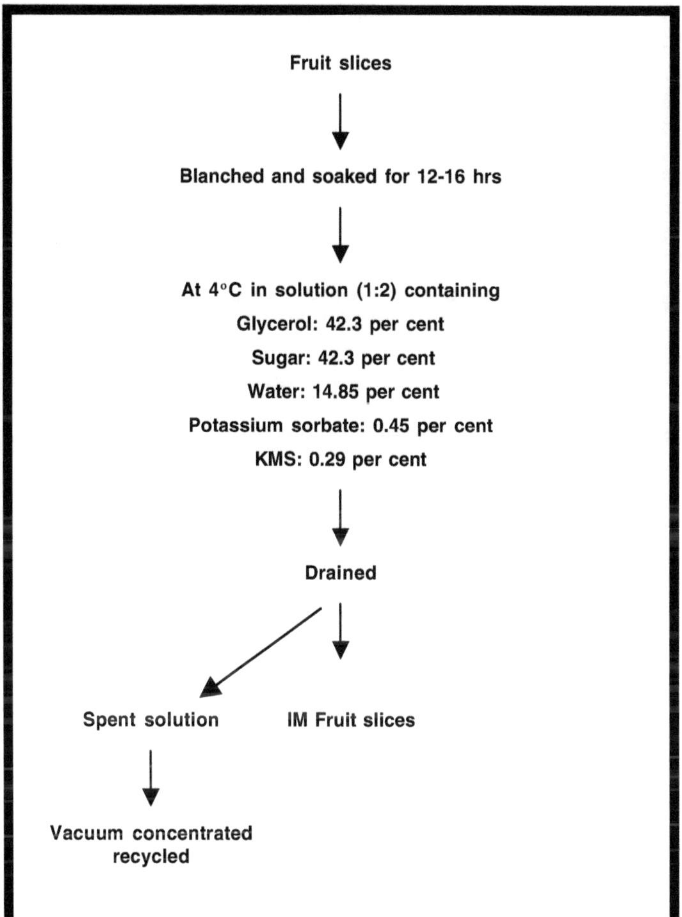

Figure 14.2: Preparation of Intermediate Moisture Fruit Slices Using Glycerol and Sugar

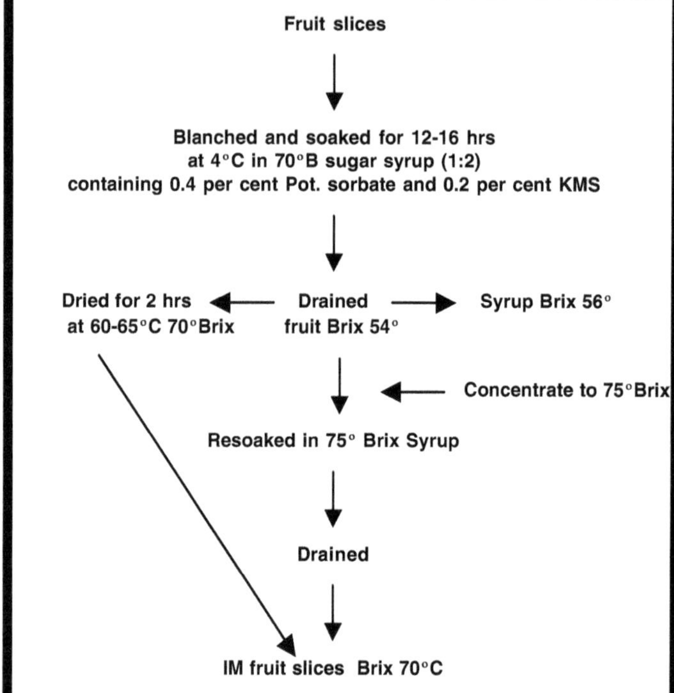

Figure 14.3: Preparation of Intermediate Moisture Fruit Slices Using Sugar only

Generally, the fruits which have been used for preparation of IMF are mango, banana, guava, pear, pineapple, aonla, bael, ber, papaya, apple, apricot, prune and strawberry.

Fruit Bars

Fruit bar is a concentrated fruit product with a good nutritive value. It can be consumed as a confectionery product. It has a good eye appeal and can be readily consumed. The product is a dehydrated one and can be packed in pouches and it possesses a good shelf-life. Fruit bars are made out of fruit pulps like mango, banana, guava, apple and others. It can be marketed as a confectionery item. Some brands of fruit bar products are already present in market.

Methods have also been standardized for the preparation of fruit bars, or slabs with a soft gel like texture suitable for direct eating from several fruits using appropriate ingredients by simple cabinet drying. The method of preparation has been explained in Figure 14.4.

These products are shelf stable for about 6 months in flexible laminate pouches and fulfil the long felt need to light weight, compact and concentrated substitutes for tinned fruits. Bars from fruits like mango, banana, guava and papaya have a great export potential.

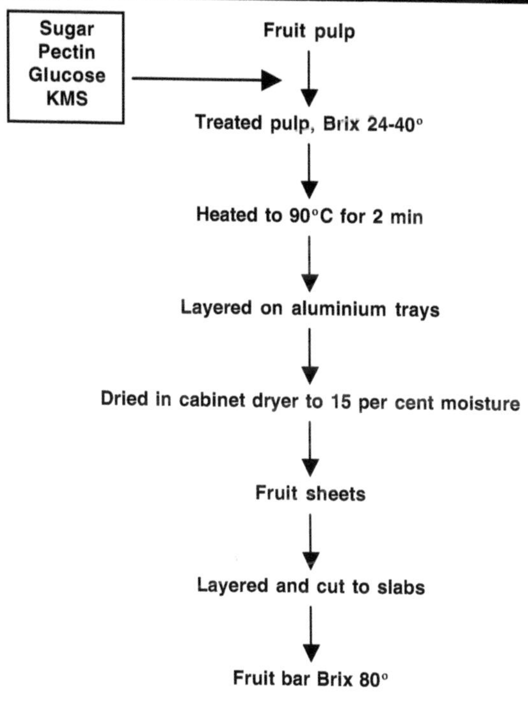

Figure 14.4: Preparation of Fruit Bars

The fruit bar prepared from various fruits showing their characteristics are given in the Table 14.1.

Table 14.1: Data on the Preparation and Characteristics of Fruit Bars (Jayaraman, 1993)

Fruit	Pulp			Bar				
	Yield (% Pulp wt.)	°Brix Initial	°Brix Final	°Brix	Moisture (%)	Acidity (%)	ERH (%)	Yield (% pulp wt.)
Mango	60	17	30	80.0	13.4	1.70	66.6	43
Banana	67	17	24	80.5	9.3	1.10	67.7	40
Guava	90	8	24	76.5	12.5	0.85	66.0	44
Papaya	70	17	30	81.5	12.5	0.70	67.0	48
Jamun	72	12	24	79.0	13.1	0.60	61.4	42
Pineapple	40	16	30	82.5	12.0	1.70	58.1	50
Jackfruit	30	28	30	76.0	13.8	0.80	66.2	33
Apple	80	9	30	80.5	13.5	1.30	68.4	50

Advantages and Disadvantages of IMF Preservation

Advantages

Intermediate moisture foods have a_w range of 0.65-0.90, and thus water activity is their primary hurdle to achieving microbial stability and safety. IMF foods are easy to prepare and store without refrigeration. They are energy efficient and relatively cheap. They are not readily subject to spoilage, even if packages have been damaged prior to opening, as with thermostabilized foods, because of low a_w. This is a plus for many developing countries, especially those in tropical climates with inadequate infrastructure for processing and storage, and offers marketing advantages for consumers all over the world.

Disadvantages

Some IMF foods contain high levels of additives (*i.e.*, nitrites, sulphites, humectants, etc.) that may cause health concerns and possible legal problems. High sugar content is also a concern because of the high calorific intake. Therefore, efforts are been made to improve the quality of such foods by decreasing sugar and salt addition, as well as by increasing the moisture content and a_w, but without sacrificing the microbial stability and safety of products if stored without refrigeration. This may be achieved by an intelligent application of hurdles (Leistner, 1994).

Fruit products from intermediate moisture foods (IMF) appear to have potential markets. However, application of this technology to produce stable products at ambient temperature is limited by the high concentration of solutes required to reduce water activities to safe levels. This usually affects the sensory properties of the food.

Chapter 15

Hurdle Technology: A Food Preservation Concept

The increasing popularity of minimally processed fruits and vegetables has resulted in greater health benefits. Furthermore, the ongoing trend has been to eat out and to consume ready-to eat foods (Alzamora *et al.*, 2000). With this increasing demand for ready-to-eat, fresh, minimally processed foods, including processed fruits and vegetables preserved by relatively mild techniques, new ecology routes for microbial growth have emerged. In order to minimize the loss of quality and to control microbial growth, and thus ensure product safety and convenience, a hurdle approach appears to be the best method (Alzamora *et al.*, 2000).

Hurdle technology is the term often applied when foods are preserved by a combination of processes. Leistner (1976) introduced the hurdle effect, to illustrate that in most traditional and novel foods, a combination of several preservation factors (hurdles) that should not be overcome by the microorganisms present account for the final microbial stability and safety of food. The hurdle effect is an illustration of the fact that in most foods, several factors (hurdles) contribute to stability and safety (Leistner, 1992). This hurdle effect is of fundamental importance for the preservation of food, since the hurdles in a stable product control microbial spoilage and food poisoning as well as undesirable fermentation.

The hurdle includes temperature, water activity, redox potential, modified atmosphere, preservatives, etc. They are present in different combinations in preservation processes. Therefore 'Hurdle Technology' foods are defined as products whole shelf life and the microbiological safety is extended by use of several factors none of which individually would be totally lethal towards spoilage or pathogenic microbes.

According to Alzamora *et al.* (2000), hurdle technology can be applied several ways in the design of preservation systems for minimally processed foods at various stages of the food chain:

1. As a "backup" measure for existing minimally processed products with short shelf life,
2. To diminish microbial pathogenic risk and/or increase shelf life (*i.e.*, use of natural antimicrobials or other stress factors, in addition to refrigeration).
3. As an important tool for improving the quality of long shelf life products without diminishing their microbial stability/safety (*i.e.*, use of heat coadjuvants to reduce the severity of thermal treatments).
4. As a synergist. According to Leistner (1994), in food preserved by hurdle technology, the possibility exists that different hurdles in a food will not just have an additive effect on stability, but could act synergistically. A synergist effect could work if the hurdle in a food hits different targets (*e.g.*, cell membrane, DNA, enzyme systems, pH, a_w, Eh) within the microbial cell, and thus disturbs the homeostasis of the microorganisms present in several aspects. Therefore, employing different hurdles in the preservation of a particular food should be an advantage, because microbial stability could be achieved with a combination of gentle hurdles. In practical terms, this could mean that it is more effective to use different preservatives in small

amounts in a food than only one preservative in large amounts, because different preservatives might hit different targets within the bacterial cell, and thus act synergistically (Leistner, 1994).

During the last decade, minimally processed high moisture fruit products (HMFP), which are ambient stable (with $a_w > 0.93$), have been developed in seven Latin American countries, under the leadership of Argentina, Mexico, and Venezuela. This novel technology was successfully applied to peach halves, pineapple slices, mango slices and purée, papaya slices, banana purée, plum, tamarind, whole figs and strawberries (Alzamora *et al.*, 1995). The methodology employed was based on combinations of mild heat treatments, such as blanching for 1-3 minutes with saturated steam, slightly reducing the a_w (0.98-0.93) by addition of glucose or sucrose, lowering the pH (4.1-3.0) by addition of 1 citric or phosphoric acid, and adding antimicrobials (1000 ppm of potassium sorbate or sodium benzoate, as well as 150 ppm of sodium sulphite or sodium bisulphite) to the product syrup. During storage of HMFP, the sorbate and sulphite levels decreased, as well as a_w levels, due to hydrolysis of glucose (Alzamora *et al.*, 1995).

The concept of hurdle technology is that for a given food the bacteria should not be able to "jump over" all of the hurdles present, and so should be inhibited. The hurdle concept illustrates that complex interactions of temperature, a_w, pH and so on are significant to the microbial stability of foods. The application of the synonymously called combined methods, combined processes, combination preservation, combination techniques, barrier technology *i.e.* hurdle technology improves product quality and identifies innovative food product possibilities.

Significance of Hurdle Technology

An intelligent application of hurdles contributes to both improvement in product quality and stability. An adjustment and calculated application of hurdles could result into saving of money, energy and several other resources.

If several hurdles are used simultaneously, a gentle preservation could be applied, which nevertheless secures stable and safe foods of high sensory and nutritional properties. This is because different hurdles in a food often have a synergistic (enhancing) or additive effect. For instance, food preserved by combined methods (hurdles) remains stable and safe even without refrigeration, and is high in sensory and nutritive value due to the gentle process applied. On the other hand, preservatives (*e.g.*, nitrite in meats) could be partially replaced by certain hurdles (such as water activity) in a food. Moreover, a hurdle could be used without affecting the integrity of food pieces (*e.g.*, fruits) or in the application of high pressure for the preservation of other foods (*e.g.*, juices). Hurdle technology is applicable both in large and small food industries. In general, hurdle technology is now widely used for food design in making new products according to the needs of processors and consumers. For instance, if energy preservation is the goal, then energy consumption hurdles such as refrigeration can be replaced by hurdles (a_w, pH, or E_h) that do not require energy and still ensure a stable and safe product.

General Description of Combined Methods for Fruits and Vegetables

Current consumer demand for foods that are less severely processed, additive-reduced, natural and fresh prompts food manufacturers to select milder preservation techniques. The hurdle technology concept can address these needs. Hence, the processes used for preservation provide shelf stability to a product by laying hurdles, which are difficult to overcome by the microorganisms. Increasing consumer demand for fresh quality products is turning processors to the so-called minimally processed products (MP), an attempt to combine freshness with convenience to the point that even the traditional whole, fresh fruit or vegetable is being packaged and marketed in ways formerly reserved for processed products (Tapia *et al.*, 1996). According to them, the widely accepted concept of MP refrigerated fruits involves the idea of living respiring tissues. Because MP refrigerated products can be raw, the cells of the vegetative tissue may be alive and respiring (as in fruits and vegetables), and biochemical reactions can take place that lead to rapid senescence and/or quality changes. In these products, the primary spoilage mechanisms are microbial growth and physiological and biochemical changes, and in most cases, minimally processed foods are more perishable than the unprocessed raw materials from which they are made.

The technology for shelf-stable high moisture fruit products (HMFP) is based on a combination of inhibiting factors to combat the deleterious effects of microorganisms in fruits, including additional factors to reduce major quality losses from reactions. In order to select a combination of factors and levels, the type of microorganism and quality loss from reactions that might occur must be anticipated (Tapia *et al.*, 1996). Minimal processing may encompass pre-cut refrigerated fruits and peeled refrigerated whole fruits which may include pre-heated vegetables and fruits, cloudy and clarified refrigerated juices, freshly squeezed juices, etc. All of these products have special packaging requirements coupled with refrigeration (Tapia *et al.*, 1996). These products, apart from special handling, preparation, and size reduction operations, might also require special distribution and utilization operations such as Controlled atmosphere/Modified atmosphere/air flow rate/vacuum storage (O_2, CO_2, N_2, CO, C_2H_2, H_2O controls), computer controlled warehousing, retailing and food service, communications network, etc. HMFP fruits are less sophisticated than MPR fruits and should be priced lower when introduced commercially (Tapia *et al.*, 1996). Careful selection of these processes should of course be made to find the appropriate methods suited to a particular rural or village situation.

An example of the hurdle technology concept is presented in Figure 15.1, in which a comparison of HMFP, IMF and MPR fruits in terms of hurdle(s) involved is made. Example A represents an intermediate moisture fruit product containing two hurdles (pH, and a_w). The microorganisms cannot overcome (jump over) these hurdles, thus the food is microbiological stable. In this case, a_w is the most relevant hurdle exerting the strongest pressure against microbial proliferation of IMF. In the preservation system of HMFP (example B), it is obvious that a_w does not represent the hurdle of highest relevance against microbial proliferation; pH is the hurdle exerting the strongest selective pressure on microflora. As in example A, HMFP does not require refrigerated storage. In example C, the mild heat treatment T(t) is applied and the chemical preservative, P, added affects the growth and survival of the flora. With these considerations in mind, it is possible to understand and anticipate the types of microorganisms that could survive, as well as their behaviour and control in such fruits.

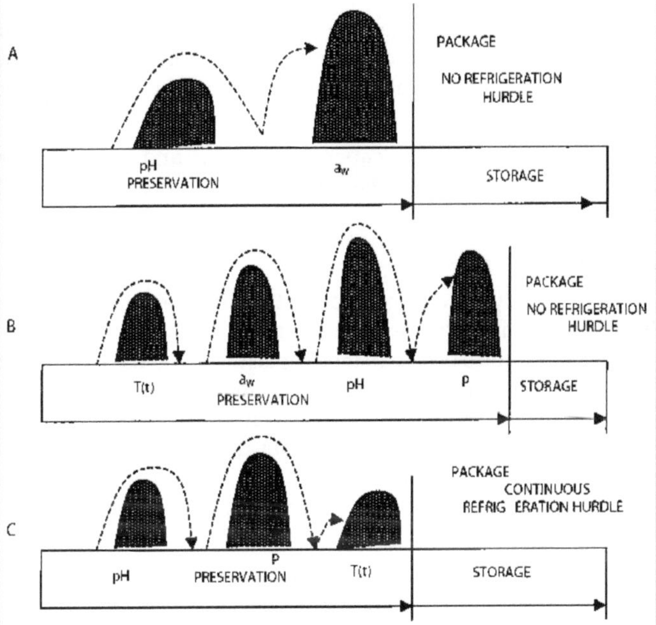

Figure 15.1. Schematic Representation of Hurdles: Water Activity (a_w), pH, Preservatives (P), and Slight Heat Treatment, T(t), Involved in Three Fruit Preservation Systems (A) an intermediate moisture fruit product; (B) a high-moisture fruit product; (C) a minimally processed refrigerated fruit product. (Adapted from Tapia *et al.*, 1996).

Recommended Substances to Reduce a_w in Fruits

Glucose

Glucose is not a very good humectant due to the lower water holding capacity, which makes it difficult to obtain the isotherm curve at low a_w.

Fructose

Fructose has a higher water activity reduction capacity and therefore is more desirable as a humectant in stabilizing food products.

Sucrose

Sucrose is one of the most studied sugars and is widely used in food systems, in the confectionary industry, both in the U.S. and Europe, but has a lower water activity reduction capacity compared to fructose.

The water reduction capacity of sugar and salts in their amorphous and anhydrous state at different a_w is presented in Table 15.1.

Other Humectants

Based solely on the water activity reduction capacity (Table 15.1), sorbitol and fructose are the most desirable humectants. Sucrose has the third best reduction capacity and lactose the poorest. The amorphous form absorbs more water at specific a_w than the corresponding crystalline form. As seen in Table 15.1, NaCl and KCl salts appear to be superior humectants at a high range of a_w. The increased a_w lowering ability exhibited by the salts may be explained by the smaller molecular weight, increasing the ability to bind or structure more water (Sloan and Labuza, 1975).

Table 15.1: Water Activity Reduction Capacities of Sugars and Salts

	Moisture Content (g H_2O/100 g Solids)							
	Anhydrous				*Amorphous*			
Sugars	a_w = 0.60	0.70	0.80	0.90	0.60	0.70	0.80	0.90
Sucrose	3.0	5.0	10.0	–	14.0	20.0	35.0	65.0
Glucose	1.0	3.5	7.5	12.5	1.0	3.5	8.0	22.0
Fructose	14.0	22.0	34.0	47.0	18.0	30.0	44.0	80.0
Lactose	0.01	0.01	0.05	0.10	4.5	4.7	4.7	–
Sorbitol (adsorption)	17.0	22.0	37.0	76.0	2 5.0	3 5.0	5 5.0	110.0
Corn syrup	–	–	–	–	1 4.0	2 0.0	3 0.0	54.0
Salts								
NaCl (adsorption)	0.1	0.1	130.0	5 85.0	–	–	–	–
NaCl (desorption)	–	-	385.0	5 90.0	–	–	–	–
KCl (adsorption)	0.1	0.1	0.1	0.1	–	–	–	–
KCl (adsorption)	–	–	0.1	580.0	–	–	–	–

Source: Sloan and Labuza (1975).

Other sugars used as humectants in food stability include lactose and sorbitol. The amorphous form absorbs more water at specific a_w than the crystalline form. Polyols are better humectants than sugars because of their greater water activity reduction capacity and are less hygroscopic than sugars. The most widely used polyols as humectants in foods are 1,3-butyleneglycol, propylene glycol, glycerol, and polyethylene glycol 400.

Recommended Substances to Reduce pH

Organic Acids

Organic acids, whether naturally present in foods due to fermentation or intentionally added during processing, have been used for many years in food preservation. Some organic acids behave primarily as fungicides or fungistats, while others tend to be more effective at inhibiting bacterial growth. The most commonly used organic acids in food preservation include citric, succinic, malic, tartaric, benzoic, lactic and propionic acids.

The mode of action of organic acids is related to the pH reduction of the substrate, acidification of internal components of cell membranes by ionization of the undissociated acid molecule, or disruption of substrate transport by alteration of cell membrane permeability. The undissociated portion of the acid molecule is primarily responsible for antimicrobial activity; therefore, effectiveness depends upon the dissociation constants (pKa) of the acid. Organic acids are generally more effective at low pH and high dissociation constants.

Citric acid is present in citrus fruits. It is more effective than acetic and lactic acids for inhibiting growth of thermophilic bacteria. Also, combinations of citric and ascorbic acids inhibit growth and toxin production *of C. botulinum* type B in vacuum-packed cooked potatoes.

Malic acid widely found in fruits and vegetables inhibits the growth of yeasts and some bacteria due to a decrease in pH.

Tartaric acid present in fruits such as grapes and pineapples exhibits the antimicrobial activity due to pH reduction.

Benzoic acid, the oldest and most commonly used preservative occurs naturally in cranberries, raspberries, plums, prunes, cinnamon and cloves. As an additive, sodium salt in benzoic acid is suitable for foods and beverages with pH below 4.5. Benzoic acid is primarily used as an antifungal agent in fruit-based and fruit beverages, fruit products, bakery products and margarine.

Lactic acid is not naturally present in foods but is formed during fermentation of foods such as sauerkraut, pickles, olives, and some meats and cheeses by lactic acid bacteria. Lactic acid inhibits the growth of spore forming bacteria at pH 5.0 but does not affect the growth of yeast and moulds.

Propionic acid occurs in foods by natural processing. It is found in Swiss cheese at concentrations up to 1 per cent produced by *Propionicbacterium shermanii*. The antimicrobial activity of propionic acid is primarily against moulds and bacteria.

Inorganic Acids

Inorganic acids including hydrochloric, sulphuric, phosphoric acid (principal acid used in fruit and vegetable processing) etc are mainly used as buffering agents, neutralizers, and cleaners.

Fermentation By-products

Fermentation by-products are formed during fermentation of fruits and vegetables, as in pickling and wine making. One by-product, lactic acid, is formed during fermentation of cabbage or cucumbers. This acid decreases the pH of fruits and vegetables, producing the characteristic flavour of sauerkraut, and acts as a controller of pathogens that may develop in the final fermented product.

Recommended Chemicals to Prevent Browning

Sulphites, Bisulphites and Metabisulphites

Sodium bisulphite is a potential browning inhibitor in fruit and vegetable products (*e.g.*, peeled potatoes and apples). This preservative when used in food production can delay or prevent undesirable changes in the colour, flavour and texture of fresh fruits and vegetables, potatoes, drinks, wine, etc. Potassium bisulphite is used in a similar way to sodium bisulphite and is used in the food industry to prevent browning reactions in fruit and vegetable products.

Sulphites, bisulphites, and metabisulphites of both sodium and potassium together with gaseous sulphur dioxides are all chemically equivalent. Sulphite levels in processed foods are expressed as SO_2 equivalents, and range from zero to about 3000 ppm in dry weight. Dehydrated, light coloured fruits (*e.g.*, apples, apricots, bleached raisins, pears and peaches) contain the greatest amounts in this range. Dehydrated vegetables and prepared soup mixes range from a few hundred to about 2000 ppm; instant potatoes contain approximately 400 ppm. The dose for wine is about 100-400 ppm and for beer about 2-8 ppm. The maximum legal sulphite level in wines permitted by the Food and Drug Administration (FDA) is 300 ppm. In the U.S. most wines have a sulphite level of 100 ppm.

Sulphites are highly effective in controlling browning in fruits and vegetables, but are subject to regulatory restrictions because of adverse effects on health. Sulphites inhibit non-enzymatic browning by reacting with carbonyl intermediates, thereby preventing further reaction. Sulphite levels in foods vary widely depending on the application. Residual levels never exceed several hundred per million but could reach 100 ppm in some fruits and vegetables.

The maximum sulphur dioxide levels in fruit juices, dehydrated potatoes and dried fruits permitted by the FDA are 300, 500, and 2000 ppm, respectively.

Recommended Additives to Inhibit Microorganisms

Potassium Sorbate

Potassium sorbate, a white crystalline powder has greater solubility in water than sorbic acid. It is used in making concentrates for dipping, spraying, or metering fruit and vegetable products. It has antimycotic actions similar to sorbic acid, but usually 25 per cent more potassium sorbate must be used than sorbic acid to secure the same protection.

The common salt of potassium sorbate was developed because of its high solubility in water, which is 58.2 per cent at 20°C (Sofos, 1989). In water, the salt hydrolysis yielded is the active form. Stock solutions of potassium sorbate in water can be concentrated up to 50 per cent, which can be mixed with liquid food products or diluted dips and sprays. Sorbates are effective in retarding the growth of many food spoilage organisms. Sorbates have many uses because of their milder taste, greater effectiveness, and broader pH range (up to 6.5), when compared to either benzoate or propianate. Thus, in foods with very low pH, sorbate levels as low as 200 ppm may give more than adequate protection. The solubility of potassium sorbate is 139 g/100 mL at 20°C; it can be applied in beverages, syrups, fruit juices, wines, jellies, jams, salads, pickles, etc.

Sodium Benzoate

The use of sodium benzoate as a food preservative has been limited to products that are acid in nature. Therefore, it is mainly used as an antimycotic agent (most yeasts and moulds are inhibited by 0.05-0.1 per cent). The benzoates and parabenzoates have been used primarily in fruit juices, chocolate syrup, candied fruit peel, pie fillings, pickled vegetables, relishes, horseradish and cheeses. Sodium benzoate is more effective in food systems where the pH is as low as 4.0 or below.

Other Additives

Other naturally antimicrobial compounds found in fruits and vegetables include:

Vanillin (4-hydroxi-3-methoxybenzaldehyde) is found primarily in vanilla beans and in the fruit of orchids *(Vanilla planifola, Vanilla pompona,* or *Vanilla tahitensis*). Vanillin is most active against moulds and non-lactic acid gram-positive bacteria. The effectiveness of vanillin against certain moulds such as *A. flavus, A. niger, A. ochraceus, or A. parasiticus* has been demonstrated in laboratory media, as well as its effectiveness against yeasts such as *Saccharomyces cerevisiae, Pichia membranaefaciens, Zygosaccharomyces bailii, Z. rouxii* and *Debaryomyces hansenii.*

Allicin is an antimicrobial present in the juice vapour of garlic. This compound is effective in inhibiting the growth of certain pathogenic bacteria such as *B. cereus, C. botulinum, E. coli, Salmonellae, Shigellae, S. aureus, A. flavus, Rhodotorula,* and *Saccharomyces.*

Cinnamon and eugenol are reported to have an inhibitory effect on the spores of *Bacillus anthracis.* Also, cinnamon was found to inhibit the growth of the aflatoxin of *A. parasiticus.* Aqueous clove infusions of 0.1 to 1.0 per cent and 0.06 per cent eugenol were reported to inhibit the growth of germinated spores of *B. subtilis* in nutrient agar.

Oregano, thyme, and rosemary have been found to have inhibitory activity against certain bacteria and moulds due to the presence of antimicrobial compounds in their essential oils (*e.g.*, terpenes, carvacol and thymol).

Recommended Thermal Treatment for Food Preservation

The Role of Heat

The main function of heat in food processing is to inactivate pathogenic and spoilage organisms, as well as enzyme inactivation to preserve foods and extend shelf life. Other advantages of heat processing include the destruction of anti-nutritional components of foods (*e.g.*, trypsin inhibitors in legumes), improving the digestibility of proteins, gelatinization of starches, and the release of niacin. Higher temperatures for shorter periods achieved the same shelf life extension as food treated at lower temperatures and longer periods and allowed retention of sensory and nutritional properties.

Hot Water

Hot water plays an important role in the sanitation of food products before processing. Some food products are treated with hot water to eliminate insects, and to inactivate microorganisms and enzymes. Foods are retained in a water blancher at 70-100°C for a specific time and then removed to a dewatering and cooling system.

Steam

Steam is a more effective means than hot water for blanching foods such as fruits and vegetables. This method is especially suitable for foods with large areas of cut surfaces. It retains more soluble compounds and requires smaller volumes of waste for disposal than those from water blanchers. This is particularly so if air-, rather than water-cooling is used. Furthermore, steam blanchers are easier to clean and sterilize.

Effects of Heat on Aerobic and Anaerobic Mesophylic Bacteria, Yeasts and Moulds

Temperatures ranging from 10 to 15°C above the optimum temperature for growth will destroy vegetative cells of bacteria, yeasts and moulds. Most vegetative cells, as well as viruses, are destroyed when subjected to temperatures of 60 to 80°C for an appropriate time. Somewhat higher temperatures may be needed for thermophilic or thermoduric microorganisms. All vegetative cells are killed in 10 min at 100°C and many spores are destroyed in 30 min at 100°C. Some spores, however, will resist heating at 100°C for several hours.

Chapter 16

Freeze Drying of Foods

Introduction

Freeze drying of foods is carried out by prefreezing in a separate freezer and by sublimation drying and secondary drying in a vaccum cabinet with heater plates.

The main physical processes of the sublimation drying are heat transfer from the heater plates to the sublimation front and water vapour transport from the sublimation front to the ice condenser. A mathematical treatment gives an insight into the process, but a true picture can only be found by laboratory tests. Such tests are also necessary to establish process parameters for each product.

The common illustration of the freeze drying process is a time-temperature diagram giving the temperature history of the product. In this context the freeze drying process is subdivided into different steps, and usually three main steps are considered : the prefreezing, the sublimation drying and the secondary drying.

In the prefreezing step the product is chilled to normal deep freeze temperature. The major part of the water content will freeze out as ice crystals, and the total water content is immobilised in its location within the product.

In sublimation drying, the ice crystals are sublimated into water vapour at the low sublimation temperature and water vapour removed from the product.

Figure 16.1 indicates the condition of a piece of food during freeze drying.

In secondary drying 'bound water,' which did not freeze out as ice crystals is reduced to the low level desired in the product by raising the temperature conditions.

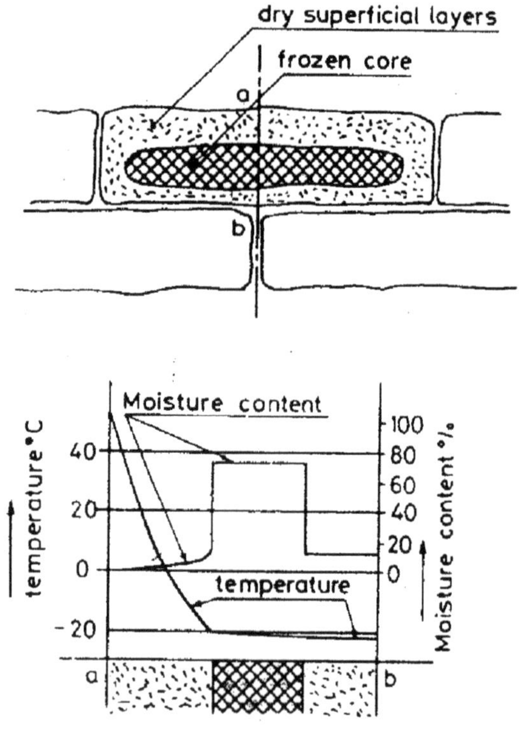

Figure 16.1: Condition of a Piece of Food During Freeze Drying

This time-temperature representation describes the freeze drying process in each microscopic element of the product. The various elements of each product piece will run through different process cycles of this kind. In macroscopic product pieces the prefreezing will normally run faster at the surface than in the interior of the product, and the sublimation drying of interior zones will overlap the secondary drying.of the superficial layers.

Let us regard one piece of food about halfway through the freeze drying process in a vacuum freeze drier. Heat radiation hits the top surface of the product piece, which is surrounded by similar produce pieces on all other surfaces. We can distinguish between the still frozen core of this piece and the superficial layers, where ice crystals not longer exist. The boundary of the frozen core is known as the sublimation front and is playing a prominent part in the freeze drying process.

The microscopic elements forming the sublimation front are running through the sublimation process step. The product elements inside the core are still in the prefreezing step, as they have not yet started the sublimation drying. The product elements of the superficial layers are now in the secondary drying period. Note that the tocal advance of the secondary drying step is linked with the local temperature rise.

At the sublimation front the major part of the water content sublimes from ice crystals to water vapour. This consumes the major part of the heat input of the process. This heat energy has to be carried to the sublimation front and the water vapour produced has to be transported away from the sublimation front. As a quantity of product ice disappears, the sublimation front moves slightly towards the product core.

These two main transport mechanisms within the product, the heat transport to the sublimation front and the vapour transport from the sublimation front, are directly coupled. In each element of the sublimation front the temperature corresponds to the water vapour pressure, the saturation pressure over ice. The driving force of the water vapour movement is the pressure differential between the water vapour at the sublimation front and the water vapour in the freeze drying cabinet at the product surface. The driving force for this heat transfer is the difference between the temperature at the surface of the product piece and the temperature at the sublimation front.

By temperature measurements at defined locations in the product during the freeze drying process, a good picture of the advancing sublimation front and of the local progress of the secondary drying can be obtained.

Another representation of the process is given by the temperatures in defined locations as a function of residual moisture.

When disregarding the secondary drying, the sublimation drying process within the product can be regarded as a combination of only two basic transfer processes, the heat conduction between the product surface and the sublimation front and the vapour flow from the sublimation front to the product surface. Process wise this is much simpler than an ordinary drying process, where the movement of a liquid phase has also to be considered.

The Figure 16.2 clearly explains about the temperatures and weight of slices of raw meat during the freeze drying process.

These processes, which external the product under freeze drying but internal in the freeze drying equipment, are indispensable parts of the total freeze drying process. They are even the most important parts to master when it comes to questions of how to design efficient equipment for freeze drying. This outside view on the freeze drying process has to be considered carefully.

Freezing of Liquid Food

Before the study of freezing of liquid food let us try to understand about the freezing of water.

The freezing of clean water normally takes place at a constant temperature of 273 K, ice crystals forming and growing from the liquid water gradually as heat is removed. To produce 1 kg of ice crystals at 273 K from water of 273 K, 334 kJ of heat has to be removed. This simple physical relation is not sufficient to understand the aspects of the prefreezing step of the freeze drying process.

In freeze drying, clean water is not considered a food. But water is an important component of all foods, and for the freeze drying process it is the all important product component. This water is normally present as part of aqueous solutions and suspensions,

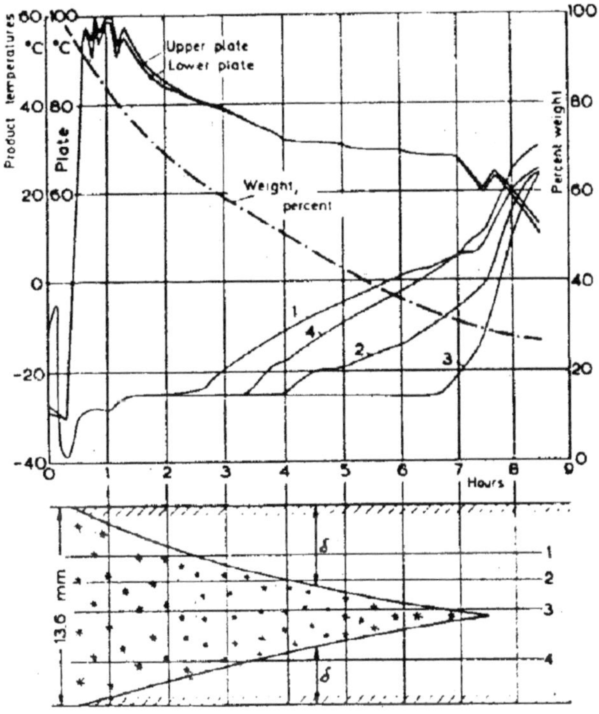

Figure 16.2: Temperatures and Weight of Slices of Raw Meat During the Freeze Drying Process

and the freezing of those is connected with phenomena such as freezing point depression, eutectic solutions, subcooling, ice crystal growth and glassy structure. We will first consider liquid foods, all of which lack an internal structure before the prefreezing step.

In the prefreezing step of a liquid food, an internal structure is created in the food material, which may be stabilised during a later sublimation drying and secondary drying. This structure is formed by a network of ice crystals incorporated in a matrix of concentrated food material. In the latter steps of the freeze drying process the ice crystals are sublimed away leaving a network of channels for the water vapour to escape to the product surfaces. The water content of the matrix reaches these channels by diffusion and also escapes to the surfaces. At the end of the freeze drying process the former liquid material is transformed to a porous solid with the concentrated solid matrix as a structural network.

When lowering the temperature of a liquid food, freezing will start at a temperature some degrees below 273 K depending on the strength and composition of the solution. Ice crystals will form leaving a stronger solution with a lower freezing point. The crystallisation will then proceed at an ever decreasing temperature and in an ever more viscous solution. The viscosity is increasing both from the lowering of the temperature and from the increasing dry solids (DS) content. At a certain temperature the remaining solution freezes to a solid mass. This may be the case of the crystallisation of a eutectic solution, but with foods it is as often the case of the formation of a glassy structure known as the vitreous state. This is the normal case with sugar-containing solutions like fruit juices. In these cases, the viscosity is so high that the material gives the impression of a solid material. The water molecules are immobilised, so that the crystallisation of ice can no longer proceed. With appropriate temperature treatment of such glass, further crystallisation may be obtained; known as devitrification.

Sugar solutions are very apt to form glass structures, which give high resistance to vapour diffusion. This may cause difficulties in the later steps of the freeze drying process.

In such cases, other means of influencing the structure of the prefrozen material may be used. One way is to freeze the liquid as a foam, so that gas bubbles are locked into the frozen matrix. This can provide extra possibilities for vapour diffusion. This foam freezing method has been extensively used in freeze drying of concentrated coffee extracts. It also serves the purpose of controlling the ultimate density of the freeze dried product.

Another example of improving the structure is in the production of freeze dried orange juice. When fruit pulp is suspended in the juice, the freeze drying process goes much easier than with a clear juice without any pulp.

When large ice crystals are formed, a structure with larger size channels will result, which will ease the water vapour transport from the product interior to the surface. Larger ice crystals are favoured by a slow freezing process.

In the prefreezing of solid foods, the structural influence is also of importance. Solid food products have a well defined natural structure, but the prefreezing may to some extent transform this structure. Slow freezing favours the formulation of large ice crystals, which can penetrate and damage cell membranes. Also ultraquick cryogenic freezing may result in drastic structure break down caused by temperature stresses. These structural changes are normally undesired in quick frozen foods. However, an open structure will favour the water vapour transport out to the product surfaces in the later steps of the freeze drying process, and it might also favour the rehydration property of the freeze dried product. For freeze dried products good compromise is often to use semiquick prefreezing.

So far we have considered the influence of the prefreezing step on the microstructure of the product. Of equally great importance is the macrostructure of the product as it is given by the size and shape of the individual product pieces. With the same microstructure, naturally the water vapour will escape more easily to the surfaces of smaller size particles than of larger size particles.

The macrostructures of solid products are usually formed before the prefreezing step in operations such as slicing and dicing. The cutting through of product membranes in such preparatory operations may be decisive for sufficient water vapour passage.

Heat Transfer and Water Vapour Transfer

In the sublimation drying step, the two main transport processes are transfer and water vapour transfer. But other phenomena are also of importance for instance chemical reactions in the material and transfer of aromatic volatiles.

In the normal vacuum freeze drier, heat is transferred to the product surface and on through the product layers to the sublimation front. The driving force of this heat transfer is the temperature difference from the heater surface to the sublimation front. The transport means are the sum of the heat conduction; the heat convection and the heat radiation. The total temperature difference is divided between the two main sections of heat transfer, the first one being from heater surface to product surface, and the other one being from product surface to sublimation front. These two sections represent a series connection of two heat transfer resistances.

The highest product temperature will be found on the product surface, where heat is transferred from the heater surface. To avoid serious quality losses it is important to keep this product temperature below a certain limit, which can be fixed individually for each kind of product.

It is the object of the heater regulation system to control the heat input to the product, so that the permissible product temperature is not exceeded but kept close to this level for process speed and economy.

The heat transfer in the vacuum chamber between the heater surface and the product surface has been the subject of numerous works. The different solutions to this task have given names to different main types of freeze driers. In these cases the sublimation drying takes place in shallow product layers, and the normal form of heater surface is the horizontal heater plate.

The first type of freeze drier was the simple contact freeze drier. Here the product is placed in a shallow layer in a metal tray, which is placed on the horizontal heater plate. The heat transfer from the heater plate to the product surface is mainly by conduction through the metal tray in contact with the heater plate and with parts of the product surface. In contact conditions are so good that an even heat distribution may be obtained.

In the 1950s the double contact freeze drying system was developed under the name of accelerated freeze drying. By establishing heat transfer to boilt sides of the product layer the process time could be drastically reduced. Here again shallow product trays were used. Beneath the product layer in the tray,, a sheet of expanded metal was placed and another sheet was placed on top of the product layer. This sandwich was placed between horizontal heater plates, which were pressed together against the interleaved product. Heat was transferred by conduction (contact) from the heater plates through the tray and through the expanded metal sheet to each of the two product sides. The expanded metal fulfilled a double purpose, It provided channels for water vapour escape between the product and the metal surfaces, and it gave an even heat distribution over the product surface. Variation in product layer thickness was compensated for by local impression of the expanded metal into the product surface.

In contrast to the contact freeze drying systems (which rely on conductive heat transfer through metal from heaters to product surfaces) are the radiation freeze drying systems with main heat transfer between heaters and product surfaces by radiation transfer, without metallic contact.

In the simple radiation freeze drying system 12 product trays with a shallow product layer are interleaved between parallel heater plates with a free space all around the heater plates. Heat is transferred by radiation through this free space from the heater plate above to the product top surface and from the heater plate below to the tray bottom, which is in contact with the product. Also in radiation freeze drying, ribbed trays have been used to increase heat transfer to the product.

The objective of the heat transfer system is to transmit a controlled heat flux evenly to the whole heat-receiving surface of the product. The control of the heat flux is affected by controlling the temperature of the heater, the even heat distribution is safeguarded by the geometry of the plates systems.

In context system, the heat transfer is mainly by heat conduction through the metal of the tray and through any vapour gaps between the tray and the heater plate. A vapour gap is an effective heat insulation compared to direct metallic contact, so minor unevenness of heater plates or tray surfaces will give detrimental unevenness of heat distribution. This is a definite weakness of the contact freeze drying system, because it is not so simple to maintain large surfaces of this kind in a geometrical flush condition. The double contact system can better adjust to practical conditions by the pressure put on to the surfaces and by the expanded metal's variable penetration into product surfaces.

Even heat distribution is maintained much more simply in the radiation freeze drying system. The radiation heat transfer between wide parallel surfaces is independent of the distance between the surfaces. This means that normal technical unevenness of trays or heater plates will not at all disturb the heat distribution.

With distances of 10 mm or more the radiation heat transfer is the dominating component. With distances of 1 mm or less the conductive heat transfer through the vapour is the dominating component. In any case the convective heat transfer component is so small that it can be disregarded, because the vacuum vapour has a very low heat carrying capacity.

To get the same heat flux to the product surface in a radiation freeze drying system may require a somewhat higher heater plate temperature than in a contact system. But even in a radiation system moderate heater temperatures (= 400 K) are sufficient for all practical purposes, so, this actually presents no complication.

The process conditions in the material itself depend on the heat flux to the product surfaces, but it is of no concern whether this heat flux is produced by 'contact' or by 'radiation'. So the heat transfer within the product material may normally be studied by itself and the heat transfer system in the freeze drier disregarded.

The heat transfer within the product material itself will differ according to the material configuration. In block form material this heat transfer can be regarded as heat conduction, and normally the heat has to pass through already dried layers of material forming good heat insulator. This heat conduction will often be rate limiting for the process. Mathematically this process is not as simple as it may seem because the heat conductivity of the material will vary as the layers gradually dry out during the secondary drying.

Several methods have been proposed to increase the efficiency of heat conduction into block material. Heat input through frozen layers is theoretically possible by freezing the product on to the tray bottom. In practice, however, it has shown to be next to impossible to obtain evenly controlled processes. Starting from the edges of the block, a dry surface layer will form and work its way successively long the tray surface and break the thermal contact between the tray metal and the frozen mass of the block.

For true freeze drying to take place, the vapour pressure at the sublimation front driving the vapour out of the product should never be higher than the saturation pressure of water at a temperature where the product is still hard frozen. Incipient melting at the sublimation front should be avoided.

The freeze drying process can also be carried out under normal atmospheric pressure. However, under those conditions the water vapour transport inside the product as a diffusion process is extremely slow, and an economic process solution has not yet been found. All freeze drying plants therefore operate as vacuum freezedriers. And even so the vapour transport out of the product may often be the rate limiting factor of the sublimation process.

In industrial freeze drying, mechanical pumping of the vast vapour volumes produced is not economical. Normally the water vapour is removed by condensation on a cold surface, known as the ice condenser or vapour trap.

The flow of water vapour from the sublimation front will flush non- condensable gases along to the vapour trap. A vacuum pumping system connected to the vapour trap removes the non-condensable gases to prevent a pressure build up. Partial pressures of gases other than water vapour are, therefore, so low in vacuum freeze driers that water vapour transport takes place as mass flow.

The flow regime of water vapour in vacuum freeze drying depends on two dimensionless numbers, the Reynolds number, Re, and the Knudsen number, Kn. Kn = $(Xm)/1$, where Xm is the mean free path of the vapour molecules and l is a characteristic dimension of the vapour flow channel.

If Kn is much larger than l, the vapour molecules bounce much more frequently against the channel walls than against each other. This situation will be normal in the pores of the product and this flow regime is known as molecular flow.

When Kn is much smaller than I, the vapour molecules bounce much more frequently against each other than against the channel walls and we have the conditions of ordinary fluid flow. Then the fluid flow regime-a turbulent flow or a laminar flow–will depend on Re. Re = $(Cl)/v$, where C is the mean vapour flow velocity, l is a characteristic dimension of the vapour flow channel and v is the kinematic viscosity of the vapour.

The kinematic viscosities of vapours are inversely proportional to the vapour pressure. Vacuum vapours consequently have relatively high kinematic viscosities, so the flow regime of the vapours outside the product is usually laminar flow.

When the flow path is known to its exact dimensions, the water vapour flow and the corresponding pressure drop can be calculated according to the physical laws for the kind of flow regimes in question.

In an idealised model product with well defined pore dimensions and heat conductivity, the whole sublimation process can be calculated exactly as simultaneous heat transfer from heater plates via product surface to the sublimation front and water vapour transfer from the sublimation front via the product surface to the vapour trap.

In real life, however, the heat conductivity of the product varies during the process, and so do the dimensions of the product pores, which will be evident as product shrinkage is observed. Therefore, laboratory test runs of the freeze drying process with the products in question will always be necessary to get exact and reliable process information.

Various Types of Freeze Drying

The secondary drying step is very similar to the latter steps of any drying process, when water is no longer mobile in the product in the liquid form and a temperature rise is necessary to release the hygroscopically bound water. Foods normally in the secondary drying can tolerate the sufficient rise in temperature over the sublimation temperature to achieve this. For some pharmaceuticals, however, a higher vacuum is necessary for the secondary drying step. For foods the normal operation is that sublimation drying and secondary drying take place simultaneously in different parts of the product and naturally in the same piece of equipment.

Theoretically the best product preservation is achieved when the residual moisture content is brought down to a monomolecular layer of absorbed water according to the BET equation. For practical purposes 20 per cent residual moisture content may serve as a guideline.

The secondary drying step is also like the sublimation drying step; a combined process of heat transfer and of water vapour transfer. Because of the complexity of this process step calculations are of very limited value. Real knowledge of this process step can be gained by laboratory testing methods.

The practical aims of these tests are to define the end point of the freeze drying process, so that the specified residual moisture content is obtained. Three different methods have been used, the pressure rise method, the temperature approach method and the weight method. In the pressure rise method the vacuum cabinet is shut off from the vapour trap, and the subsequent rise in the cabinet pressure is observed. At specified process conditions, the rate of this pressure rise is characteristic of the residual moisture content. This rate naturally depends also on product quantity and vacuum cabinet size and this makes it difficult to use that method in practice.

In the temperature approach method, product temperatures are observed during the conclusion of the freeze drying operation. Product temperatures tend to approach the heater plates temperature as the product dries out. The temperature difference between heater plates and product will be characteristic of the moisture content. The measurement of exact product temperatures during the process is a delicate task, but when standardised measuring techniques are used, the temperature difference readings will serve the purpose nicely. As it is easy to standardise temperature measurements, this method has found wide application.

When using the weight method a product sample is weighed continuously or at regular intervals during the process. The rate of weight reduction per gramme of product is then a measure of the residual moisture content of the sample. This method can well be used in laboratory testing equipment. In industrial operations, however, the weighing of the same sample is not always practicable. In laboratory testing this method is often used together with the temperature approach method. The correct temperature approach can be established quickly and can be used for measurements in industrial operations.

In the freezing step the product volume increases because ice crystals have about 10 per cent higher volume than liquid water. In the sublimation step, the product volume will remain fairly constant. Even when the ice crystals have sublimed away, the remaining product matrix is rigid enough to keep its shape at the sublimation temperature.

However, with the temperature rise in the secondary drying step, the rigidity of the matrix diminishes, so a volume shrinkage will often occur, although not as obviously as in hot air drying of products. The extent of this shrinkage will depend largely on product characteristics but also on process conditions. Many products can be freeze dried without significant shrinkage, but in some cases as much as 30 per cent volume reduction may be observed.

The other two types of freeze drying which we can study are, continuous freeze drying and batch freeze drying. Let us study them one after another.

Continuous Freeze Drying

In continuous freeze drying driers, the vacuum is kept in the freeze drying cabinet throughout the operation. At short time intervals small portions of the prefrozen material are sliced into the cabinet through a vacuum lock. Each product portion is carried through the cabinet past the heater plate system and ultimately sluiced out of the chamber as finished product for final packaging.

Here the heaters are kept at constant temperature in each of several zones along the path of the product through the cabinet. In this way each portion of the product runs through the same process as in batch freeze drying. The temperature programme for continuous freeze drying is established the same way as in batch freeze drying by laboratory tests with the products in question.

Batch Freeze Drying

In batch freeze drying, a whole charge of prefrozen products is prepared in the trays. Awaiting the sublimation drying step this charge has to be kept under low temperature.

The full charge is then brought quickly into the vacuum cabinet with each tray in the correct position in relation to the heaters. The cabinet is closed and evacuated down to the correct freeze drying pressure. When the charging and evcuating operations are effected fast enough, there is no need to place the vacuum cabinet or part of it in a refrigerated room. The heaters in the cabinet should be chilled to ambient temperature before the charging operation. If parts of the product have picked up some heat during the charging operation, the product temperature will, shortly after the evacuation, be brought back to normal by slight evaporation from warmer zones.

The heating is then turned on and the heater temperature is controlled to a constant level, for instance 400 K, over the entire heater surface. The heater temperature is kept constant over a certain period of drying cycle. Then with the progress of the sublimation and secondary drying of the product, the heater temperature is gradually reduced to an end point temperature of, for instance 325 K, where it may be kept constant again for a period of time.

This temperature-time programme has to be decided for each individual product type and each tray loading by laboratory tests with the products. Normally tray loadings are chosen giving freeze drying cycle times of 6–8 h.

When the freeze drying process is finished the cabinet vacuum is broken by opening, allowing the entrance of atmospheric air, or in special cases dry nitrogen. When the cabinet is opened the product charge is withdrawn for packaging.

In batch freeze drying the total cycle time is the sum of the freeze drying process time and the cabinet operation times. So the cabinet operation times should be as short as possible. This means that the product must be brought in and out without delay, that evacuation and vacuum breaking must be carried out quickly, that plate cooling and plate heating are affected rapidly and that deicing of the vapour trap can take place during the freeze drying operation. In an efficient plant the whole cabinet operation time of the cycle should be down to 15 min.

Freezers and Freeze Driers

For the prefreezing process normally the same types of freezers are used as in the quick frozen foods industry. The choice of freezer type depends largely on the type of product but to some extent also on the type of freeze drier used. For freeze driers with straight trays, the products may often be prefrozen in the same trays carried through a conventional air blast freezing tunnel. Particularly for batch freeze driers, this may give a simple and practical installation.

When a free-flowing product of relatively small but uniform particle size is produced, a freezer giving a free-flowing prefrozen product, *e.g.* IQF–freezer, is favourable because such free-flowing prefrozen material can easily be charged into freeze drier trays. A typical freezer of this kind is the fluidised bed freezer. IQF-freezing is most suitable for continuous freeze drying.

With liquid raw material a free-flowing prefrozen granulate is often preferred. Prefreezing in a thin layer on a rotating drum has shown great advantages. The prefrozen slices that are released from the freezing drum are then dropped into a mill situated in a cold room to produce a granulate of the desired particle size.

Prefreezing of liquids in blocks or lumps followed by crushing in mills in a cold room has also been used, but the technological difficulties increase sharply with the block size.

Also prefreezing of liquid products by dripping into a cold bath of liquid nitrogen or other refrigerant has been proposed.

The various batch freeze driers are mainly characterised by the method of heat transfer between heater plates and product trays and by the shape of the product trays and by the design and arrangement of the ice condenser system.

Taking a look at a popular type of batch freeze drier characteristics will be discussed.

An even surface temperature of the heater plates safeguards in the simplest possible way an even heat distribution to the main product surfaces.

Likewise this arrangement safeguards the easy escape of the water vapour from the product bed. The straight tray is also most versatile with regard to product type and product shape. It is important that the product is distributed in an even layer in the product tray.

An alternative heating system is the contact heating system with ribbed trays. The practical difficulties with this system to achieve even heat distribution has been mentioned earlier. Here the product layer thickness is fixed *i.e.* the distance between the tray ribs. However, some product forms will not be suitable for such fixed layer thickness.

The purpose of the ribbed tray is to achieve a higher heat flux. This higher heat flux is linked with a proportional water vapour flux out of the product bed. In freeze drying this vapour flux has very definite limitations otherwise product particles may be thrown out of the bed. To channel the water vapour away from the product, more complex ribbed trays have been introduced.

Heat transfer systems with moving product in direct contact with heater plates have been advocated. This may give extremely high heat fluxes but the water vapour fluxes inevitably carry material away, which has to be collected and processed separately. This system is also problematic from the hygienic point of view.

The ice condenser of the batch freeze drier shown is built into the vacuum cabinet with good flow conditions for the water vapour. The deicing procedure operates automatically, so that only thin ice layer is allowed to build up before being melted off. The condensing surface therefore operates at full capacity at any time.

Other freeze driers operate with ice condensers in separate vessels connected through pipelines with valves, normally designed for deicing at longer intervals, usually only between batches. This gives rise to pressure drops in the vapour path and to temperature differentials through heavy ice layers. This necessitates lower refrigerant temperatures for satisfactory operation.

The major part of the whole energy consumption on freeze drying is spent on refrigerating the ice condenser. Thus, the thermal efficiency of the ice condenser is vital to an economic operation of the freeze drying production.

In a continuous freeze drier the product is carried as a shallow stationary bed in straight trays.

The trays enter the long vacuum cabinet one by one through a narrow vacuum lock seen to the right in front. The vacuum lock can be actively sealed off from the atmosphere as well as from the vacuum cabinet. After a tray is brought in, the lock is sealed and evacuated by a separate vacuum pump down to freeze drying pressure. Then the connection to the vacuum cabinet is opened and the tray transferred. In the entrance end of the cabinet, the trays are successively collected to a stack (15 trays), which in one movement is pushed one step forward through the cabinet, each tray in a pair of guide rails.

Each pair of rails guides a continuous line of trays in correct radiation position between the stationary heater plates through the whole length of the cabinet. In the exit end of the cabinet a stack of 15 trays is pushed out when the stack is pushed forward in the entrance end. The trays of the exit stack have each moved down between the whole line of heater plates during 6 h, so by this time the freeze drying process is finished. This stack of trays is transferred one by one out through an exit vacuum lock of the same design as the entrance vacuum lock, which is also operated by means of a separate vacuum pump. The empty exit lock must be evacuated down to the freeze drying pressure, before it is opened to the vacuum chamber for acceptance of a tray with finished product.

Chapter 17

Freezing of Foods

Frozen food products have assumed considerable importance during the past two decades or more. According to the recent figures compiled by the Food Department (FPO) as well as GOI, India produced 1294 tons of frozen fruits pulps as compared to only 678 tons of frozen vegetables. The frozen fruits and vegetables are valued at Rs. 1.41 crores and 0.62 crores, respectively. Thus, frozen fruit pulps accounted for nearly two thirds of total frozen fruits and vegetables in India, quite in contrast to USA and Europian countries where frozen fruits accounted for less than 10 per cent as compared to 26 to 59 per cent of frozen vegetables. Figure 17.1 shows the rising trend over the years in production of frozen fruits and vegetables.

Freezing is a quick, convenient, and popular way to preserve fruits and vegetables. Its purpose is:

1. To lower the temperature of food (act of cooling)
2. To transform one state of a substance to another (as water into ice)
3. To maintain the substances in a desired healthy state (as in the storage of ice for the preservation of food).

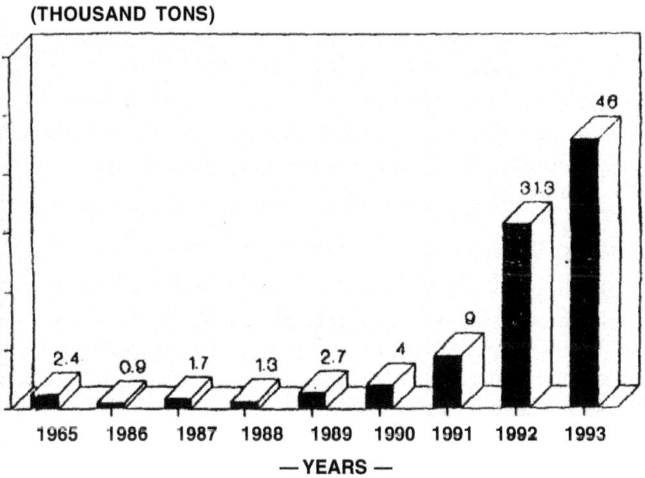

Figure 17.1: Trends in the Production of Frozen Fruits and Vegetables in India. ***(In thousand tonnes)***
Source**: APEDA, Govt. of India, 1995.**

The advantages of freezing foods are as under:

1. Freezing is not usually as economical as canning, but it preserves more nutrients in the food if properly done.
2. Freezing retards the growth of bacteria, moulds, and yeasts. Many microorganisms, such as bacteria, yeasts and moulds cause spoilage of foods. At normal temperature, these microorganisms are active, causing food decomposition to varying degrees, the rate depending on the type and the condition of the food product as well as the extent and nature of microbial load. If the temperature of these foods is reduced sufficiently, the action of these microorganisms is retarded greatly making it possible to keep food longer. The preservation of food is therefore another highly important use of refrigeration. Once the food is thawed, microorganisms may continue to grow.
3. Natural enzymes in foods cause changes in flavor, color, texture and nutritive value. Freezing slows this activity

but does not stop it. To prevent further enzyme activity, vegetables need to be blanched in boiling water or steamed before freezing. Some nutrient loss occurs when vegetables are blanched before freezing. By comparison, however, the nutrient losses from enzymatic activity are greater if vegetables are not blanched. Enzymatic browning in light colored fruits can be prevented by using ascorbic acid mixtures or other substances.

Freezing Basics

Freezing is one of the simplest and least time-consuming ways to preserve foods at home. Freezing does not sterilize food; the extreme cold simply retards growth of microorganisms and slows down changes that affect quality or cause spoilage in foods. Freezing cannot improve the flavor or texture of any food, but when properly done it can preserve most of the quality of the fresh product. Select only the best-quality fruits and vegetables at their peak of maturity for freezing. Frozen fruits and vegetables of high quality and maximum nutritional value can be produced if the directions below are followed. These directions are based on:

1. The chemical and physical reactions which take place during the freezing process;
2. Scientific knowledge of the effect of freezing on the tissues of fruits and vegetables; and
3. Food microbiology.

Different Types of Freezing Processes

Quick Freezing of Foods

Quick freezing process is 'Ultra Rapid' freezing to a very low temperature (-30° to -40°C). It is designed to bring the inside of the product to a temperature of -18°C as quickly as possible. Rapid lowering of the temperature also makes it possible to pass quickly below the critical temperature of 0-5°C at which intracellular liquids freeze. When quick freezing process is used, these liquids solidify to form extremely small crystals of ice and the cellular structure is left intact, whereas in ordinary freezing process, where low temperature is reached more slowly, the texture of the product is altered, liquid forming large ice crystals which lacerate the tissues, which is responsible for poor texture of the frozen food after its thawing. Quick freezing is done by a process of lowering temperature food to much below their freezing point in a short period of time, say to –20°C in 3-30 min. The process involves direct immersion or indirect contact of foods with the refrigerant, and air blast over the foods being frozen.

Individual Quick Freezing (IQF)

'Individual Quick Freezing' refers to quick freezing of solid food/pieces/grains like green peas, cauliflower pieces, cut beans, meat, fish etc. while 'Quick Freezing' relates mostly to liquid, pulpy or semi liquid products like fruit juices, purees, mango/papaya pulp/juice concentrates etc. In IQF, food pieces remain separate and there is no clumping together of pieces or grains, hence the term IQF is designated. The main advantages of such a technology over the conventional slow or deep freezing technology are:

1. There is less damage to the texture of food as the ice crystals formed are very small in size.
2. Due to shorter freezing period, less diffusion of salts occurs. Water is not separated in the form of water.
3. There is less microbial decomposition of food as the lower temperature is attained very quickly by the food. This temperature is much below the one at which microbial growth can occur.
4. The inherent speed and greater output of Individual Quick Freezing and hence higher capacity for commercial freezing plants with the resultant cost reduction is the major advantage over conventional freezing.

Steps involved in Quick Freezing of foods are as follow:

Selection of Varieties

Not all varieties of fruits and vegetables are suitable for freezing. Peaches, apricots, berries, pineapple, apples and grapefruit respond well to freezing, but bananas, avocados, watermelons and grapes do not respond so satisfactorily. Even different varieties within the same fruit respond differently. The variety of fruits and vegetables suitable for quick freezing should have the following desired quality characteristics:

1. Low browning tendency
2. Firm flesh
3. Rich and characteristic flavour
4. Attractive colour
5. Good yield
6. Resistant to insects and diseases and
7. Uniformity in maturity and ripening

Handling, Washing and Cleaning

Initial Sorting and Grading

Pre-Freezing treatments

Some fruits require peeling which is usually done by lye-dipping followed by gentle rubbing and washing off the peel. Some products can have the peel loosened by gentle warming or heating or freezing the surface. Most fruits are pitted by means by halving and pit removal (*e.g.* peaches and apricots)

1. Slicing, cutting, cubing or dicing
2. Blanching

Tabe 17.1: Commercial Freezing Methods

Air Freezing	*Indirect Contact Freezing*	*Immersion Freezing*
Still air 'Sharp' freezer	Single plate	Heat exchange fluid
Blast freezer	Double plate	Compressed gas
Fluid bed freezer	Pressure plate	Refrigerant spray
	Slush freezer	

In IQF method, industrial freezing methods are used. In industrial freezers, so as to remove heat, the product is brought into contact with a cold medium which can be cold air in a blast freezer, cryogenic liquids or gases in a cryogenic freezer, or cold surfaces in a plate freezer. The heat transfer mechanism in cold air is through convection. This can be assisted by blowing the air over the product. Cryogenic liquids have a higher thermal density and remove heat more rapidly still due to an improved heat transfer capability. Heat transfer to a cold surface tends to be by conduction. Therefore, the choice of appropriate freezer is in part, governed by the product size and in part, governed by the required freezing rate. Small fruits can be frozen by IQF method while fruits packed in large cans or drums will not freeze quickly due to heat transfer limitations and so are frozen in a cold room. Retail packs of approximately rectangular brick shape are often frozen in plate (conduction) freezers.

Chemical Changes During Freezing

Fresh fruits and vegetables, when harvested, continue to undergo chemical changes which can cause spoilage and deterioration of the product. This is why these products should be frozen as soon after harvest as possible and at their peak degree of ripeness.

Fresh produce contains chemical compounds called enzymes which cause the loss of color, loss of nutrients, flavor changes and color changes in frozen fruits and vegetables. These enzymes must be inactivated to prevent such reactions from taking place.

Enzymes in vegetables are inactivated by the blanching process. Blanching is the exposure of the vegetables to boiling water or steam for a brief period of time. The vegetable must then be rapidly cooled in ice water to prevent it from cooking. Blanching also helps to destroy microorganisms on the surface of the vegetable and to make some vegetables, such as broccoli and spinach, more compact.

The major problem associated with enzymes in fruits is the development of brown colors and loss of vitamin C. Because fruits are usually served raw, they are not blanched like vegetables. Instead, enzymes in frozen fruit are controlled by using chemical compounds which interfere with deteriorative chemical reactions. The most common control chemical is ascorbic acid (vitamin C). Ascorbic acid may be used in its pure form or in commercial mixtures with sugars.

Some directions for freezing fruits also include temporary measures to control enzyme-activated browning. Such temporary measures include soaking the fruit in dilute vinegar solutions or coating the fruit with sugar and lemon juice. However, these latter methods do not prevent browning as effectively as treatment with ascorbic acid.

Another group of chemical changes that can take place in frozen products is the development of rancid oxidative flavors through contact of the frozen product with air. This problem can be controlled by using a wrapping material which does not permit air to pass into the product. It is also advisable to remove as much air as possible from the freezer bag or container to reduce the amount of air in contact with the product.

Textural Changes During Freezing

Water makes up over 90 per cent of the weight of most fruits and vegetables. This water and other chemical substances are held within the fairly rigid cell walls which give support structure, and texture to the fruit or vegetable. Freezing fruits and vegetables actually consists of freezing the water contained in the plant cells.

The water in fruits and vegetables expands during freezing and breaks the cell walls. That's why thawed foods sometimes have a soft and mushy texture. This textural difference is especially noticeable in products which are usually consumed raw. For example, when a frozen tomato is thawed, it becomes mushy and watery. This explains why celery, lettuce, and tomatoes are not usually frozen and is the reason for the suggestion that frozen fruits, usually consumed raw, be served before they have completely thawed. In the partially thawed state, the effect of freezing on the fruit tissue is less noticeable.

Textural changes due to freezing are not as apparent in products which are cooked before eating because cooking also softens cell walls. These changes are also less noticeable in high starch vegetables, such as peas, corn and lima beans.

To maintain as much of the crisp texture as possible, fruits and vegetables should be frozen at the lowest possible setting on the freezer. Freeze no more than can be solidly frozen in 24 hours. This is usually 2 to 3 pounds of food for each cubic foot of available storage space. Store frozen foods at 0°F or lower.

Rate of Freezing

The extent of cell wall rupture can be controlled by freezing produce as quickly as possible. In rapid freezing, a large number of small ice crystals are formed. These small ice crystals produce less cell wall rupture than slow freezing which produces only a few large ice crystals. This is why it is recommend that the temperature of the freezer be set at the coldest setting several hours before foods are placed in the freezer. About 2 to 3 pounds of vegetable per cubic foot of freezer space per 24 hours is recommended. Overloading the freezer with unfrozen products will result in a long, slow freeze and a poor quality product.

Changes Caused by Fluctuating Temperature

To maintain top quality, frozen fruits and vegetables should be stored at 0°F or lower. This temperature is attainable in separate freezer units and in some combination refrigerator-freezers. A freezer thermometer can help to determine the actual temperature of the freezer. Storing frozen foods at temperatures higher than 0°F increases the rate at which deteriorative reactions can take place and can shorten the shelf life of frozen foods. While some nutrients are lost during storage, the loss is lower when stored at 0°F or below than when stored at higher temperatures. Constant storage temperature helps in retention of quality. Fluctuating temperatures in the freezer can cause the migration of water vapor from the product to the surface of the container. Fluctuating temperature is the texture of frozen fruits and vegetables as the ice melts and then refreezes in the cells.

This defect is sometimes found in commercially frozen foods which are improperly handled.

Moisture Loss

If fruits and vegetables are not properly packaged, air can cause changes that affect flavor. Moisture loss, or ice crystals evaporating from the surface area of a product, produces freezer burn–a grainy, brownish spot where the tissues become dry and tough and off-flavors develop. To prevent this "freezer burn," use moisture vapor-proof or resistant packaging, such as "can or freeze" glass jars, plastic freezing containers, heavy weight aluminum foil, plastic coated freezer paper, and plastic wraps. Be sure to use sealing tape that is designated for freezer use; other tapes will not adhere at freezing temperatures. Freezer bags are popular and can be placed within rigid cartons or containers for easy stacking. Plastic sandwich bags and waxed paper are not suitable. Freeze and cook bags are expensive but convenient. A heat sealer is needed for closing them.

Storage Times for Frozen Foods and Vegetables

Most frozen fruits maintain high quality for 8 to 12 months. Unsweetened fruits lose quality faster than those packed in sugar or sugar syrups. Most vegetables maintain high quality for 12 to 18 months at 0°F or lower. Longer storage of fruits and vegetables than those recommended above will not make the food unfit for use, but will decrease its quality.

Steps in Freezing Foods

Vegetables

1. Choose young, tender vegetables at their peak of flavor and texture. Freeze as soon after picking as possible. Refrigerate vegetables if they cannot be frozen soon.
2. Work in small quantities.
3. Blanch in boiling water or steam (Table 17.2). Start counting blanching time as soon as the vegetable is placed in boiling water or in steamer.

Table 17.2: Preparing Vegetables for Freezing

Vegetables	*Preparation and Blanching in Boiling Water*
Asparagus	Wash and sort small, medium, and large stalks. Leave whole or cut in 1- or 2-inch lengths. Blanch small stalks 2 minutes. Blanch medium stalks 3 minutes, large stalks (1/2- to 3/4-inch diameter) 4 minutes. Chill in ice water for same time.
Beans (green and yellow podded)	Wash, snip off tips, and sort for size. Cut or break into suitable pieces or freeze small beans whole. Blanch 3 minutes. Chill in ice water 3 minutes.
Beans, lima	Wash, shell, and sort. Blanch small beans 2 minutes; medium beans, 3 minutes; large beans 4 minutes. If desired, blanch in the pod and shell after cooling. Chill in ice water for same time.
Beets	Remove all but 2 inches of top; wash. Cook until tender. Chill. Remove skins. Slice or dice large beets. Since beets are completely cooked, blanching is unnecessary.
Broccoli	Remove tough leaves and woody butt ends. Cut through stalks lengthwise, leaving heads 1 inch in diameter. Soak 1/2 hour in salt brine (1/2 cup salt to 1 quart water) to drive out small insects. Rinse and drain. Blanch 3 minutes in water or steam-blanch 5 minutes. Chill in ice water.

Contd...

Table 17.2–Contd...

Vegetables	*Preparation and Blanching in Boiling Water*
Beans and green soybeans	Harvest when beans have filled the pod and in the green stage; wash. Boil beans in the pods for 5 minutes. Cool promptly in ice water. Squeeze beans out of the pods.
Brussels sprouts	Wash and trim. Soak 1/2 hour in salt brine (see Broccoli). Rinse and drain. Blanch small heads 2 minutes; medium heads 4 minutes; large heads 5 minutes. Chill in ice water.
Cabbage	Wash and discard coarse outer leaves. Cut into wedges or shred coarsely. Blanch wedges 3 minutes and shredded cabbage 11/2 minutes. Chill in ice water.
Carrots	Trim, wash, and scrape. Dice or slice 1/4-inch thick. Blanch 2 minutes. Chill in ice water.
Cauliflower	Trim and wash. Slit heads into individual pieces 1 inch in diameter. Soak 1/2 hour in salt brine. Rinse and drain. Blanch 3 minutes. Chill in ice water.
Sweet corn on-the-cob	Husk, remove silks, and trim ends. Use a large kettle for blanching. Blanch small ears (1, 1/4-inch diameter) 7 minutes; medium ears (1, 1/4-11/2-inch diameter) 9 minutes; large ears (over 1, 1/2-inch diameter) 11 minutes. Chill in ice water until cool. Corn that is not thoroughly cooled may become mushy. Cooling corn-on-the-cob will take longer than blanching time.
Sweet corn, cut whole kernel	Husk, remove silks, and trim ends. Use a large kettle. Blanch whole kernel corn to be cut from the cob 4, 1/2 minutes. Chill in ice water. Cut from cob after cooling.
Corn, cream style	Prepare as for whole kernel corn except cut only the kernel tips; then scrape the cobs with the back of a knife to form juice and remove the heart of the kernel.
Herbs	Wash and drain, but do not blanch leaves. Wrap a few sprigs or leaves in foil or seal in plastic bags. Store in carton or glass jar in freezer. Wash and drain, but do not blanch leaves. Wrap a few sprigs or leaves in foil or seal in plastic bags. Store in carton or glass jar in freezer.
Mushrooms	Wash and remove stem base. Freeze small mushrooms whole; cut large ones into 4 or more pieces. When blanching mushrooms, add 1 teaspoon citric acid, or 1 tablespoon lemon juice, or 1/2 teaspoon ascorbic acid per quart of water to prevent darkening. Blanch medium or small whole mushrooms 5 minutes; cut pieces 3 minutes. Chill in ice water.

Contd...

Table 17.2–Contd...

Vegetables	*Preparation and Blanching in Boiling Water*
Okra	Wash. Cut off stems without opening seed cells. Blanch small pods 3 minutes; large pods 4 minutes. Chill in ice water.
Peas (green, english, blackeyed)	Wash. Shell small amount at a time. Blanch green and english for 1, 1/2 minutes. Blanch blackeyed peas 2 minutes. Chill in ice water.
Peas (edible, podded or Chinese)	Wash. Remove stems, blossom ends, and any string. Leave whole. Blanch 1, 1/2 to 3 minutes. Chill in ice water.
Pumpkin	Cut or break into fairly uniform pieces. Remove seeds. Bake at 350°F, or steam until tender. Cool, scoop pulp from rind, and mash or put through ricer. Spoon into moisture-proof containers.
Potatoes	Wash; peel; remove deep eyes, bruises, and green surface coloring. Cut in 1/4- to 1/2-inch cubes. Blanch 5 minutes. Cool. For hash browns—cook unpeeled until almost done; peel and grate; form in desired shapes; freeze. For french fries—peel and cut in thin strips; fry in deep fat until very light golden brown; drain and cool; complete browning before serving.
Potatoes, new	Choose those of uniform size; scrub vigorously to remove tender skin. For 3/4-inch diameter potatoes, blanch 4 minutes. For 1-inch diameter, blanch 6 minutes. For 1, 1/2-inch diameter, blanch 7 minutes. For those larger than 11/2-inch diameter, blanch 8 to 10 minutes. Chill potatoes in ice water 3 to 5 mintues; drain well. Use within a month for best quality.
Spinach and other greens	Sort and remove tough stems. Blanch most leafy greens 2 minutes. Blanch collards and stem portions of Swiss chard 3 to 4 minutes. Blanch very tender spinach 1, 1/ 2 minutes. Chill in ice water.
Tomatoes, cooked	Select firm, sound, ripe tomatoes. Wash, core, cut, and cook until soft.
Tomato juice and purée	Select firm, sound, ripe tomatoes. Wash, core, and cut into pieces. Simmer about 5 minutes and put through food mill for juice. Cool. For purée, cook juice until concentrated to about half its volume.

(a) *To blanch in boiling water*: Blanch no more than 1 pound of vegetables per gallon of boiling water at a time. Immerse wire basket or mesh bag with vegetable in boiling water. Cover and blanch for required time.

(b) *To steam blanch*: Place one layer of vegetables in a wire basket suspended above several inches of rapidly boiling water in steamer. Cover and steam for about 1, 1/2 times longer than suggested time for blanching in Table 17.2. Generally, water blanching takes less time and fuel. However, the nutrient losses may be somewhat less for steam blanching.

(c) *Microwave oven blanching*: Because microwave ovens do not have standardized power levels, it is impossible to publish a timetable that can be used with all ovens. Follow directions given by the oven manufacturer. Little research has been published on the effectiveness of blanching in the microwave oven. Some studies have shown microwave-blanched vegetables to be lower in vitamin C and color than steam- or water blanched vegetables.

4. Cool immediately in ice water for the same time as for blanching. Drain thoroughly.
5. Pack tightly in freezer bags or rigid containers. Squeeze as much air as possible from bags before sealing. Leave 1/2-inch headspace for expansion in containers. There are two basic methods for packing vegetables for freezing, the tray pack and the dry pack.

(a) *Dry pack*: This is the term used to describe the packing of blanched and drained vegetables into containers or freezer bags. Pack the vegetables tightly to cut down on the amount of air in the container. If the vegetables are packed in freezer bags, press air out of the unfilled part of the bag. When packing cauliflower or broccoli, alternate the heads and stems. Allow ½ inch headspace (except for loose packing vegetables such as asparagus and broccoli that do not require headspace). Seal.

(b) *Tray Pack*: This is the method of freezing individual pieces of blanched and drained vegetables on a tray or shallow pan, then packing the frozen pieces into a freezer bag or container. This method produces a product similar to commercially frozen plastic bags of individual vegetable pieces and is particularly good for peas, corn and beans. Pack the frozen pieces into a bag or container as soon as they are frozen. Long exposure will result in loss of moisture. Do not leave headspace. Seal.

6. Label with product and date. Freeze at lowest possible setting.
8. Store at 0°F or lower. Use within 8 to 12 months for best quality and nutrient retention.

Using Frozen Vegetables

Most frozen vegetables can be cooked without thawing. However, corn on the cob should be partially defrosted. Cook vegetables in a small amount of water, 1/2 cup or less. Cook until just tender–about half as long as if the vegetable were fresh.

Fruits

1. Do not use underripe or overripe fruit. The flavor and texture will be undesirable.
2. Wash and sort fruit according to size.
3. Pare and remove pits, seeds and blemishes. Leave whole, slice, or puree according to instructions given in Table 17.3.
4. Use an anti-darkening agent on fruits that turn brown. Ascorbic acid preparations or sodium bisulfite are the most effective.
5. Use dry sugar, syrup, or unsweetened pack. The flavor, color and texture of most fruits is retained best when some sweetening is used.

   (a) *Dry Sugar*: A 3 to 1 ratio is usually used. However, less sugar can be used. Sprinkle sugar over fruit. Gently stir until pieces are coated with sugar and juice. About 48 calories is added to each half cup serving when using these proportions.

   (b) *Syrup*: To make sugar syrup, dissolve the needed amount of sugar in water, mixing until the solution is clear. Chill syrup before using. Use just enough cold syrup to cover the fruit. A 30 to 40 per cent syrup is used for most fruits (see Syrup Proportions Chart below). Dissolve sugar in water. Heavy syrups will require heating to dissolve the sugar. Cool and use just enough to cover the fruit. Each pint will require 1/2 to 2/3 cup syrup. Keep fruit under syrup with a small piece of crumpled, water-resistant wrapping material on top of the fruit. Then close and seal the container.

Table 17.3: Preparing Fruits for Freezing

Fruits	*Preparation*
Apples	Wash in cold water, peel, core, and cut into pie slices. To prevent darkening, use one of these methods: (*a*) Soak slices in sodium bisulfite solution (1 teaspoon in 1 gallon water) for 5 minutes. Mix solution in glass, earthenware, stainless steel, or enamel container. Drain. Pack using 1 cup sugar for 10-12 cups apples. (*b*) Soak apple slices in brine solution (1/2 teaspoon ascorbic acid to 1 quart water). (*c*) Scald in boiling water 2 minutes. Chill in ice water. Pack using 1 cup sugar for 10-12 cups apples.
Apricots	Quality is better when apricots are canned rather than frozen. Wash in cold water and sort. Dip 6 fully-ripened apricots into boiling water until skins loosen, 15 to 20 seconds. Chill. If desired, peel, halve, and remove pits. Fill containers one-third full of syrup (3 cups sugar to 1 quart water with 1/2 teaspoon ascorbic acid). Pack apricots in syrup. Or: Halve soft ripe fruit, steam 4 minutes, crush and pack using 1 cup sugar for 8-9 cups fruit.
Sweet cherries	Wash in cold water and sort. Pack in syrup (2 cups sugar to 1 quart water plus 1/2 teaspoon ascorbic acid) and either 1 teaspoon citric acid or 4 teaspoons lemon juice for flavor.
Cranberries, currants	Wash in cold water, sort, and pack without sugar.
Gooseberries	Wash in cold water and sort. Pack without sugar or syrup, or mix berries and sugar as called for in pie recipe, using 1 to 1, 1/2 cups sugar for 4 cups berries.
Grape juice	Choose fully ripe, firm, sweet grapes. Sort, stem, and wash. (*a*) For juice as a beverage: Cover with water. Heat slowly to simmering and cook until grapes are soft. Strain through bag and sweeten if desired. Common proportion is about 1/2 cup sugar per quart juice. (*b*) For juice for jelly: Add just enough water to keep grapes from sticking. Cook at simmering until grapes are soft. Strain through jelly bag. Let juice stand in refrigerator overnight to remove tartrate crystals. Strain again.

Contd...

Table 17.3–Contd...

Fruits	*Preparation*
Melons	Wash in cold water. Peel and cut flesh into 1/2- to 3/4-inch cubes or balls. Cover with syrup (2 cups sugar to 1 quart water). Flavor with lime juice and/or add whole seedless grapes, if desired. Serve partially frozen.
Peaches	Wash in cold water and sort. Dip 3 or 4 peaches into boiling water until skins loosen. Chill. Peel and slice peaches into containers one-third full of syrup (3 cups sugar to 1 quart water with 1/2 teaspoon ascorbic acid).
Pears	Select from full-flavored pears, not mealy in texture. Wash, peel and core. Slice. Heat in a boiling syrup in the proportion of 3 cups sugar to 4 cups water for 1-2 minutes. Drain and cool. Add 3/4 teaspoon ascorbic acid to 1 quart syrup.
Pineapple	Peel and core. Dice, slice, or cut into wedges, removing center core. Cover with syrup (3 cups sugar to 1 quart water). Or pack using 1 cup sugar for 8-9 cups fruit. Do not use uncooked pineapple in gelatin mixtures.
Plums	Select firm ripe fruit soft enough to yield to slight pressure. Sort, wash, halve, and pit. For dry sugar pack, use 1/2-3/4 cup sugar per quart fruit. For syrup pack, cover fruit with a 40 per cent syrup (3 cups sugar to 1 quart water). For best quality, add ascorbic acid to the cold syrup.
Raspberries	Pack raspberries for jam without sweetening. Wash in cold water and sort. Pack raspberries in syrup (3 cups sugar to 1 quart water). Or pack using 1 cup sugar for 7-8 cups fruit.
Strawberries	Wash in cold water, sort, and stem. Pack whole, sliced (preferred), or crushed berries with 1 cup sugar for 7-8 cup of fruit. Or pack whole berries in syrup (3-4 cups sugar to 1 quart water).

(c) *Unsweetened*: Use with frozen fruit that will be used later in jams, jellies, pies, or special diets. Fruit also may be packed in its own juice, other fruit juices, or water to which an anti-darkening agent is added. Gooseberries, currants, cranberries and blueberries freeze well without sugar. Be sure to treat sugarless packs with an anti-darkening agent if the fruit is light colored.

Simply spread a single layer of prepared fruit on shallow trays and freeze. To prevent freezer burn, package the fruit as soon as it is frozen and return to the freezer. Sugar substitutes may be used in any of the unsweetened packs. Both saccharin and aspartame work well in frozen products, or they can be added to the fruit just before serving.

6. Pack in freezer bags or rigid containers. Squeeze as much air as possible from bags before sealing. Leave 1/2-inch headspace for expansion in containers.
7. Label with product and date. Freeze at lowest possible setting.
8. Whole berries may be placed in a single layer on a tray and frozen until nearly solid. Transfer to freezer bag or containers.
9. Store at 0°F or lower. Use within 8 to 12 months for best quality. Unsweetened fruits lose quality faster than those sweetened with sugar or syrup.

Table 17.4: Syrups for Processed Fruits

Syrup	*Sugar*[1] *(Cups)*	*Water (Cups)*	*Ascorbic Acid*[2] *(Teaspoons)*	*Yield (cups)*	*Uses*
Light (30%)	2	4	¾	5	Small, soft fruits
Medium (40%)	3	4	¾	5½	Peaches, apples, pears, sour berries
Heavy (50%)	4¾	4	¾	6½	Sour fruits

To make syrup: Dissolve sugar in water. Refrigerate syrup until you use it. You will need approximately 1 to 1½ cup(s) syrup for each quart of fruit.

[1] Light corn syrup or light, mild-flavored honey can replace ½ of the sugar.

[2] Add to cold syrup, if you are preserving peaches, apples, pears, plums, nectarines or sweet cherries. If you use a commercial mixture, follow the manufacturer's instructions.

Preventing Discoloration in Fruits

Some fruits, such as peaches, apples, pears and apricots, darken quickly when exposed to air and during freezing. They may also lose flavor when thawed. There are several ways to prevent darkening of fruit and flavor loss.

Ascorbic Acid (Vitamin C)

Ascorbic acid or vitamin C is effective in preventing discoloration in most fruits. Not only does it preserve natural color and flavor of fruits, but it adds nutritive value as well. Ascorbic acid in powdered form is available at some drugstores or where freezing supplies are sold. Ascorbic acid tablets may be more readily available and less expensive, but are more difficult to dissolve. They do need to be finely crushed before use. Fillers in the tablets may make the syrup cloudy, but they are not harmful. One-half teaspoon powdered ascorbic acid equals 1500 mg. Ascorbic acid may be added by the following methods when freezing fruits.

Syrup or Liquid Packs

Add powdered or crushed ascorbic acid to cold syrup shortly before using. Stir it gently so you do not stir in air. Keep syrup refrigerated until use.

Sugar or Dry Packs

Dissolve the ascorbic acid in two to three tablespoons of cold water and sprinkle dissolved ascorbic acid over fruit just before adding sugar.

Crushed Fruits, Fruit Pulses and Fruit Juices

Add ascorbic acid to prepare fruit and stir well.

Ascorbic Acid Mixture

Ascorbic acid mixtures are special anti-darkening preparations, usually made of ascorbic acid mixed with sugar or with sugar and citric acid. The important active ingredient in these mixtures is ascorbic acid. Follow the manufacturer's directions for use.

Citric Acid and Lemon Juice

Citric acid or lemon juice are sometimes used in place of ascorbic acid but are not as effective as ascorbic acid. When used in large quantities, they often mask natural fruit flavors.

Using Frozen Fruits

Thaw fruit at room temperature in its original package. If faster defrosting is required submerge in lukewarm water or partially defrost in microwave oven. Serve with a few ice crystals still remaining. Completely thawed fruits will be limp or mushy.

Spices and Seasonings

When using seasonings and spices, season lightly before freezing, and add additional seasonings when reheating or serving. Pepper, cloves, garlic, celery seasonings, green pepper, imitation vanilla and some herbs tend to get stronger and bitter. Onion, paprika and curry change flavor during freezing. Salt loses flavor and has the tendency to increase rancidity of any item containing fat.

Freezing Prepared Foods

Preparing food ahead of time and freezing it can save time, energy and money. Frozen cooked foods also add variety to the menu, offer quick meals for unexpected company and provide nutritious choices for busy days. When one is preparing a main dish, it takes only a little more effort and time to make enough for several meals. One can freeze all of the prepared food in meal-size packages, or serve part of the food immediately and freeze the rest. It is more economical to make one's own supply of prepared dishes than to purchase commercially prepared foods. Consider freezing for the following:

1. Leftovers that cannot be used immediately.
2. Foods that take a long time to prepare.
3. Seasonal fruits and vegetables.
4. Foods which one can prepare in quantity.
5. Foods that still taste good after storage.

Problems in Prepared Frozen Foods

1. Unbaked biscuits will be smaller and less tender. Unbaked muffins are likely to have poor texture.
2. Custards and cream pies soak into the crust.
3. Meringue on meringue pies toughens and sticks to the wrapper.
4. Milk sauces sometimes curdle and separate. Stirring while reheating helps keep them smooth. Using waxy rice flour or waxy corn flour as the thickener also helps.
5. Gravy tends to separate and curdle when thawed. It is better to make gravy just before serving. Or use waxy rice flour or waxy corn flour as the thickener.

Table 17.5: Freezing Recommendations for Commonly Prepared Foods

Food	*Preparation*	*Serving*	*Storage Time*
COMBINATION DISHES	Foods should be undercooked.		
Cooked bean dishes; rice; chow mein; soups; seasonal vegetable: stuffed capsicum etc.	Omit potatoes in soups. Cook until nearly done, and cool by setting pan or oven/ freezer- tempered baking dish in a shallow pan of ice water. Stir occasionally. Wrap package compactly to eliminate air space and freeze promptly.	Partially thaw in refrigerator to prevent overcooking and heat in top of double boiler, or at 400°F for 30-45 minutes.	4-6 months
POTATOES			
Mashed	Store leftovers in a straight-sided container; or add a beaten egg and shape into patties. Shape and store with double layer of paper between patties.	Thaw just enough to slip potatoes into top of double boiler; heat over boiling water. Dip patties in flour and fry slowly without thawing.	2-4 weeks
Stuffed (baked)	Cut freshly baked potatoes in half lengthwise. Scoop out potato, leaving skin unbroken. Mash scooped-out potato and season. Pile into shells. Cool and wrap.	Remove from package or wrap. Place on baking sheet. Reheat at 325 to 350°F until piping hot. Add cheese if desired.	1 month
French Fries	Use a high-starch content potato for frying. Fry in hot fat until a light brown.	Spread on cookie sheet and place in 400°F oven until thawed and crisp.	1-2 months
Sweet Potatoes	Dip boiled, peeled potatoes in lemon juice or package with cooking water around them. Freeze candied or mashed potatoes in a casserole. Leave skin on baked potatoes.	Reheat in 325 to 350°F oven until hot throughout.	3 months

Contd...

Table 17.5–Contd...

Food	*Preparation*	*Serving*	*Storage Time*
BREADS			
Biscuits, muffins, cake, fruit bread	Prepare as usual, cool. Freeze in pan and cover tightly or package in aluminum foil for reheating.	Thaw in wrapping at room temperature or heat in pan or foil in 325 to 350 °F oven until hot.	Biscuits, quick breads: 2-3 months; muffins: 6-12 months
Waffles	Cook to a light brown. Separate in layers in plastic container or freeze separately and then package.	Heat without thawing in the toaster until hot and crispy.	1-2 months
Yeast: rolls, sweet rolls, loaves, Soft rolls	Dough may be shaped, placed in pans and frozen; or baked before freezing. Cover uncooked bread tightly and freeze before product rises. For baked bread, cool, then cover pan tightly or place product in foil.	For unbaked bread, uncover and let bread rise before baking. Bake at 350 to 375 °F. For baked bread, reheat in pan or foil at 300 °F for 15 to 20 minutes.	3-6 months
Sandwiches	Use day-old bread, spread to edges with soft butter or margarine. Add filling and spread evenly. Omit crisp vegetables, hard-cooked egg white, tomato, jellies and jams. Mayonnaise tends to separate. Use salad dressing or home-cooked dressing. Package.	Thaw at room temperature in wrappings, 3 to 4 hours. Sandwiches for grilling, such as cheese, should be thawed before grilling. Frozen sandwiches in a lunch box will keep other foods cool as they thaw.	Cheese, 3-4 weeks; Others: 3-6 months

Contd...

Table 17.5–Contd...

Food	*Preparation*	*Serving*	*Storage Time*
CAKES			
Shortened cakes	Prepare and bake as usual in layers or loaf pan. Cool. Remove from pans and wrap tightly. Loaf cake may be cut to family-size pieces before wrapping. For best results, freeze cake and frosting separately. See note below.*	Thaw cake in wrapping at room temperature. Ice and serve.	2-4 months
Angel food, chiffon, sponge cakes	Bake thoroughly; cool. Frosted: Freeze before wrapping. Do not use egg white frosting. Unfrosted: Wrap and freeze. Place in container that will prevent crushing. *Note: Fudge frosting and powdered sugar icings made with fat freeze well. Cooked candy-type frostings may stay soft and creamy between layers, but often crack and crumble on outside of cake.	Frosted or filled: Unwrap and thaw in refrigerator. Unfrosted: Thaw in wrap on rack 1 to 2 hours at room temperature. If wrapped in aluminum foil, thaw at 300°F. for 15 to 20 minutes.	Egg-white cakes: 6 months Whole-egg cakes: 4-6 months Egg-yolk cakes: 2 months
COOKIES Baked	Prepare and bake as stated in recipe. Cool thoroughly. Package in foil or rigid container with two layers of waxed paper between cookies.	Thaw in containers at room temperature. Remove from containers and serve.	6 months

Contd...

Table 17.5–Contd...

Food	Preparation	Serving	Storage Time
Unbaked	Refrigerator cookies: Form dough into roll. Slice if desired. Package in moisture- and vapor-resistant paper. Drop cookies: Drop on sheet or just package bulk dough.	Thaw dough in refrigerator. Firm cookie dough may be sliced before completely thawed and baked.	6 months
PIES			
Baked fruit, mince, nut pies	Make as usual. Cool rapidly. Freeze before packaging. Pies are easier to wrap after freezing.	Let stand at room temperature about 15 minutes, then heat in 350 °F oven until warm, about 30 min.	3-4 months
Unbaked fruit, mince, nut pies (unbaked fruit pies have a better fresh-fruit flavor than frozen baked pies, but bottom crust tends to get soggy)	Make as usual except add 1 extra tablespoon flour or tapioca or ½ tablespoon cornstarch to juicy fillings to prevent boiling over when pies are later baked. Do not cut vents in top crust. Steam and cool light fruits before making pies. Freeze in pan. Package.	Unwrap and cut vent holes in upper crust. Put pan on cookie sheet. Bake without thawing at 450°F, 15 to 20 minutes. Reduce heat to 375°F for 20 to 30 minutes or until top crust is brown.	Fruit pies: 3-4 months Mince pies: 6-8 months Nut pies: 3-4 months

6. Cooked, creamed vegetables tend to lose flavor rapidly and should only be stored for a few days. Cook vegetables. Cool quickly. Add sauce. Package. Leave headspace.
7. Lettuce, other greens and raw tomatoes lose crispness and become soggy.
8. Ham and other cured meats may lose color when frozen and become rancid more quickly than other meats.

Packaging and Labeling of Frozen Foods

Foods for the freezer must have proper packaging materials to protect their flavor, color, moisture content and nutritive value from the dry climate of the freezer. Packaging materials must:

1. Be moisture- and vapor-resistant
2. Be durable and leak-proof
3. Not become brittle and crack at low temperature
4. Be resistant to oil, grease or water
5. Protect foods from absorption of off-flavors or odors
6. Be easy to seal and easy to mark

There are usually two types of packaging materials used for domestic purposes *i.e.* rigid containers and flexible bags or wrappings.

Rigid Containers

Rigid containers made of plastic or glass are suitable for all packs and are especially good for liquid packs. Straight sides on rigid containers make frozen food much easier to get out. Rigid containers are often reusable and make the stacking of foods in the freezer easier. Cardboard cartons for cottage cheese, ice cream and milk are not sufficiently moisture and vapor-resistant to be suitable for long-term freezer storage, unless they are lined with a freezer bag or wrap. Covers for rigid containers should fit tightly or the seal should be reinforced with freezer tape, which is especially designed to stick at freezing temperature.

Regular glass jars break easily at freezer temperatures. If using glass jars, choose wide-mouth dual-purpose jars made for freezing and canning. These jars have been tempered to withstand extremes in temperature. The wide mouth allows easy removal of partially thawed foods. If standard canning jars with narrow mouths are used for freezing, leave extra headspace to allow for expansion of foods during freezing. Expansion of the liquid could cause the jars to break at the neck. Some foods will need to be thawed completely before removal from the jar.

Flexible Bags or Wrappings

Bags and sheets of moisture and vapor-resistant materials, heavy-duty aluminum foil, and laminated papers are suitable for dry-packed vegetables and fruits, meats, fish, or poultry. Bags can also be used for liquid packs. Laminated papers are sometimes used as protective overwraps. Protective cardboard cartons may be used to protect bags and sheets against tearing and to make stacking easier. Lightweight aluminum foil, wax paper and bread wrappers are not sufficiently moisture and vapor-resistant to be suitable for long-term freezer storage.

"Freeze-and-Cook" Bags

These bags withstand temperatures from below 0°F to above the boiling point and are suitable for both freezing and cooking the product. They come in 1½ pint and quart sizes and also as large rolls of plastic so they can be made the size desired. A heat sealer is necessary for closing these bags. These products are more expensive but convenient since one can cook in them.

Some of the practical tips to be considered while packaging of frozen foods are:

1. Quickly cool all foods and syrup before packaging. This speeds up freezing and helps retain the natural color, flavor and texture of food.
2. Pack foods in quantities that will be used for a single meal. Do not freeze fruits and vegetables in containers with a capacity over one-half gallon. Foods in larger containers freeze too slowly to result in a satisfactory product.
3. Follow directions for individual foods to determine which can be packed dry and which need added liquid. Some loose foods such as blueberries may be "tray packed" (frozen first on a tray before packaging).
4. Pack food tightly, leaving as little air as possible in the package.

5. Most foods require headspace between the packed food and closure to allow for expansion of the food as it freezes. Foods that are exceptions and do not need headspace include loose packing vegetables such as asparagus and broccoli, bony pieces of meat, tray-packed foods and bread.
6. Seal rigid containers carefully. Use a tight lid and keep the sealing edge free from moisture or food to ensure a good closure. Secure loose-fitting covers with freezer tape.
7. Label each package, including the name of the product, any added ingredients, packaging date, the number of servings or amount, and the form of the food, such as whole, sliced, etc. Use freezer tape, marking pens or crayons, or gummed labels that are made specifically for freezer use.
8. Transportation plays an important role in the frozen food distribution system. The success of frozen food industry depends on a safe and dependable refrigerated transportation system.

Foods That Do Not Freeze Well

1. Cabbage, celery, cress, cucumbers, lettuce, parsley, radishe
2. Cheese or crumb toppings
3. Cream or custard fillings, milk sauces, sour cream
4. Cooked egg whites, icings made from egg whites
5. Fried foods
6. Fruit jelly, gelatin
7. Cooked macaroni, spaghetti or rice

Remedies for Freezing Problems

These are mentioned in Table 17.6.

Table 17.6: Remedies for Freezing Problems

Problems	*Cause*	*Prevention*
Freezer burn-(Surface of food light colored and dried out; food is tough, dry and less flavorful)	1. Torn or unsealed packages	1. Be sure all packages are tightly sealed so no air can get in. Handle carefully to avoid tears.
	2. Packaging not moisture- and vapor-resistant	2. Use only packaging approved for use in freezing.
	3. Too much air in package	3. Always press out all air in wrapped foods. Use just the right size containers for the amount of food and proper headspace.
Gummy liquid in fruits	Fruits frozen too slowly or freezer temperatures too warm or fluctuating	Freeze foods at 0°F or below immediately after packaging and maintain that temperature throughout storage. Do not freeze more than 2 to 3 pounds of food per cubic foot of storage space at one time.
Mushy food	Large ice crystals form in the food, breaking the cell structure	
"Grassy" flavors in vegetables Green vegetables turn olive brown colors	Freezing unblanched vegetables	Blanch all vegetables as directed.
Rancid flavors - (strong oily-like flavor in fats)	Spoilage of the fat in a product	Blanch all vegetables as directed. Package food correctly and exclude all unnecessary air. Do not store for longer than recommended.
Maroon-colored bones or pink meat in frozen poultry after it is cooked.	Hemoglobin (coloring matter) in bones.	Natural occurrence. There's nothing wrong with the meat.

Chapter 18

Food Irradiation

Food Irradiation–What is it?

Many processing methods have been developed to help control food spoilage and improve safety. The traditional methods of preservation, such as drying, smoking and salting, have been supplemented with pasteurisation, canning, freezing, refrigeration and chemical preservatives. Food irradiation is another technology that can be added to this list. However, it should never be used as a substitute for good manufacturing practices.

Just a mention of the word irradiation usually sparks strong objections, usually based on misunderstanding of the process involved and its effects on foods. Very simply, food irradiation is a process of exposing foods, either prepackaged or in bulk, to very high-energy, invisible lightwaves (radiation). The process controls the amount of radiation the food absorbs.

Radiation is broadly defined as energy moving through space in invisible waves. Radiant energy has differing wavelengths and degrees of power. Light, infrared heat and microwaves are forms of radiant energy. So are the waves that bring radio and television broad-casts into our homes. Broiling and toasting use low-level radiant energy to cook food. The radiation of interest in food preservation is ionizing radiation, also known as irradiation. These shorter wavelengths are capable of damaging microorganisms such as those that contaminate food or cause food spoilage and deterioration. That capability plus the fact that much of our food supply is lost due to spoilage by insects each year- is why scientists have been experimenting with irradiation as a method of food preservation since 1950. They have found irradiation to be a controlled and very predictable process.

History of the Irradiation Preservation of Food

Actually, the first patent application for food preservation by irradiation was filed in 1905! But it has only been during the last 40 years that irradiation has been in use. Early in the 1920s, a French scientist discovered that irradiation could be used to preserve food. This technology was not adopted in the U.S. until World War II. At this time there was a need to feed millions of men and women in uniform. The U.S. Army sponsored a series of experiments with fruits, vegetables, dairy products, fish and meats. Since World War II, extensive research on radiation processing has been sponsored by the U.S. Army and studied by scientists. Although it has been a pioneer in research on food irradiation, the U.S. has not been involved in food irradiation on a commercial scale except in the case of spices. Only a very small percentage of spices are irradiated, however. In 1963, the U.S. saw its first approval of food irradiation when FDA approved its use to control insects in wheat and wheat flour. In 1964, additional approval was given to inhibit the development of sprouts in white potatoes.

In 1983, approval was granted to kill insects and control microorganisms in a specific list of herbs, spices and vegetable seasonings. Then, in 1985, treatment of pork to control trichinosis was added to the list of approvals. In the same year, approval was granted to control insects and microorganisms in dry enzyme preparations used in fermentation type processes. In 1986, approval was granted to control insects and inhibit growth and ripening in such foods as fruits, vegetables and grains.

In the United States, the Food and Drug Administration (FDA) has approved irradiation for eliminating insects from wheat, potatoes, flour, spices, tea, fruits and vegetables. Irradiation also can be used to control sprouting and ripening. In the U.S., irradiated foods have been used by astronauts, the military and hospital patients, but the adoption of the technology for other purposes has been slow despite approvals from the Food and Drug Administration (FDA). For example, the process to control trichinosis in pork, approved in 1985, is still not used. Using irradiation to control Salmonella and other harmful bacteria in chicken, turkey, and other fresh and frozen uncooked poultry was approved in May 1990. FDA supports it as an effective control of microorganisms responsible for a major portion of foodborne illness, including *Salmonella, Yersinia,* and *Campylobacter*. The process reduces but does not eliminate all bacteria. Therefore, processed poultry are safe longer than unprocessed poultry, but still require refrigeration.

In December 1997, FDA approved the use of irradiation to control pathogens (disease causing micro-organisms such as *E. coli* and *Salmonella* species) in fresh and frozen red meats such as beef, lamb, and pork.

To date, more than 40 countries have given approval for over 60 food products to be irradiated. Currently regulations on food irradiation in the European Union are not harmonised. Some countries, such as Belgium, France, The Netherlands, Portugal and the UK are in favour of a Directive to allow the irradiation of specified foods, whereas other countries such as Denmark, Germany and Luxembourg remain opposed. Within the UK seven categories of foods are cleared for irradiation to specified doses. The regulations also make provision for labelling to ensure that consumers are fully informed whether foods or ingredients within them have been irradiated.

Sources of Radiation (Ionizing Energy)

Food irradiation is a food processing technology that exposes certain types of food to a source of ionising energy. Two things are needed for the irradiation process.

1. A source of radiant energy
2. A way to confine that energy

The food is passed through a field of ionising energy from either machine-generated electron beams or gamma rays from cobalt-60. There are three types of energy that can be used for irradiation of food: X-rays, electron beams and gamma rays. Machines called "electron accelerators" are used to produce beams of electrons to which food can be exposed. If the electrons from the accelerator are stopped by a metal target (such as tungsten foil) first, X-rays are produced and the food is exposed to X-rays. The amount of energy produced by electron beams can be adjusted.

Gamma rays are produced by radioactive substances (called radioisotopes) that continuously emit the high energy gamma rays. The approved sources of gamma rays for food irradiation are cobalt-60 (the most common) and cesium-137.

Irradiated food doesn' t become radioactive. Irradiation is known as a cold process. It does not significantly increase the temperature or change the physical or sensory characteristics of most foods. An irradiated apple, for example, will still be crisp and juicy. Fresh or frozen meat can be irradiated without cooking it.

During irradiation, the energy waves affect unwanted organisms but are not retained in the food. Similarly, food cooked in a microwave oven, or teeth and bones that have been X-rayed do not retain those energy waves.

The ionising radiation passes through the food generating large numbers of short-lived free radicals. These can kill micro-organisms, such as *Salmonella*, and inhibit many processes, such as those that cause sprouting and ripening. The radiation energy used is able to cause changes in molecules, such as breaking chemical bonds. At small irradiation doses, properties of the food, such as sprouting and ripening can be modified (usually inhibited). Higher doses can alter molecules in microorganisms so they can no longer cause spoilage or human illness. The microorganisms may be killed or genetically altered so they can't reproduce. It takes lower doses to damage microorganisms and insects than to alter enough molecules in the food to damage it.

At no time during the irradiation process does the food come into contact with the radiation source and, by using cobalt-60 or electron beams up to 10 MeV, it is not possible to induce radioactivity in the food. When the treatment stops, energy does not remain in the food. Where a radioactive Cobalt 60 source is used, the gamma rays do not have enough energy to make food radioactive. Also, the food does not come into contact with the energy source, so it cannot become contaminated by radioactive material. The length of time the food is exposed to the ionising energy and the strength of the source determine the irradiation dose, measured in kiloGrays (kGy), the food receives.

Process of Food Irradiation

The process of irradiating food is quite simple. The food is exposed to ionising radiation, either from gamma rays or a high energy electron beam or powerful X-rays. Gamma rays and X-rays are a form of radiation that shares some characteristics with microwaves, but is of much higher energy and penetration. The rays pass through the food just like microwaves in a microwave oven, but the food doesn't heat up to any significant extent. Exposure to gamma rays does not make food radioactive. Electron beams and X-rays are produced using electricity, which can be switched on or off, and do not need the use of radioactive material. In both cases, organisms that are responsible for spoiling foods such as insects, moulds and bacteria including some important food poisoning bacteria can be killed. Food irradiation cannot kill viruses.

Large cases of food or food products can be loaded in pallets onto a conveyor system which will carry them into the irradiation chamber, past the irradiation source, and back out to an unloading station. The speed at which the conveyor moves determines the dose the food product receives and is controlled by computers. The irradiation source (usually cobalt-60) is contained in slender pencil-like stainless steel casings about 18 inches long by 3/8 inch in diameter. At the heart of the facility, the casings are contained in a lead-lined chamber surrounded by 6-1/2 feet thick concrete walls.

Because waves of irradiation energy are so very short, they penetrate foods or food packaging easily. The energy level is so high, it can even penetrate and break up molecules. Exposure of the food to irradiation energy is controlled so that the dose only causes physical and chemical changes in targeted cellular components, not nuclear changes in the atoms that make up food. Living microorganisms may be killed or their genetic material may be damaged to prevent reproduction, but molecules are not changed to the extent that they become radioactive. Gamma irradiation may be compared to turning on a light, illuminating a room, and turning it off again. Gamma rays pass into foods, affect the food or target organism, and leave the food. How the rays affect the food depends on the food, the dose and the organisms in the food.

Irradiated Foods are not Radioactive

There is a common misconception that irradiated food is radioactive. The radiation used to process foods is very different from the radioactive fallout that occurs after, for example, a nuclear accident.

In food processing, the radioactive source never contacts the food and the sources used emit only gamma rays and no neutrons. That means the food cannot become radioactive. The World Health Organization (WHO), the American Dietetic Association and the Scientific Committee of the European Union are three internationally recognised bodies which support food irradiation.

Application of Food Irradiation

Irradiation is most useful in four areas:

Preservation

More traditional preservation methods are used for the same reason. They include drying, smoking, salting, curing, canning, freezing or pasteurisation and these processes keep food safe to eat by delaying the natural decomposition processes that result from bacterial action. Different preservation techniques are used with different types and forms of food. Food can only be irradiated if there is no other safe method available.

Food is irradiated to destroy parasites, bacteria, moulds and yeasts which cause food decomposition and food poisoning. When radiation energy is absorbed by food, it causes a variety of chemical and physical reactions. The amount of energy the food absorbs is controlled so the changes produced have desirable food preservation effects while maintaining the safety, quality, and wholesomeness of the food. The food itself does not become radioactive. Hence, food irradiation is a food preservation process and a quarantine measure that is used to extend product shelf life.

Irradiation can be compared to pasteurization. As in the heat pasteurization of milk, the irradiation process greatly reduces but does not eliminate all bacteria. Irradiated poultry, for example, still requires refrigeration, but would be safe longer than untreated poultry. Strawberries that have been irradiated will last two to three weeks in the refrigerator compared to only a few days for untreated berries. Irradiation complements, but does not replace, the need for proper food handling practices by producers, processors and consumers.

Irradiation can be used to treat herbs and spices in place of chemical treatments. It can also be used to destroy the unwanted pests that can accompany food when it is traded regionally or imported from other countries. It is an energy-efficient food preservation method that has several advantages over traditional canning. The resulting products are closer to the fresh state in texture, flavor and color

Using irradiation to preserve foods requires no additional liquid, nor does it cause the loss of natural juices. Both large and small containers can be used and food can be irradiated after being packaged or frozen.

Perishable foods treated with all but high doses, however, must still be refrigerated. Refrigerated storage life is extended, but the need for cold storage is not replaced by irradiation. Microorganisms are destroyed by radiation more easily than enzymes which lead to food deterioration (color, flavor and texture changes, for example). Many enzymes survive the current levels of radiation processing, although refrigeration can slow down many enzymatic changes. In addition, not all microorganisms are destroyed. The process does not protect the food against reinfestation or contamination. Therefore, irradiated perishable foods are still considered perishable.

Sterilization

Foods that are sterilized by irradiation can be stored for years without refrigeration just like canned foods. With irradiation it will be possible to develop new shelf-stable products. Sterilized food is useful in hospitals for patients with severely impaired immune systems, such as some patients with cancer or AIDS. These foods can be used by the military and for space flights.

Control Sprouting, Ripening, and Insect Damage

In this role, irradiation offers an alternative to chemicals for use with potatoes, tropical and citrus fruits, grains, spices, and seasonings. However, since no residue is left in the food, irradiation does not protect against reinfestation like insect sprays and fumigants do.

Table 18.1: Potential Food Irradiation Uses

Type of Food	*Effect of Irradiation*
Meat, poultry, fish	Destroys pathogenic organisms, such as *Salmonella*, *Clostridium botulinum*, and Trichinae
Perishable foods	Delays spoilage; retards mold growth; reduces number of microorganisms
Grain, fruit, vegetables, dehydrated fruit, spices and seasonings	Controls insect infestation
Onions, carrots, potatoes, garlic, ginger	Inhibits sprouting
Bananas, mangoes, avocados, papayas, guavas, other non-citrus fruits	Delays ripening
Grain, dehydrated vegetables	Reduces rehydration time

Control Foodborne Illness

Irradiation can be used to effectively eliminate those pathogens that cause foodborne illness, such as Salmonella, Campylobacter etc.

Benefits of Food Irradiation

Some of the benefits of this food processing technique include:

1. Extended shelf life of some products.
2. Less food spoilage.
3. Reduced risk of food-borne diseases caused by micro-organisms, such as *Campylobacter, Salmonella, E coli* and *Listeria* (especially in meat, poultry and fish).
4. Less need for pesticides.
5. Less need for some additives, such as preservatives and antioxidants.
6. Lower risk of importing or exporting insect pests hidden inside food products.
7. Reduced need for toxic chemical treatments, such as those used to kill bacteria found in some spices.

8. As an alternative to current treatment for disinfecting imported fruits, grains and vegetables, which uses an ozone-depleting gas.
9. Reduced sprouting in potatoes and onions.

Dosing of Food Irradiation

In food irradiation, the radiation "dose" that a food receives should not be confused with something added to the food. The "dose" for food irradiation is the amount of radiation absorbed by the food and it is not the same as the level of energy transmitted from the radiation sources. The dose is controlled by the intensity of radiation and the length of time the food is exposed.

Terms used to describe this "dose" or amount of absorbed radiation, are unfamiliar and confusing to most people. In the past, the term used was rad, short for "radiation absorbed dose," which is 100 ergs absorbed by 1 gram of matter. The rad has been replaced by the gray (Gy). One gray is equal to 100 rads or 0.00024 Calorie (*i.e.*, diet calorie) per kilogram of food (0.00024 Calorie per kilogram equals 0.0001 Calorie per pound).

Strictly from the scientific point of view, no ceiling should be set for food irradiated with doses greater than the currently recommended upper level of 10 kGy by the Codex Alimentarius Commission. The food irradiation technology itself is safe to such a degree that as long as sensory qualities of food are retained and harmful microorganisms are destroyed, the actual amount of ionizing radiation applied is of secondary consideration. That was the main conclusion of a week-long meeting on high dose irradiation (Geneva, 15-20 September 1997) organized jointly by the World Health Organization (WHO), the United Nations Food and Agriculture Organization (FAO) and the International Atomic Energy Agency (IAEA).

The knowledge of what can and does occur chemically in high dose irradiated foods which derives from over 50 years of research tells us that one can go as high as 75 kGy, as has already been done in some countries, and the result is the same–food is safe and wholesome and nutritionally adequate", commented Dr Fritz Kaferstein, Director of the WHO Programme of Food Safety and Food Aid.

The participants reviewed all relevant data related to the toxicological, microbiological, nutritional, radiation chemical and physical aspects of food exposed to doses greater than 10 kGy and came to the unanimous conclusion that the food is safe for consumption. The focus of the meeting was on considering the generic wholesomeness of foods, appropriately treated and packaged, that are irradiated in the range of 10-100 kGy to eliminate all spoilage and pathogenic microbial contaminants.

According to Dr Terry Roberts, Chairman of the meeting, former Head of Microbiology at the Institute of Food Research, Reading Laboratory, United Kingdom, food irradiation is perhaps the most thoroughly investigated food processing technology. He is quite satisfied with the existing scientific evidence that higher doses of radiation can provide wholesome, nutritious and safe foods. Similar to thermal sterilization of food, the dose should be sufficient to produce a shelf-stable and microbiologically safe product depending on the type of food and specific consumer requirements.

The presence in food of harmful microorganisms such as *Salmonella* species, *Escherichia coli* O157:H7, *Listeria monocytogenes*, or *Yersinia enterocolitica* is a problem of growing concern to public health authorities all over the world. In an attempt to reduce or eliminate the resulting risks, national regulations on food safety are being tightened up in many countries. In the Unites States of America, for example, the Department of Agriculture had issued regulations for meat and poultry requiring that testing for *Escherichia coli* began in January 1997 and that raw meat and poultry processed by large firms be virtually free of *Salmonella* began in January 1998. For some of these products, food irradiation may well be the best method to ensure the absence of these microorganisms. However, in some instances, an upper limit of 10 kGy for the overall average dose could preclude the effective use of the technology.

In the case of irradiation of spices, this need for a greater average dose has already been recognized in several countries. France permits an average dose of 11 kGy for the irradiation of spices and dry aromatic substances, whereas Argentina and the United States of America permit a maximum dose of 30 kGy for this purpose.

Still higher doses are required for sterilization of food, for instance for immuno-compromized hospital patients. For this purpose, the Netherlands permits an average dose of 75 kGy which implies that some parts of the food are exposed to doses over 100 kGy. Some other countries, such as the United Kingdom, permit the use of this technology without specifying an irradiation dose limit for this application. South Africa has permitted the marketing of shelf-stable meat products irradiated to an average dose of 45 kGy. For over 20 years, US and Soviet astronauts have been enjoying their irradiated foods, indeed choosing them over other food preservation technologies.

The FDA's regulations describe radiation levels in terms of the kilogray (kGy), equal to 1000 Gy.

The dose (number of kGy) permitted varies according to the type of food and the desired action. Treatment levels have been approved by FDA as follows:

"Low" Doses–(up to 1 kGy) Designed to

1. Control insects in grains
2. Inhibit sprouting in white potatoes
3. Control trichinae in pork
4. Inhibit decay and control insects in fruits and vegetables

"Medium" Doses–(1-10 kGy) Designed to

1. Control *Salmonella, Shigella, Campylobacter and Yersinia* in meat, poultry and fish
2. Delay mold growth on strawberries and other fruits

"High" Doses–(greater than 10 kGy) Designed to

1. Kill microorganisms and insects in spices
2. Commercially sterilize foods, destroying all microorganisms of public health concern (*e.g.*, to sterilize food to the same degree as if they were thermally sterilized (canned); however, no commercial applications of this dosage have been approved by FDA for use in the U.S. with the exception of some special hospital diets for immune-deficient patients.)

"For high dose food irradiation, as with other methods of food production, it is important to use raw materials of good quality, to provide adequate packaging, to follow proper processing procedures and to have sound record-keeping, to follow good personal hygiene and sanitation practices, and to handle the processed foods appropriately during distribution", explained Dr Roberts. The group came to the following overall conclusions.

Doses greater than 10 kGy:

1. Will not lead to changes in the composition of the food that, from a toxicological point of view, would have an adverse effect on human health;
2. Will greatly reduce potential microbiological risk to the consumer;
3. Will not lead to nutrient losses to an extent that would have an adverse effect on the nutritional status of individuals or populations.

Therefore, foods treated with doses greater than 10 kGy can be considered safe and nutritionally adequate when produced under established Good Manufacturing Practice.

"Given these reassuring conclusions, the World Health Organization hopes that food irradiation will now become more acceptable as a means for the improvement of food safety which remains one the Organization's priorities", said Dr Fernando Antezana, WHO Assistant Director-General.

Food Irradiation Applications and their Dosing in U.S.A.

These are as under:

1. Irradiation of poultry and poultry products, including mechanically recovered meat, to reduce numbers of *Salmonella, Campylobacter* and other food poisoning bacteria. Doses of up to 3 kGy (fresh) and up to 7 kGy (frozen) have been recommended.
2. Irradiation of red meats, including hamburger meat, to reduce numbers of Escherichia coli O157:H7 and other food poisoning bacteria. Doses of up to 4.5 kGy (fresh) and up to 7 kGy (frozen) have been recommended.
3. Irradiation of dried herbs and spices to reduce levels of contaminating micro-organisms generally and reduce the

chances of survival of food poisoning bacteria in particular. Doses up to 10 kGy have been recommended.

4. Irradiation of some imported seafood, in particular warm water prawns and other shellfish, to improve their microbiological safety. Doses up to 3 kGy have been recommended.
5. Irradiation of certain fruit and vegetables in order to reduce the numbers of micro-organisms, particularly those which cause spoilage. Doses up to 2 kGy have been recommended.
6. Irradiation of bulbs and tubers, such as potatoes and onions, to prevent sprouting. Doses of 1 kGy have been recommended.
7. Irradiation of cereals, grains and certain fruits, as a quarantine measure, to kill insects. Doses of 1 kGy are recommended.
8. High-dose irradiation to produce sterile foods, such as ready meals. These foods are irradiated by recommended doses of 45 kGy under frozen conditions to lessen adverse sensory effects, but are not stored frozen.

Changes in Irradiated Food

As with other food pasteurization and sterilization technologies involving thermal, mechanical, or photonic energy input, the objective of processing with ionizing radiation is to destroy pathogenic and spoilage microorganisms without compromizing safety, nutrition and sensory quality. Common to all these processes are subsequent physical and chemical changes, the extent of which differs significantly among them. Many of the resulting compounds are the same as those formed when food is cooked or preserved in more traditional ways. In comparison to thermally sterilized foods, the amount of chemical change in a radiation sterilized food is uniform and relatively small. Irradiation changes the chemical composition of the food to a minimum extent. In the case of irradiation, different doses have different effects. At low doses, irradiation lengthens the shelf life of fruits like strawberries by destroying moulds and inhibits sprouting in vegetables like potatoes. At higher doses, irradiation helps to kill the bacteria and pathogens that cause food poisoning.

All food preservation methods change the taste, appearance, texture, composition and nutritional value of the food to varying extents. Food irradiation is sometimes called a "cold" process; it achieves its effect with little rise in the temperature of the food. There is little if any change in the physical appearance of irradiated foods as they do not undergo the changes in texture and/or color as foods preserved by heat pasteurization, canning or freezing. Food remains close to its original state. Problems that have occurred are some off-flavors in meat and excess tissue softening that has been documented in fresh peaches and nectarines. Some foods, such as dairy foods and eggs, can't be irradiated because it causes changes in flavour or texture. Fruits, vegetables, grain foods, spices and meats (such as chicken) can be irradiated. Therefore, irradiation process is not suitable for all products. Foods with high fat contents, such as fatty fish and some dairy products, develop off-odours and tastes due to the acceleration of rancidity, even at relatively low doses. Also, foods with a high protein content, such as meat and poultry, can suffer from changes in flavour and odour after irradiation. These effects can be minimised by irradiating at chill or frozen temperatures.

At the radiation energy levels used in food processing, only chemical changes are possible, not nuclear changes that would make the food itself radioactive. Over 35 years of research suggest that the chemical by-products of radiation ("radiolytic products") are mostly the same as by-products of conventional cooking or other preservation methods. When high energy particles strike matter, electrons are lost from atoms and ions are formed. Newly formed radiolytic products may then interact to create new compounds in the food that were not present before treatment, a few of which could produce off-flavors. (In meat, this can be partly controlled by maintaining low product temperatures during the irradiation process.) The most common chemical reaction during food irradiation is the conversion of water to hydrogen peroxide. Reactions like these occur in all types of food preservation, and those few reactions unique to irradiation are not harmful.

The FDA concluded that "very few of these radiolytic products are unique to irradiated foods; approximately 90 per cent of the radiolytic products are known to be natural components of food" (FDA, 1986). Some of these are fatty acids just like those that result from the breakdown of triglycerides, amino acids that make up proteins, and compounds (hydrocarbons) commonly found in the waxy coverings of fruits like apples, pears and berries. Others are fatty compounds just like those found from cooking meat by common methods such as grilling. The other 10 per cent of radiolytic compounds are chemically very similar to natural components in food. The chemistry of irradiation is very predictable and the products of an individual component such as proteins are not affected by the type of food or other food components present. Radiolytic products have been critically tested for toxicity and no evidence of hazard has been found.

Animal feeding studies show no toxic, teratogenic or mutagenic effects from irradiated foods. In addition, irradiation leaves no chemical residues in food.

Nutritional Quality of Irradiated Foods

Scientists believe that irradiation produces no greater nutrient loss than what occurs in other processing methods, such as canning. Irradiation does not cause any significant loss of macronutrients. Proteins, fats and carbohydrates undergo little change during irradiation even with doses over 10 kGy. Similarly, the essential amino acids, essential fatty acids, minerals and trace elements suffer no significant losses. There can be a decrease in certain vitamins but these are of the same order as occur in canning and drying.

Radiation does not impair activity of certain nutrients, but overall nutrient retention in irradiated foods is similar to retention with other preservation methods. The process alters the nutrient content of some foods because it reduces the level of some of the B-group vitamins, but this loss is similar to other accepted food processing techniques such as canning or blanching.

Vitamin C (ascorbic acid) reduction has been reported, but it is attributed to a shift from ascorbic acid to dehydroascorbic acid, a change that is mostly insignificant from a nutritional standpoint. Tocopherol, which have Vitamin E activity, appear to be very sensitive to irradiation in the presence of oxygen. Vitamin K seems relatively stable. These adverse effects of irradiation on vitamins can be reduced by excluding oxygen and light, keeping the food at a low temperature, and using the lowest dose need for treating (processing) food.

Table 18.2: Thiamine Retention Comparison

Meat	*Percent in Canned Sample*	*Percent in Irradiated Sample*
Beef	21	44
Chicken	22	66
Pork	12	57

Reference: Journal of Food Science 46:8, 1981.

Table 18.3: Vitamin Content Comparison of Cooked Chicken Amounts are for 2.2 pounds (1 kilogram) cooked chicken

Vitamin	*Non-irradiated Sample*	*Irradiated Sample*
Vitamin A, international units	2200	2450
Vitamin E, milligrams	3.3	2.15
Thiamin, milligrams	0.58	0.42
Riboflavin, milligrams	2.10	2.25
Niacin, milligrams	58.0	55.5
Vitamin B6, milligrams	1.22	1.35
Vitamin B 12, milligrams	21	28
Pantothenic acid, milligrams	13	17
Folacin, milligrams	0.23	0.18

Reference: *Journal of Food Processing and Preservation* 2:229, 1978

Adoption of the Technology

The food industry is slow to invest in this technology because of uncertainty over the public acceptance of food. More than forty countries allow the use of irradiation for food safety reasons. As examples: most of the frogs legs sold in France are irradiated; most of the herbs and spices sold in South Africa have been irradiated; and in Thailand there is growing demand for irradiated Nham, a fermented pork sausage that is usually eaten raw. To date, in Australia and New Zealand, only herbs and spices and some tropical fruits have been approved to be irradiated.

Although some food radiation treatments have been approved in the U.S. since 1963, only irradiated spices are found on the market today. The only two times that irradiated food other than spices has been sold in the U.S. (as of June 1989) were research test markets. In 1986, 2 metric tons of mangoes irradiated in Puerto Rico were test marketed in Miami, FL. In 1987, irradiated Hawaiian papayas were tested in Anaheim and Irvine, CA. These test market studies as well as others in France, Argentina and South Africa indicated good consumer acceptance and willingness to buy.

Worldwide, over 30 countries approve some form of irradiation and many groups of consumers readily accept these products. As of 1986, Japan irradiated over 10,000 pounds of potatoes annually. The Netherlands irradiated 2 tons of food daily, and Belgium irradiated 1 ton daily. South Africa routinely irradiates mangoes, papaya, and other vegetables. Canada has a facility dedicated to irradiating potatoes.

In the UK, the Advisory Committee on Novel and Irradiated Foods approved irradiation in 1986 as a safe and satisfactory method of food processing. This opinion was reaffirmed in 1987 after receiving submissions from industry, consumer groups and interested parties. In 1991, the Food (Control of Irradiation) Regulations in the UK cleared 7 categories of foods for irradiation to specified overall average doses: Fruits (2.0 kGy), Vegetables (1.0 kGy), Cereals (1.0 kGy), Bulbs and tubers (0.2 kGy), Spices and condiments (10 kGy), Fish and shellfish (3.0 kGy) and Poultry (7.0 kGy). The regulations also make provision for labelling to ensure that consumers would be fully informed whether foods or ingredients within them had been irradiated. Irradiation facilities must also meet specified criteria before they can be licensed to process foods. To date, the only licences granted within the UK have been for the treatment of certain spices.

Labelling of Irradiated Foods

The labelling requirements vary, however, from country to country. Some like the European Community require the labelling of any food that contains an irradiated ingredient, however small the percentage of that product. This is also the case in Australia and New Zealand. In the United States, labelling applies only where the whole food item is treated. Under the Food Labelling Regulations (1996), irradiated foods and ingredients have to be identified with the words "irradiated" or "treated with ionising radiation" in UK.

The FDA approved initial labeling requirements in April, 1986. The regulations required that all produce be labeled at the packing/ wholesale and retail levels. Suggested wording and an international symbol (logo) were provided. At the retail level, the label must bear the symbol plus one of these statements: "treated with radiation" or "treated by irradiation." In addition, the manufacturer may add a phrase which truthfully describes the primary purpose of the treatment, such as "treated with radiation to control spoilage." For unpackaged fruits and vegetables, labels may be on each piece, on the shipping container placed in prominent consumer view, or on a sign near the commodities identifying the use of the treatment. For irradiated foods sold at the wholesale level, the logo and the wording are still required. In addition, however, they must be accompanied by the caution "do not irradiate again." Irradiated spices sold in the U.S. do not need to carry the logo if they are in packaged foods where they constitute one small ingredient. Most packaged spices in stores are usually treated by chemical fumigation, not by irradiation.

Since 1986, all irradiated products must carry the international symbol called a radura, which resembles a stylized flower.

Safety of Irradiated Foods

In 1981, the Food and Agricultural Organization (FAO) of the United Nations, the International Atomic Energy Agency (IAEA), and the World Health Organization concluded that "any food irradiated up to an average dose of 1 Mrad or less is wholesome for humans and therefore should be approved without further testing" (WHO, 1981 in IFT, 1987). A Mrad equals 10 kGy. Also in 1981, the U.S. FDA concluded that food irradiated with up to 1 kGy is "wholesome and safe for human consumption, even where the food that is irradiated may constitute a substantial portion of the diet," and that food "compromising no more than 0.01 per cent of the daily diet [ex., spices] and irradiated at 50 kGy or less also can be considered safe for human consumption" (FDA, 1981).

Before approving low dose irradiation of foods, the FDA reviewed over 400 toxicity studies, which included animal feeding studies. The U.S. Army Medical Department fed irradiated foods to 41 human volunteers in 15-day tests in 1955-58 and showed no unfavorable effects. More recently, in 1986, the People's Republic of

China reported on eight experiments with 439 human volunteers. Their diets included 60 per cent of ingredients irradiated at levels from 0.1-8 kGy. During and after these trials, there were no significant differences in the clinical, physiological and biochemical evaluations between the subjects receiving the irradiated foods and those receiving comparable food without such processing.

Community Concerns and Acceptance

Many surveys and market trials have been carried out to assess consumer attitudes to food irradiation. Results have consistently shown that many consumers have misconceptions about the technology, such as that it makes food radioactive. When consumers are given information on the process and they are given a chance to try irradiated products for themselves, as in market trials, they are much more likely to accept the technology. One of the most successful market trials was carried out in 1991 in a small food store in Chicago, USA. Irradiated strawberries, oranges and grapefruits outsold the non-irradiated fruits by a 9:1 ratio. In the following season irradiated strawberries became the number one best seller in that store with the ratio expanding to 20:1 over the non-irradiated product. This positive experience has encouraged approximately 60 stores in Indiana, Illinois and Ohio to sell a variety of irradiated foods.

The upsurge in support of food irradiation in the US has also been reflected in position statements from a number of professional bodies such as the American Dietetic Association and the American Medical Association. These bodies support the use of irradiation to enhance the safety and quality of the food supply and see their role as assisting in the education of consumers about the technology. Many sectors of the food industry are speaking in favour of food irradiation, such as the Grocery Manufacturers of America. The American media has also become increasingly positive towards the technology with newspapers such as The New York Times and the Wall Street Journal running stories in support of the process. In most cases, the main benefit of irradiation is that it is regarded as making food safer to eat.

Although the technology appears to be gaining increasing acceptance in the US, it is still regarded with suspicion within many countries within Europe, including the UK. There is obviously a role for respected professional bodies to inform consumers about the benefits and limitations of food irradiation so that they can make informed choices.

People have expressed a number of concerns in relation to food irradiation. They include:

Radioactivity

A belief that irradiated food is radioactive and therefore harmful to eat. Irradiated foods do not become radioactive. Extensive testing has demonstrated that irradiating food is as safe as canning, pasteurising and freezing. Facilities for irradiating food are similar to those in operation for sterilizing medical equipment and do not resemble nuclear reactors in any way. There are no explosives or materials that could cause widespread dissemination of radioactive material. Facilities using cobalt do not use materials from nuclear defense industries. Facilities used, however, must comply with plant and worker safety requirements of the Nuclear Regulatory Commission and the Occupational Safety and Health Administration. At an irradiation facility, the radiation source (usually the cobalt-60) is contained in slender pencil-like stainless steel casing about 18 inches long by 3/8 inch diameter. The casings, in turn, are contained in a lead-lined chamber. Packaged food travels in pallets on a conveyor between 6-1/2 foot thick concrete walls into and through a chamber where it is exposed to the radiation source (gamma rays if cobalt-60 is used). Pallets may be turned to allow uniform exposure over the route. Radiation dosage is controlled by a computerized rate of passage (conveyor speed) through the chamber.

Lack of Choice

Some consumers have indicated they want to be able to choose between irradiated and non-irradiated foods. There are currently limited ways to determine whether or not irradiation has been used, because there is no radiation left behind in the food after the irradiation process. Labelling is compulsory on irradiated foods in many countries. Irradiated foods are marked by a symbol called a radura, which is the international symbol for irradiation.

Effects on Food Hygiene

There are concerns that the extensive use of this method of food preservation may result in less stringent food hygiene and handling practices. Irradiation is not a substitute for good hygienic practices. Proper food handling is still needed. Food irradiation does not replace the need for correct food handling practices in industry and in the home. For instance, a few bacteria may survive the irradiation of meat. If the meat is left unrefrigerated, these bacteria could still multiply and cause food poisoning.

Natural Warning Signs Eliminated

Concern that irradiation will make it difficult to tell if foods are old or have gone 'off', because normal indicators like smell or mould are destroyed by irradiation.

Not for Consumer's Benefit

Concern that food irradiation is driven purely by market needs, rather than consumer demand.

Studies on consumer acceptance have been done with few irradiated products. Bruhn, Schutz and Sommer (1986) examined change in consumer attitudes toward food irradiation as the result of opportunities to read about and discuss the technology. They found that consumers in general showed a higher level of concern for preservatives and sprays than for food irradiation. "Alternative" consumers who were a vocal segment of ecologically sensitive consumers, were more skeptical about the safety and advantages of the process. They also were initially more aware of food irradiation and their concern increased after educational efforts. After educational efforts, conventional consumers adopted a minor concerned attitude toward irradiation, but a discussion leader knowledgeable in food irradiation reduced concerns of conventional consumers compared with the impact of a non-expert leader. These researchers concluded that the attitudes of conventional consumers can be positively influenced by an educational effort and the influence is most effective when the consumer can interact with someone knowledgeable about irradiation.

Wiese Research Associates (1984) found that less than 25 per cent of a national telephone sample were aware of food irradiation technologies. Among those aware, 75 per cent were concerned, but in general, consumers were more concerned about chemical sprays and preservatives than irradiation. Those with fewer concerns had more formal education and higher levels of awareness.

A market survey of the effects of irradiation on demand for papaya in California in the 1960s suggested that demand would be reduced by 23 per cent if the fruit were treated with irradiation (Huang, et.al., 1987). Acceptance of irradiated produce is quite high in South Africa. Reportedly, 90 per cent of consumers reacted positively to its introduction in the late 1970s. However, the test marketing was accompanied by an extensive educational campaign that involved mass media and supportive consumer organizations (Bruhn, et.al., 1986).

Even though some test marketing has shown some hope for consumer acceptance, irradiated foods have caused picketing and violent protests at other areas. Consumer studies seem to suggest that people are more interested in the safety and wholesomeness of the foods rather than benefits of the process for any specific foods. These and other findings also seem to indicate that as knowledge and awareness of the technologies increase, so might its acceptance. History indicates that consumer acceptance will continue to influence the extent to which irradiation gets adopted as an alternative food processing technology.

There is no doubt that food irradiation involves many complex issues. If properly applied, irradiation can be an effective method of reducing organisms which cause food spoilage and human illness. The potential benefits of food irradiation may come closer to realization now that regulatory barriers are beginning to be removed, interest in alternatives to pesticides is high and some studies are demonstrating consumer acceptance of products. Food irradiation appears to be gaining consumer acceptance in the US but it is slow to gain support within many parts of Europe, including the UK. There is a role for respected professional bodies to inform consumers of the advantages and limitations of the technology so that they can make an informed decision on buying and eating irradiated foods.

Chapter 19

Microwave Applications to Food

Introduction

Cooking by microwaves is now relatively well established and the flavour of a number of foods heated by microwave energy is discussed. Investigations have been carried out regarding the feasibility of the application of microwave energy for enzyme inactivation.

The interaction of food and microwave energy is complex. To assist understanding and enable predictions to near optimum methods to be made further knowledge of the dielectric properties of foods, actual temperature profiles within foods during cooking or processing and other data which determine food quality, *e.g.* sensory parameters, moisture contents, etc., is required. Computers are beginning to be used to assist in these studies and in the future, more optimisation studies on the application of microwaves to food should occur.

In 1971 Bengtsson and Risman stated that "Knowledge of the dielectric properties of food materials is important for proper understanding of the heating pattern during microwave heating of foods; such as in the cooking of foods or in reheating of precooked foods from frozen or refrigerated condition. Unfortunately such data are missing to a considerable extent, inspite of a few noteworthy publications over the last decade."

In the same paper, Bengtsson and Risman presented measurements of the dielectric properties of a number of foods at 3 GHz using the cavity perturbation technique. In subsequent years the following work appeared:

1972–Roebuck and co-workers investigated the dielectric properties of carbohydrate-water mixtures at microwave frequencies of 1 and 3 GHz over a wide compositional range.

1974–Tinga and Nelson prepared a comprehensive table of the dielectric properties of materials for microwave processing.

All the evidences indicate that the inactivation of enzymes by microwaves is due to a thermal effect. Conventional heat sources are frequently used in food processing to inactivate enzymes. Investigations on a number of foods have been carried out to determine whether microwave energy can be used to replace conventional methods.

Capson described a method and apparatus for inactivating enzymes in fruit juice in particular pectin-methylesterase which is present in raw fruit juice should be inactivated. Pectin-methylesterase catalyses the hydrolytic removal of methoxyl groups from the pectin molecule to form low ester pectins which results in the separation of the liquids and solids in the juice, *i.e.*, the cloudy appearance of the fruit juice is lost. Microwave energy was applied at 2450 MHz so that fruit juice reached temperatures in the range of 60–80°C. Flavour tests on orange juice concentrate processed in this way showed that the natural flavour and quality of the juice was not impaired.

A number of studies on the blanching of broccoli for freezing have been reported. The percentage retention of ascorbic acid by different blanching methods has been reported by a number of workers. Microwave blanching and steam blanching gave the highest retention of ascorbic acid. From Noble and Gordon's experiments, the steam blanching method showed a greater retention of ascorbic acid *i.e.* 79 per cent (21 per cent loss) than water blanching indicating

57 per cent retention (43 per cent loss). However, this difference disappeared when the samples were cooked, the average retention being approximately 29 per cent of the original content.

Eheart also studied the effect of microwave blanching and water blanching on total acid, pH and the chlorophyll content of broccoli. Total acids (wet basis) and percentage retention of chlorophyll *a* contents were higher in microwave blanched broccoli than those in water blanched broccoli. After frozen storage and cooking, microwave blanched samples were lower in pH, higher in total acids and in ascorbic acid (wet basis) but lower in total chlorophyll than broccoli blanched in 100°C water. The higher chlorophyll losses in microwave blanched broccoli were thought to be due to the lower pH of these samples. However, prior to freezing the total chlorophyll retention and ascorbic acid content were the highest for microwave blanching. Retention of chlorophyll is of some importance because Sweeney *et al.* showed that percentage retention of chlorophyll in cooked broccoli correlates directly to sensory panel scores for colour.

The results of these studies clearly indicate that when evaluating a blanching method, more than one parameter should be measured to give an overall evaluation of the quality of the product.

A comparison of three blanching methods *i.e.* microwave, conventional and a combination method was carried out by Dietrich *et al.* After blanching the sprouts were frozen and the main aim was to achieve flavour stability without noticeable colour deterioration. For good colour retention a chlorophyll conversion to pheophytin of between 0–25 per cent is satisfactory. This can easily be exceeded because a long blanching time is often required for sprouts to achieve flavour stability. The minimum blanching time by each method was therefore found and the methods compared with respect to enzyme inactivation, heat penetration, chlorophyll, flavour and ascorbic acid retention. Storage stability at temperatures of – 29°C, – 18°C and – 7°C was compared.

After a 6-min microwave blanch, peroxidase in the centre of the sprouts was inactivated and no pink centres were seen but there was pink discoloration and peroxidase in the outer leaves. Excessive dehydration also occurred and off-flavours developed on storage. Steam blanching for 6–8 min at approximately 100°C was adequate for flavour stability both initially and after storage at – 18°C for six months and – 7°C for two months. A combination of blanching method using microwave energy and steam or water to inactivate peroxidase produced sprouts which were flavour stable at – 29°C, – 18°C and – 7°C. Chlorophyll and ascorbic acid retention was as good or better than conventionally processed samples. It was concluded that concurrent application of steam and microwave heating for about 2 min may achieve an adequate blanch with only 15–20 per cent chlorophyll conversion.

A combined method using steam and microwave for blanching Brussels sprouts has been described by Meredith. At the moment in the UK microwaves are not being used industrially for blanching, however a large-scale trial using a prototype 30 kW machine (Magnetronics Ltd, Leicester, UK) will soon be complete. For economic and technical reasons, a combination method is recommended. A steam tunnel to raise the temperature of the product to 60°C followed by microwave heating is the most economic process. The quality of the treated sprouts so far has been described as being as good as conventionally blanched sprouts and in some instances better, particularly when lower grade sprouts are used.

A patent which contains details of equipment design and methods of blanching vegetables has been described. Four stages are recommended:

1. Preliminary surface blanch using hot water or steam at atmospheric pressure
2. Application of microwave energy (9.15 MHz)
3. Hot water blanch
4. Quenching in cold water.

Solid Food Sterilisation

By sterilisation of solid food the organisms like bacteria, *clostridium botulinum* can be killed easily.

The destruction is a time/temperature relationship; *Clostridium botulinum* spores are normally destroyed by 3 min at 121°C but with conventional heating of food in cans it may take 1½h or more for the centre of the food to reach this temperature. Meanwhile, the surface will have been at this temperature for some time and the food will be overcooked and the organoleptic properties of the product may be impaired.

Figure 19.1 illustrates the time/temperature history for steam sterilisation of food in a can and in a pouch; it also shows microwave sterilisation of food in a pouch.

Heat is conducted slowly from the surface to the centre during the sterilisation of cans by the conventional method, hence the long processing time. However, by reducing the thickness of the sample, *e.g.* by sterilising the food in an aluminium foil pouch, very much faster heat transfer can be achieved and the quality of the food is improved as there is less thermal damage. As can be seen, an even faster rate or heating obtained by microwave sterilisation. The thickness of the two pouches is identical.

Since temperatures in the order of 121°C or greater are obtained within the pouch during microwave processing, internal pouch vapour pressure corresponding to temperature will exceed the rupture point of flexible pouches. An external air counter pressure (range 30–45 psig) is therefore provided to maintain pouch integrity during heating.

Further work done in the US Army Natick Laboratories showed that the temperature spread between the centre and the surface during the sterilisation of beef slices was quite large. They used paper thermometers which were sealed in polypropylene envelopes, the

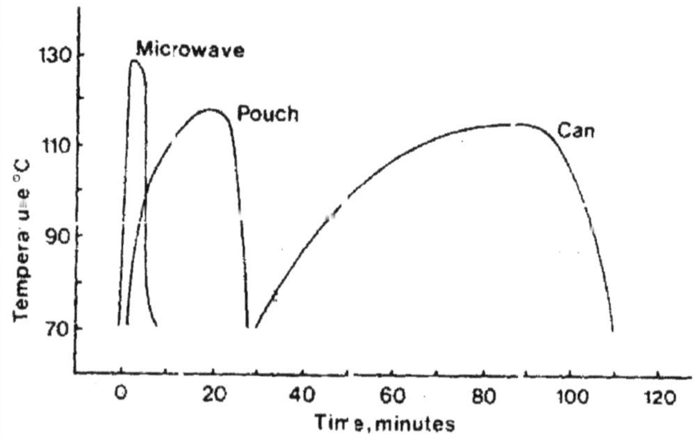

Figure 19.1: Time/Temperature History for Steam Sterilisation

resulting thermogram covered the rectangular slices of beef and were placed between the two 0.5 cm slices, inside each pouch. The work showed that the slowest heating point was at or near the geometric centre of the product where a temperature of 121°C was reported, whereas at the outer edges the temperature recorded was 138°C.

Similar results have been reported elsewhere. To achieve low temperature spread between the corner edges and the central parts, Stenstrom and Ohlsson stressed the necessity of immersing the food to be sterilised in a liquid during microwave heating.

Equipment has been designed to carry out this process and an almost uniform temperature distribution has been achieved; for example a variation of ± 2 or 3 °C just after microwave heating has been reported. A comparison of beans and peas sterilised by this process and by conventional canning, using a sensory panel, showed the microwave product to be superior in quality.

O'Meara *et al.* agree that organoleptic evaluation of several foods demonstrated a quality advantage in favour of a microwave/hot water sterilising process rather than hot water heating alone. No data however to substantiate this statement were presented by these authors.

Sterilisation is the availability of a suitable packaging material which will allow the passage of microwaves but prevent the absorption of oxygen and water. Hecht and Haskell reported that a film consisting of a thin coating of an inorganic metallic phosphate, which is sandwiched between a polyester film and a heat-sealable layer, are compatible with microwave heating and have barrier properties necessary to provide shelf-stable packaging. A test run was carried out at the US Army Natick Laboratories; beef was packaged in a laminate containing this inorganic coating and sterilised in a continuous microwave unit. The product showed no sign of spoilage after 10 months at room conditions.

Sterilisation of Liquid Food Stuffs

Milk can be sterilised by microwave heating. The whole process of heating, holding and cooling can be carried out with a free jet of milk, not touching the walls of the equipment.

Milk has been pasteurised by a batch and a continuous methods. Bacteriological tests indicated that for the eight samples taken, the pasteurisation efficiency varied and this was attributed to the fact that it was difficult to maintain the processing milk temperatures at 82°C throughout the run. Improved equipment should overcome this but as with the sterilisation of milk the high cost of equipment is a major drawback to the application of this technique on a large scale.

A method for pasteurising oysters and similar products has been patented. Freshly shucked oyster meat was fed through a microwave oven (range 2450–2500 MHz) for approximately 1½ min., the temperature of the product was raised to about 60°C and this was sufficient to pasteurise the oysters but maintain their firmness and taste when compared to freshly shucked oysters.

Use of Microwave Cooking in Food Dehydration

The dehydration of pasta has also been studied quite extensively. It has been reported that conventional drying of pasta takes 5–10 h. Both temperature and humidity are carefully controlled so that the moisture is slowly driven out and core hardening prevented. Microwave processing has been shown to be an ideal way of overcoming this potential problem and a commercial process has proved very successful. The pasta is first dried in a shaker drier to remove some water and so prevent clumping. This is followed by a 10 min. exposure through a microwave chamber (915 MHz). Hot air is forced through the chamber to carry off the moisture liberated. Cooling, holding and packaging complete the process. The advantages are increased production and energy savings, also a reduced floor area is required for the equipment. The time of rehydration and cooking of the product is reduced possibly because as the moisture escapes from the product during microwave drying, tiny tunnels are formed throughout the pasta. Total plate counts of bacteria are said to be much lower in the microwave product when compared to conventionally dried pasta.

Experiments have been carried out using microwave energy (2450 MHz and 915 MHz) and untreated air to dry corn. Conventional hot air drying causes core hardening of the kernels and the other disadvantage is the low thermal conductivity of the grain. Although the microwave process is feasible, there is a limit of the speed at which the corn may be dried and beyond this limit the swelling of the kernels would lower the market grade. It is thought that this may preclude the application of microwave power for drying corn.

Tomato paste foam has been dried by both hot air and microwave energy. A disadvantage of drying a foamed product by hot air is the poor thermal conductivities of foams. Both methods yield a product of high quality but the microwave method (using microwave energy at 2450 MHz) reduces the drying time considerably.

Application of Microwaves in the Freeze Drying

The application of microwaves in the freeze drying process has been discussed in a number of papers, the main advantage being a reduction in the overall time of the process. A number of problems have been highlighted, one of them being electrical breakdown (corona discharge) which has a deleterious effect on the product being dried.

Decareau stressed the merits of a combined conventional/microwave freeze drying process where conventional freeze drying is used to remove the bulk of the moisture followed by microwave energy to remove the residual moisture. The problem is the design of the combination process.

A process has however been patented which involves partially freeze-vacuum drying a food in particulate form to an average moisture content of 10–35 per cent. The frozen, partially freeze-vacuum dehydrated food is then subjected to microwave: energy and the remaining core moisture is redistributed. The particles are then subjected to pressure to form a compressed food. The food processed could be uncooked vegetables or cooked meat.

Reports state that many of the problems of microwave freeze drying have been overcome, but the process has not been adopted commercially.

Use of Microwaves in Meat Thawing

Now let us try to understand about partial thawing of meat. Dicing, cutting and mincing meat is usually carried out at a temperature of between –3 to –2°C. A microwave tempering tunnel enables blocks of meat to be taken straight from the cold store so that it enters the microwave unit at –20°C, and after 0.5–2 min leaves the microwave unit at a temperature of approximately –3°C. The tempering process, conventionally, is carried out in conditioning rooms and takes 48–72 h. In the UK, microwave equipment at a frequency of 896 MHz ± 10 MHz is used and there are various types of tunnels (ABR : Food Machinery International Ltd. 'Bletchley, Milton Keynes, UK) ranging from a throughput of 1000 kg/h to 3350 kg/h which can be used for tempering meat, fish and poultry for medium- to large-scale processes.

During thawing, as the temperature of beef reaches 1°C, major changes occur in the dielectric properties of the beef resulting in a marked decrease in the penetration of microwave energy which often results in 'runaway heating'. For this reason experiments have been carried out on the microwave thawing of tuna fish, beef and pork using a surface cooling process–electrostatic spraying of cryogenic liquids.

The Use of Microwave Cookers

The power available for cooking in the majority of microwave cookers ranges from 72 to 720 W. The roof of the interior of the cooking space of the microwave cooker usually incorporates a mode stirrer, this may be covered by a spatter shield. The function of the stirrer is to attempt to distribute evenly within the cooking space the microwaves from the magnetron. Some cookers have a turntable which replaces the stirrer although the Belling MW1 has both. Experiments to improve the heating uniformity in microwave ovens are continually being carried out. Kashyap and Wyslonzil described a multi-slot wave guide oven feed suitable for domestic microwave ovens. Other experiments they described involve sweeping the frequency over a certain range.

The turntables are usually driven by a motor which is coupled by gears to a rotary shaft which extends into the cooking space through the bottom plate. A patent exists for a rotary table with a magnetic driving system, thus eliminating the necessity of the rotary shaft extending into the cooking space. A portable battery driven turntable suitable for use in microwave cookers has also been developed.

In spite of these attempts to distribute the energy evenly in the cavity in practice the energy distribution, even when the cooker is empty, is not uniform. Many tests designed to measure the electric field distribution in a microwave cooker have been described. The aim of these experiments being if possible, to enable the user to utilise the hot and cold spots appropriately. The lack of uniformity in the distribution of energy in the cooking space is a major problem in the utilisation of microwaves for heating food.

Many cookers can also cycle the power on and off. The ratio of the on : off time can be controlled by a simple switching arrangement. This facility is extremely useful for defrosting frozen food and also for showing down the cooking process.

The Microwave Cooked Beef Samples

MacLeod and Coppock investigated the flavour of beef cooked conventionally and by microwave energy. The odour volatiles were analysed by using gas chromatography/mass spectrometry. The separated odour volatiles were then evaluated by sensory analysis. For both conventional and microwave heating methods samples were analysed in the presence and absence of water. Their results indicated that for both the boiled and roast series of samples, cooked to an acceptable and equivalent degree of doneness by conventional and microwave cooking produced different amounts of volatiles. The microvave cooked beef samples liberated approximately one half of the total volatiles of the conventional product.

There was an initial increased rate of production of volatile flavour components, due to the more rapid rise in temperature in the microwave heated samples but this was not sufficient to make up for the shorter cooking time when the total production of volatiles was considered.

In further work on meat in which 100 volatile components were isolated it was found that in general the classes of components which represent the greater proportion of isolates for the conventional boiled beef were the benzenoids, aldehydes and furans, whereas the classes of compounds which in general occur in greater amounts in the microwave boiled beef isolates were the alkanes, alkenes, alcohols and pyrazines.

It is of interest to note that several of the alkanes, alkenes and alcohols present in high concentrations in the microwave boiled beef aroma were associated with relatively undesirable odour assessments such as sour, harsh, pungent, unpleasant, dull cardboard. The conventionally boiled beef aroma was described in more desirable terms such as pleasant, solvent, fruity and meaty. More evidence was presented by MacLeod and Coppock substantiating that relatively undesirable odour qualities are present in microwave samples cooked for the normal time required for cooking the meat by either boiling or roasting. The same authors point out that chemical reactions in general are strongly temperature dependent and the rapid initial temperature rise in microwave cooking has to be contrasted with the slower conventional process of conduction where a high temperature gradient initially exists throughout the food. The compounds present in higher concentrations in the microwave samples are probably representative of compounds formed by reactions which have relatively high rates of reaction and are thus formed in significant amounts in the relatively short microwave cooking time. They concluded from this work that time appears to be an important factor as far as desirable beef flavour production is concerned. The microwave samples were cooked by a continuous microwave method (2450 MHz).

Starrack commented on the improvement of roast meat when an intermittent method of microwave cooking is used, details of methods for roasting are also given.

Meat roasting has been a subject of research at Natick Development Center for many years. Decareau has recommended that because of the temperature pattern in beef roast cooked by microwaves, the roast should be removed from the cooker before the internal recommended temperature was reached. Time should then be allowed for conduction to occur and the temperature to equilibrate. The timing devices on some cookers which control intermittent microwave cooking do allow some conduction to take place during the cooking cycle.

Recent work at Natick aims to optimise the process of the roasting meat in terms of yield, quality and cooking time. Uniform, cylindrical pieces or meat are being used (12 cm by 20 cm). First of all conventional roasting studies were carried out carefully monitoring changes in weight with time and also collecting and measuring drip losses so that a figure for evaporative cooking could be worked out. Dielectric properties of beef as a function or temperature and microwave frequency were obtained and a computer program was developed which can provide a graphical print-out or the temperature in a cylindrical roast of beef given specific cooking conditions.

To verify the computer studies an experimental oven was built which has variable microwave power at both 915 MHz and 2450 MHz to approximately 2 kW, controlled oven temperature from 93 to 260°C, controlled steam pressure at 5, 10 and 15 psig and an IBM card reader to permit programmes to be run in an automatic mode.

The Microwave Cooked Vegetables and Cereals

The effect of microwave cooking on vegetables has shown that both wet and dry microwave cooking resembled each other but were both different from the conventional method, *i.e.*, the ideal. One of the differences was a marked increase in methyl alcohol during microwave heating of vegetables. This increase occurred sooner during dry microwave heating as opposed to wet microwave heating. A similar effect was also observed for allyl isothiocyanate, thus showing the more rapid action of microwaves when no added water is present.

Non-volatile organic acids (lactic, succinic, malic and citric) have been identified and quantified in frozen peas cooked by microwaves. Gas-liquid chromatography was used for this study. The concentration of these acids increased after cooking and in particular samples cooked by microwaves without water at 115 V and 550 W cooking power. Malic acid was identified in frozen carrots cooked in the same way but in this vegetable concentration was not affected by cooking method. The non-volatile organic acids found in both peas and carrots were not correlated with sensory scores for flavour.

Their findings showed that the sensory scores for the flavour of peas cooked with and without water in a consumer microwave cooker (115 V, 550 W) were not significantly different from each other and that all the scores were in the acceptable range. A difference was found, however, between peas cooked in microwaves at 115V (550W cooking power) and those cooked at 220 V (1150 W cooking power). Significantly higher scores for flavour were obtained by the 115 V, 550 W method. Higher flavour scores were obtained for carrots cooked by microwaves without water when compared to the other methods.

The flavour scores for microwave cooked peas and carrots were either not significantly different from the conventionally cooked counterpart or were rated significantly higher. The authors stress, however, that caution must be exercised in generalising as to the effects of microwave cooking on vegetables. They noted differences in trends for peas and carrots. They also commented on biological variation within and between lots of vegetables.

These results, however, do indicate that cookers in which the power call be varied could be beneficial for the flavour of certain products.

The poor flavour scores for the microwave baked whole-wheat bread were not unexpected and were attributed to the lack of caramelisation and development of flavour compounds which occur during conventional baking. Some of the panel members commented that the microwave bread had a very mild flavour or less flavour than the bread baked in the convection oven.

With the rye bread however the panels did not detect any difference in flavour between the two baking methods.

The continued expansion and improvement of the application of microwaves to food requires integration of the basic knowledge of the composition and heating parameters of food with the technology of food processing. The complexity of many foods and biological variation do not assist the application of microwaves. The collection of further data on conventional, as well as microwave processing and the use of computer may help to solve some of these problems.

Even distribution of microwave energy and the standardisation of setting on microwave cookers would greatly assist both the food manufacturers and the users of microwave cookers. It seems a contradiction that the microwave cooker, an appliance of a scientific age, often requires the development of an art to use it! Standardisation of the cookers followed by the standardisation of cooking procedures in a scientific manner would enhance consumer acceptance for those without inclination to develop the art. Improvements are being continually made to achieve these aims and the sophistication of the recent cookers enhance their versatility.

Chapter 20

Use of Extrusion Technology in Food Industries

Introduction

Extrusion cooking is an increasingly important and widespread processing technique in modern food and feed industries. New applications are constantly appearing for this continuous, space and energy efficient technology. Practical industrial applications are in this case ahead of research findings and theory development. The advantages of the technique are, however, large enough so that the art of extrusion daily gains new practitioners.

Harper has recently published a review on food extrusion. General, descriptive articles on the technology have been written by Smith, Smith and Ben-Gera. Clark, Horn and Bronikowski and Williams *et al.* The classical article on the modelling aspects in food extrusion is by Rossen and Miller. This art is also presented in a series of articles in *J. Food Process Eng.*, where problems arising in mathematical modelling of extrusion have also been discussed.

This chapter is not intended to be a review or a general description of extrusion cooking, nor does it concentrate on the general aspects of economics and lines of equipment used when applying this technology in practice. Instead it discusses the basic phenomena (denaturation and plastification of biopolymers, such as starch), which take place in HTST cooking extrusion in thermodynamically polytropic machines that are most frequently used in food and feed industries. The central mechanical processing equipment, the extrusion cooker, is discussed as a general screw reactor which creates the physical environment where natural biopolymers, such as starch, proteins and cellulose, change their physical properties in a 'reaction analogous' manner, and chemical reactions take place.

The extrusion cooker is a high temperature, short time, medium to high pressure, medium to high shear screw reactor for biopolymer modification and conversion. This approach presupposes knowledge about biopolymer reaction kinetics and mathematical models that can be applied to the design and optimisation of the reactor. No comprehensive theory or models on the reactions and changes in biopolymer structures, however, exists at present for the physical and chemical conditions encountered in extrusion cooking. This is partly due to the fact that most pasting and gelatinisation studies, have been performed with much higher water contents than normally encountered and at atmospheric pressure, while the water contents in extrusion range from 5 to 40 per cent (w/w) and pressures range from 0.2 to 50 M Pa. Attempts to model mathematically the basic changes in biopolymer properties have been published. Remsen and Clark have considered the basic phenomena taking place in extrusion cooking. The heuristic phenomena level approach, which is given below, is a further development of this kind of thinking.

In the opinion of the author, some kind of theoretical model is essential, if the changes that take place in a cooking extruder are to be understood and mastered. Following equation gives a heuristic model for the changes in material states or properties in thermal processing in an extruder.

$$y = f(RM, I, MC; \qquad)/X_4 \qquad (1)$$

where y is a dependent variable such as mass viscosity in a given point in the extruder, mass pressure or temperature before the die, water absorption or solubility index of the product after extrusion, product density, cold water, consistency of ground, water suspended extrudate, etc. It can also be a combination of several such product quality indicators, which amounts to a multi-dimensional quality vector Y. *RM* is the raw material composition of the processed recipe such as wheat flour, soy protein, etc. I is the ingredients added such as monoglycerides, sugar, salt, etc. *MC* is the moisture content in the extruder. It is separated from other recipe components due to its prime importance on the basic phenomena. is the pressure profile in the extruder from feed point to the die exit. is the temperature profile in the extruder from feed point to the die exit. is the residence time profile in the extruder from feed point to the die exit. is the shear profile in the extruder from the feed point to the die exit. *f* is some mathematical function.

The state of the mass in the extruder and in the ready processed stage, as the dependent variables indicate, is thus a result of the temperature and shear action under pressure while the mass resides in the extruder. The temperature is a resultant of added or removed heat by heat exchanging and by mechanical energy dissipation in the extruder. In some modern well designed plastics extruders the temperature profiles are regulated within ± 0.5°C around the desired values. Such accuracy is but a dream in most food extruders today.

Hypothetical Model for Changes in Starch

The critical factor in food and feed thermal processing is the change in biopolymer structures of nature's products. Let us build a hypothetical model for changes in starch as it is subjected to the (   ) profile actions in extrusion. We can here use a reaction analogy.

Starch is a biopolymer of α-D glucose units (primary structure), which are α-1,4 bonded in amylose, and α-1,4 and α-1,6 bonded in amylopectin (secondary structure). Amyloses at least are thought to have a helical tertiary structure. Aggregation of amylose and possibly some parts of amylopectin molecules by hydrogen bonding and/or possibly by the action of other bonding material yields a partly crystalline quaternary structure. Native, dry starch is a powder, which in extrusion is converted to a coherent and, in the cool dry state, continuous mass.

If only changes in the physical structures of the aggregates take place, the phenomenon is not a proper reaction; it is a thermal operation. This is why it is described as being 'a reaction ana1ogous change in biopolymer aggregate structures.' Biopolymers are always processed in the presence of water. Thus water is considered to be the second 'reactant'.

When a starch-water slurry is heated, starch first reversibly absorbs water until the so-called gelatinisation temperature is reached. This temperature is connected with abrupt changes in the physical properties of starch; this is due to the changes that take place at least in the quaternary structure of starch crystalline aggregates. Donovan has called this change in nature's biopolymer aggregates 'denaturation.' Denaturation in this chapter denotes the initial, irreversible change in starch and other biopolymer aggregate structures. Water, the second 'reactant,' is at least an energy transfer agent in this process, and evidently, at the initial denaturation temperature, water has reached high enough 'reactivity' so that hydrogen and other possible bonds that hold the natural aggregate structure together are at least partly ruptured, Although other molecules than starch and water may be involved, we can simplify the inspection of the basic phenomena by considering a starch-water system.

Denaturation of biopolymers is a necessary prerequisite for the dough or malt development and texturisation in extrusion. Mercier and co-workers have demonstrated by X-ray techniques that changes in crystalline state occur at this stage. Once denaturation has started, we can assume that it follows first-order kinetics, and that the rate constants depend on temperature according to the Arrhenius equation. The limiting factor in the 'reaction' is the reaction rate below 110°C and water diffusion rate at temperatures above 110°C.

We can assume that starch continuously forms new aggregate structures, which in turn are ruptured by the action of heat and shear as processing continues. In extrusion cooking this denaturation-plastification takes place under pressure and at elevated temperatures. Thus, high water 'reactivities' are reached even if the moisture content is relatively low. The effective mixing and shear actions overcome the limiting factor of water diffusion into the denaturing, plastifying mass. The ground, extruded, expanded starchy products show a glass- or plastic-like appearance under the light microscope. The mass looks as if it were amorphous, but Mercier and co-workers have shown that with some starches there are crystalline structures, which clearly deviate from the structure of natural starch crystalline aggregates.

One way to observe the 'progress of denaturation and subsequent plastification is to investigate the pasting behaviour of ground, water suspended starch or starchy material, and to compare it with scanning electron micrographs (SEM), and some other indicators such as the water solubility index (WSI) and water absorption index (WAI). The series of experiments, (discussed below), were performed in order to demonstrate the aggregate level changes.

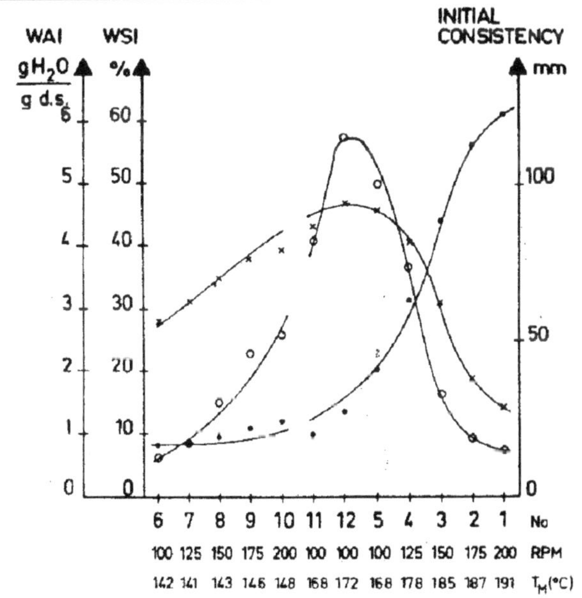

Figure 20.1: Water Absorption Index. ×, water solubility index, ●, and initial 30°C consistency. O, for experimental points given in Table 20.1. Screw speed and observed mass temperatures are given for each data point.

Table 20.1: Processing Data in a Series of Experiments Used to Demonstrate WAI and Consistency Maximum Existence in a Twin-Screw Extrusion Cooking of Wheat Flour

Sample	Processing Variable Levels			Observed Process Data			Functional Properties Data			
	Screw	Feed Rate (g/min)	Screw Speed (rpm)	Mass Temperature (°C)	Current (A)	WAI gH_2O/gds	WSI % ds	Initial (mm)	Consistency Peak (mm)	Cooling (mm)
RM	–	–	–	–	–	0.90	7.5	2	40	35
RM+0.1% Hg_2Cl	–	–	–	–	–	0.90	7.5	2	200	230
6	C	200	100	142	8.5	2.82	8.1	12	163	158
7	C	200	125	141	8.0	3.12	8.3	17	160	153
8	C	200	150	143	8.0	3.52	9.5	30	150	150
9	C	200	17.5	146	9.0	3.81	11.2	46	127	121
10	C	200	200	148	9.0	3.95	12.1	52	1[illegible]9	110
11	R	600	100	168	21	4.34	10.0	82	107	102
12	R	400	100	172	16	4.70	13.5	11.5	96	67
5	R	200	100	168	16	4.58	20.5	100	39	36
4	R	200	125	178	14	4.06	31.7	74	20	22
3	R	200	150	185	13	3.13	44.2	33	7	9
2	R	200	175	187	11.5	1.90	61.5	19	4	7
1	R	200	200	191	10	1.44	66.2	15	5	7

For the experiments wheat flour was used; it contains proteins, lipids, glycoproteins and other constituents that can be involved in the aggregate structure formation. However, it is the aggregate structures that are important, and the author considers the basic phenomena inspection to be relevant to this series of experiments.

There is a maximum in WAI and initial consistency values at around a mass temperature of 170°C. This maximum also exists for rye, barley and oat. The WSI shows a continuous increase that accelerates when the WAI and initial consistency peaks have been reached. Anderson *et al.*, Conway, Lawton *et al.*, and Mercier and Feillet have reported similar phenomenal behaviour maxima for different starchy materials and mixtures of flours and soya concentrates.

The consistency developments of the raw material used, and of ground, water suspended extrudates of samples 6, 12 and 1 representing the mildest, intermediate and most severe treatments in extrusion, respectively are given in Figures 20.2 & 20.5. Scanning electron micrographs (SEM) are displayed with their respective pasting curves. Figure 20.2 shows the consistency curves of the raw material. The solid curve shows the pasting behaviour without enzyme.

The dotted curve shows the pasting behaviour when 0.1 per cent mercury chloride ($HgCl_2$) has been added in order to eliminate the effect of amylolytic enzymes on the raw material pasting behaviour. The latter curve is the proper reference for extruded products as no amylolytic enzyme activity survives in the conditions used in this experimental series.

When wheat flour, or other starchy raw materials, undergo heat induced denaturation-plastification during extrusion cooking, the initial cold water consistency of ground, water suspended products starts to rise, and the hot area peak consistency and 50°C consistency upon cooling are reduced when compared with the raw material pasting curve with enzyme deactivation. The cold water suspended, freeze dried SEM pictures show that some starch particles have survived the mildest treatment, but that some extracellular (amylose) aggregate networks have already formed. The WAI value for the mildest treatment also shows an increase from the raw material WAI.

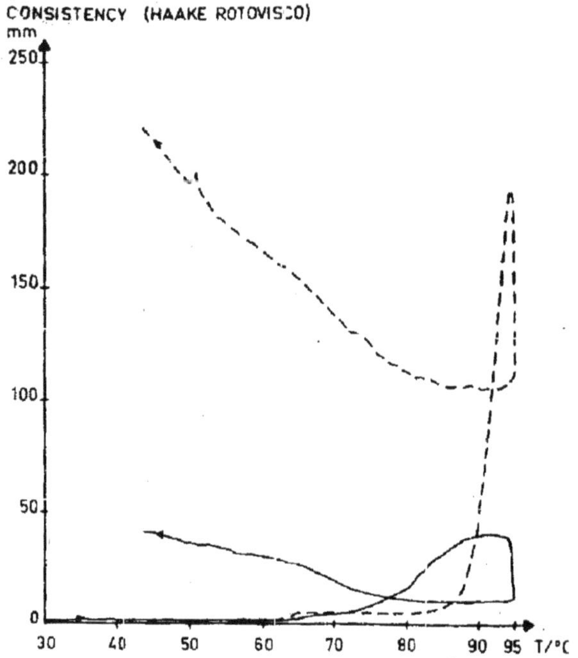

Figure 20.2: Relative Consistency in a Pasting Study as a Function of Temperature. The solid curve indicates the raw material pasting curve when amylolytic enzymes are not denatured. The dotted curve shows the pasting behaviour when amylolytic enzymes have been deactivated by using 0.1 per cent $HgCl_2$ and is the proper reference curve for extruded materials.

When the 'hardness of treatment' increases, the initial cold water consistency of products increases further, and the hot area peak and 50°C consistency upon cooling are further reduced until the maxima concurring with the WAI maximum are reached. Cold water suspended, freeze dried extrudate SEM pictures show that this maximum area is connected with a sponge-like typical aggregate structure with few if any starch granules present. Even before the maximum area, and especially in areas further on the 'hardness of

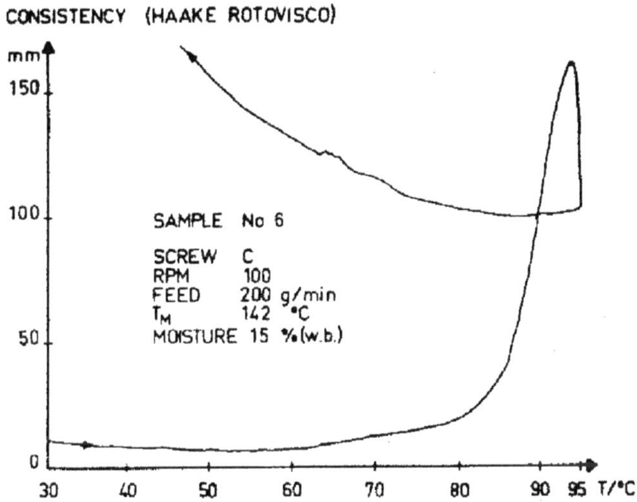

Figure 20.3: Consistency Curves, and SEM Pictures (magnification 300 ×) for the Mildest Treated Sample No. 6 in Table 20.1. The SEM picture on the left shows the dry, ground extrudate. The SEM picture on the right shows the water suspended, freeze dried extrudate.

treatment', the ground, dry extrudates look like pieces of transparent plastic or glass in optical microscope and SEM observations.

When the 'hardness of treatment' further increases after the WAI maximum area has been reached, the initial cold water pasting consistency of ground extrudates starts to decrease while the hot area consistency and 50°C consistency upon cooling continue to decrease, until a pasting curve with low consistency throughout is obtained. The low consistency curve is accompanied by the disappearance or the spongy aggregate structures in the SEM pictures and high initial consistencies at the maximum. The consistencies at the initial cold water state, hot area peak and 50°C upon cooling during pasting are given in Figure 20.6.

The SEM pictures that are prepared by suspending the material in water and freeze drying afterwards do not show the real structure of the material. They are biased because of ice-crystal formation and

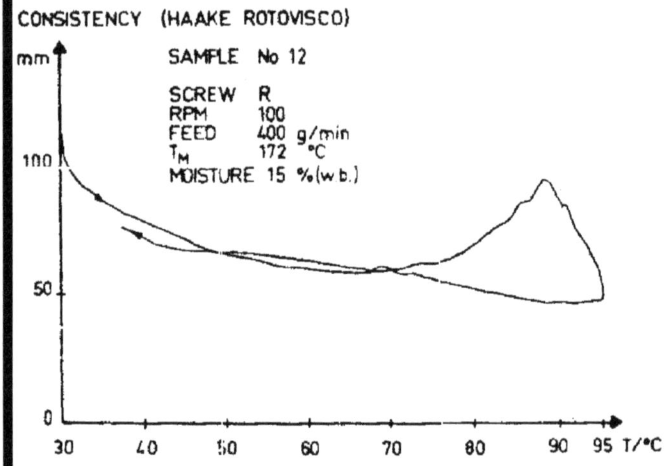

Figure 20.4: Consistency Curves and SEM Pictures (Magnification 300 ×) for the Maximal Initial Consistency Sample

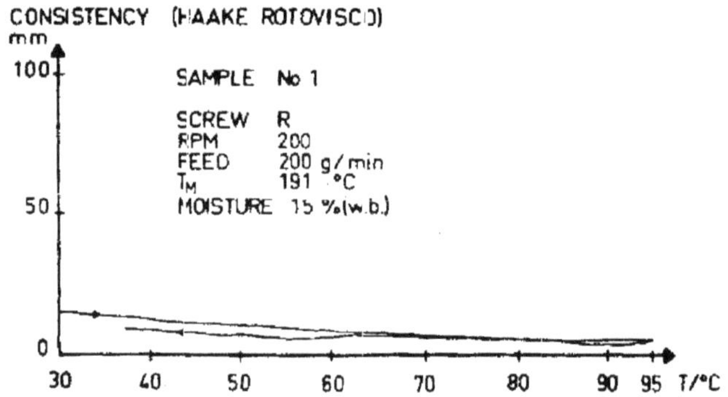

Figure 20.5: Consistency Curves and SEM Pictures (Magnification 300 ×) for the most Severe Treated Sample

subsequent changes during drying. However, the SEM observations in the above series, and in other published and unpublished data, reflect changes in the aggregate structures which are so large that

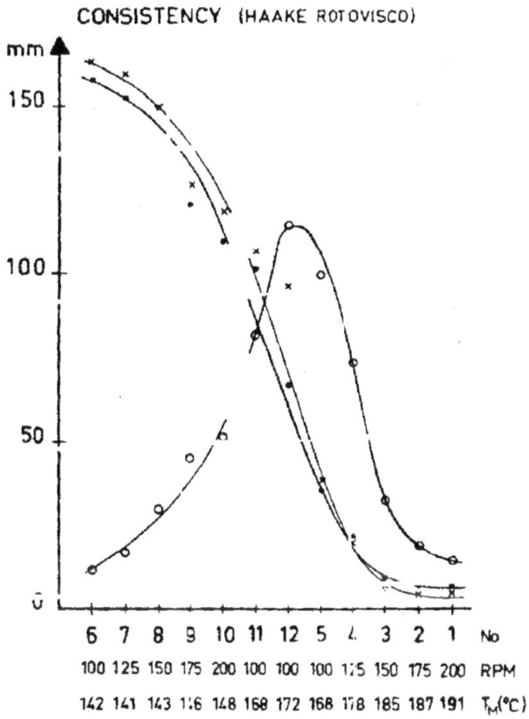

Figure 8.6: Initial 30°C. O, Hot Area Peak, ×, and 50°C, ●, Consistencies for the Experimental Series in Table 20.1. Samples 6–10 are extruded with a screw combination having compression screws only. Samples 11, 12 and 5–1 have been extruded with a counter element in the screw composition.

the bias is small in its magnitude compared with the changes due to progress of the treatment. So the aggregate level changes can be considered to hold the key to understanding the progress of denaturation and subsequent plastification of starches and other biopolymers.

The microstructure of ground, dry extrudates is like that of plastic. If similar biopolymer products are dried to low moisture contents close to 0 per cent water, they display plastic-like behaviour. For example, they have a softening or melting temperature zone.

Here, the word 'melt' is a proper one to be used in connection with the plastifying mass in the extruder. This terminology suggests that the biopolymer-water mass is first formed into a dough, which upon denaturation and plastification is transformed into a melt during its passage through the extruder. At this point we can conclude that in our phenomena model the denaturation of biopolymers is a necessary prerequisite for plastification and texturisation of starch, proteins and cellulose alike.

The fact that at the aggregate level, structural changes are responsible for the changes in functional properties needs further elaboration. Since the WAI experiments were performed at 30°C, the water can be adsorbed by the molecular biopolymer structures and/or physically entrapped by the three-dimensional aggregate networks. What then causes the increased water solubility, decreased water absorption, and decrease in cold water consistency connected with the hard thermal processing in the extruder, or long cooking of starch under atmospheric conditions while shear is applied? In one laboratory, only insignificant or nonexistent hydrolysis of starch to glucose or oligosaccharides has been found in enzyme denaturated extrudates when applying twin-screw extrusion on cereal materials. There are indications that some depolymerisation of starch to the magnitude of 100–500 d.p. might occur, but this does not seem to explain the observed facts.

Changes in Biopolymers

When biopolymers are denatured, in that their quaternary structure is changed, the hydrogen bonds holding together the original aggregates are ruptured, and the macromolecules are released to form new hydrogen bonds. There is a possibility that chemical reactions proper also take place and their course also has an effect on the aggregate network formation. In starchy materials, for example, glycoproteins may be a part of dough formation and subsequent aggregate formation. In proteins the disappearance and reappearance of S-S covalent bonds may also be a factor to be considered.

At mass temperatures around 170°C something seems to happen in low water extrusion processing of starch and cellulose, and possibly of proteins as well. One can speculate that at this temperature changes in the tertiary structures of biopolymers start to take place predominantly, which leads to a loss of aggregate forming ability, and thus to changes in functional properties. Physically modified starches tolerate extrusion cooking better than native starches or flours. One should, however, remember that the physical conditions in extrusion are those of a 'reactor.' Thus the modification agents can immediately, after their release from the original cross-bonding, rebond with starch which may well lead to lowered denaturation of tertiary structures by protecting the helices associated with these structures.

The dough formation-denaturation-plastification phenomena have been elaborated above, since the author thinks them to be a necessary basis for the understanding of extrusion cooking. 'Reaction analogous' in this connection refers to the fact that true chemical reactions such as hydrolysis are not the predominant mechanisms. Instead changes on the aggregate level hold the key. However, the above hypothesis gives a perspective into the understanding of the effects of heat, shear and ingredients such as monoglycerides, fat and sugar on biopolymer denaturation and plastification.

If consider the biopolymer, in this case starch, and water to be two 'reactants,' we can say that ingredients do affect the 'reaction' of aggregate structure changes by:

1. Influencing the state of water, or its 'reactivity' as sugar does,
2. Complexing immediately the aggregate forming molecules (amylose) being released by denaturation as monoglycerides or fatty acids of proper dimensions do, or
3. Affecting both the biopolymer and water as salt does.

Thus the 'reaction analogy' gives us a way to evaluate the effects of the recipe and ingredients, and their effect on the functional properties of products.

Chemical reactions proper, such as vitamin destruction, Maillard reactions between amino acids and carbohydrates, and to some extent dextrinisation or higher level depolymerisation of starch, also take place under the conditions encountered in extrusion cooking. Fortunately, their reaction rates seem to be lower than those of the aggregate physical changes, and so extrusion cooking yields a good retention of nutrients when applied in its normal food and feed processing conditions. Aggregate level changes, *i.e.* denaturation and plastification of biopolymers, and chemical reactions under these conditions are in principle irreversible. This means that the physical properties of products can be obtained from successive, irreversible changes taking place in the different zones of the extruder barrel, in the die and upon expansion at the die exit.

The Food Processing System

The food processing system (FPS) consists of raw materials and ingredients, unit operations and processes, and the necessary mechanical equipment needed to produce a food item for the consumer. It takes into account the composition of materials used and the effects of processing conditions on defined quality criteria as they are evaluated by the consumer.

There are three major subsystems involved in the FPS. Subsystem 1, the phenomena level subsystem, relates the consumer quality to the raw materials and ingredient selection through the impact of basic changes during processing and storage. Subsystem 2, the mechanical equipment subsystem, consists of the hardware for the production and handling of a food item. It also forms the physical environment in which the basic changes during processing take place. Subsystem 3, the information and automation subsystem, collects, processes and transmits information in a FPS, and executes the necessary control measures.

It integrates the phenomena level subsystems and the mechanical equipment subsystems through instrumental measurements, information processing and process variable manipulation.

In asmuch as the FPA essentially combines raw materials, the basic changes they undergo, unit operations and the properties of the product at the time of consumption, a special problem for the food industry is to predict and control consumer sensory quality factors, nutritional quality and safety already during processing, since the actual consumption may take place after an extended storage period.

Quality formation in processing is further redefined and simplified from the FPS concept.

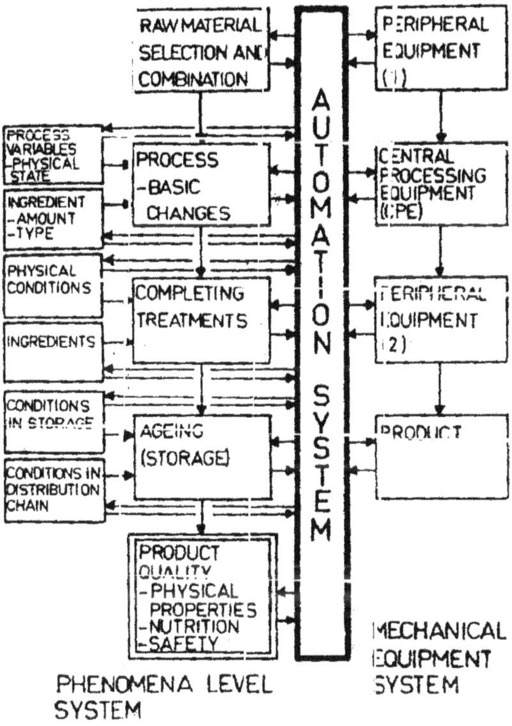

Figure 20.7: Schematic Presentation of a Food Processing System (FPS)

The key to analytical and consumer quality formation is the interaction of product bases, or the raw material combination and processing conditions, on the basic phenomena. As the basic phenomena taking place in extrusion cooking are irreversible, the problem can be transformed to a general sequential optimisation problem and handled according to Bellman's principle of optimality for sequential systems which states: 'An optimal policy has the property that, whatever the initial state and the initial decisions are, the remaining decisions must constitute an optimal policy with

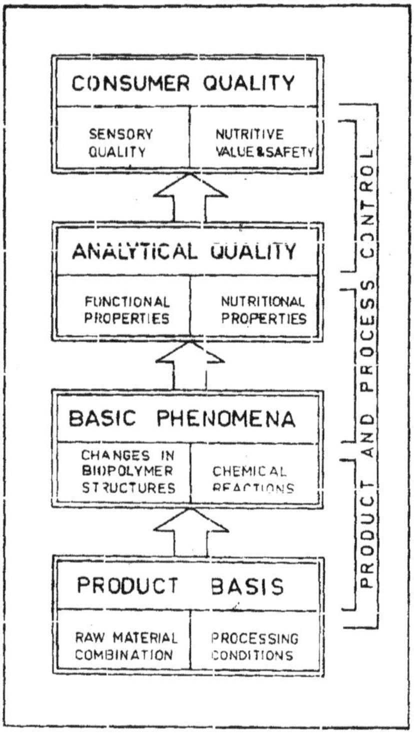

Figure 20.8: Schematic Presentation of Food Product Quality Formation in Heat Processing

regard to the state resulting from the first decision'. We shall now apply the sequential principle, which is the basis for the so-called dynamic programming, as a heuristic model for extrusion cooking.

Single-Screw Extruder

The mechanical central processing equipment (CPE in Figure 20.9) in the FPS consists of a feeding device, a barrel that can be heated or cooled if the need arises, a screw rotating inside of the barrel, and a die through which the plastified mass re-enters the atmosphere.

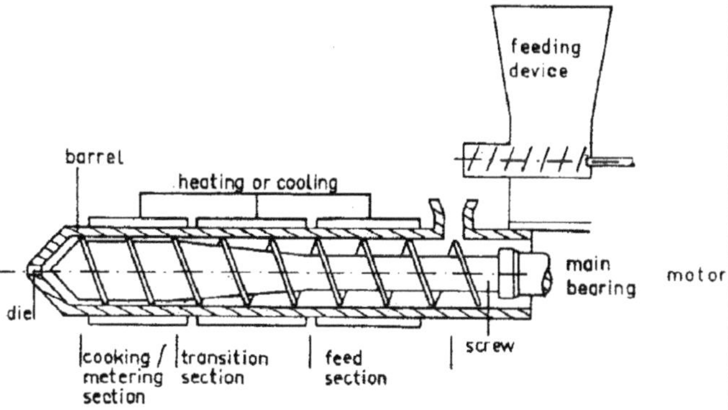

Figure 20.9: Mechanical Parts and Functional Sections of a Single-Screw Extruder

Thermodynamically polytropic screw reactors which create the necessary physical environment, where the 'reaction analogous' basic changes in biopolymer aggregates and chemical reactions take place. The extruder is a high temperature, short residence time, relatively high shear, moderate to high pressure screw reactor.

Equation (1) describes the emergence of the overall functional and/or sensory properties in extrusion cooking as a function of raw materials, ingredients, moisture content and the effect of complete pressure, temperature, residence time and shear profiles in the CPE. This overall effect is, however, a combination of the effects in the different functional sections of the extruder.

Figure 20.10 explains a heuristic sequential model of the functional sectors for the single screw extruder.

$Y_0, \ldots . Y_6$ indicate the state of the quality vector at different stages of extrusion processing. RPM_1 indicates the rotation speed of the feeding device controlling the reed rate (*F*). *RPM* indicates the rotation of the extruder screw. MEC_2, MEC_3, MEC_4 indicate the mechanical construction of the screw and the barrel in the extruder feed, transition and metering sections, respectively. Q_2, Q_3, Q_4 indicate the external heat added or removed in the extruder feed, transition and metering sections, respectively. MEC_5 indicates the

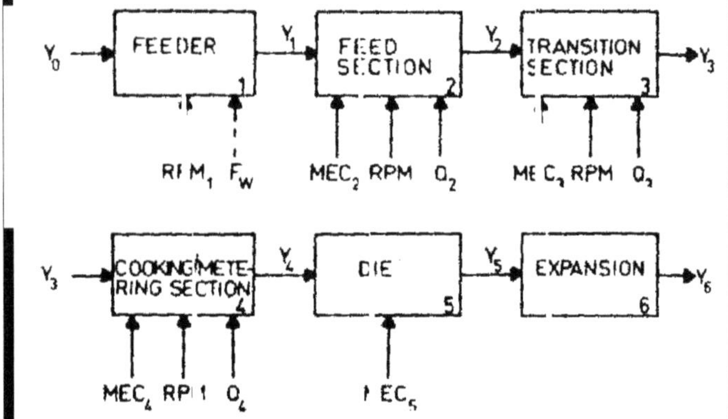

Figure 20.10: Heuristic Sequential Model of the Functional Sectors for the Single Screw Extruder

die geometry including its diameter, periphery and length. F_w indicates the water feed rate, if water is separately added to the feed section.

The quality property states in the sequential analysis of the extrusion cooker are described below:

$$Y_0 = f_0(RM, I, MC_0) \tag{2}$$

Y_0 is the raw materials and ingredients mixture at the moisture content it holds in the feeding device (MC_0). This implies knowledge of the type of raw materials and ingredients, and their ratios.

$$Y_1 = f_1(F + Y_0) = f_1[(F + f_0(RM, I, MC_0)] \tag{3}$$

Y_1 is the amount of raw materials and ingredient mixture per unit time being fed into the extruder feed section. This amount is regulated in normal praxis by feeding device actions such as rpm of a feeder screw. In practical applications most extruders operate with partly starved screws, which means that the feed sections of extruders are not completely filled. The same applies for extruders equipped with flow restrictions such as steam locks also in other sections of the machine. Twin-screw extruders are always run in a partly starved condition. Thus, the mass feed rate (F) is an independent process variable.

If water is added into the feed section separately, as often is the case, there is another independent variable, the water feed rate (F_w). In this case the total water content coming into the feed section is $MC = MC_0 + MC_1$ or the sum of moisture in the feed mixture and the moisture added separately into the feed section.

$$Y_2 = f_2[(Y_1 + f_{PC2}(\widetilde{P}_2, \widetilde{T}_2, \widetilde{R}_2, \widetilde{S}_2)] \quad (4)$$

Y_2 is the state of the mass leaving the feed section. It is a function (f_2) of the entering mass, its moisture content, and reed rate, and including a function (f_{PC2}) of the changes that the pressure, temperature, residence time and shear profiles induce in this section. It is important to notice that the incoming mass and the extruder operation variable interact in the emergence of the $\widetilde{P}, \widetilde{T}, \widetilde{R}, \widetilde{S}$ profiles.

MEC_2 denotes the mechanical construction variables; the screw design inclusive flight depth, flight pitch, flight edge thickness, flight clearance and barrel surface construction (smooth or rifled). *RPM* is the screw rotation speed, and Q_2 is the amount of heat added or removed from this section, which in operational conditions is controlled by the barrel temperature. Feed sections are frequently cooled so that steam is not generated, since any steam in the feed section would result in uneven material flow and thus instability of the operation.

The mass that enters the extruder at feed rate *F*, is either a prepared dough or dry granular powder. If it is a granular mass it will at least be partly worked into a dough by the mixing and shearing actions of the screw and the temperature rise, given that sufficient moisture for dough formation is fed in. The entering mass has thus flow properties, which are changed by the dough formation. It also has a specific heat and heat transfer properties that are largely a function of the moisture content. Moisture also acts as a lubricant by reducing the viscous heat dissipation in the mass from the mechanical energy input through shearing action into the forming dough.

The temperature in the feed section can rise for either of two reasons. The first one is the viscous heat dissipation as the dough, if formed, is a highly viscous mass. Some ingredients such as bran material, and undissolved crystals such as sugar and salt increase the mass viscosity. They behave like glass beads in a solution and need material to cover them.

The other mechanism leading to a temperature rise can be mechanical friction. Dry ingredients, if their particle size is so small that it allows them to enter the clearance in between the screw and the barrel, and if they are forced into this clearance can create mechanical friction whereby more torque is needed and the mechanical energy is transformed into heat energy leading to a temperature rise. If sufficient moisture is present this friction-creating mass either gelatinises or water lubricates the surfaces thus solving the problem.

Without any mass entering the extruder through the feed section there is no formation of $\widetilde{P}, \widetilde{T}, \widetilde{R}, \widetilde{S}$ profiles. In summary the mass, its initial physical properties, and changes in the physical properties due to the mass-machine interaction create the final state Y_2 of the mass leaving the feed section. At this stage the conditions or the state is seldom such that the initial denaturation temperature has been reached, and if it has been reached plastification has not yet substantially progressed. The normal maximal state of the mass in the feed section can be described to be a dough much like that in baking. The Y_2 state is thus a mass with a set of physical properties leaving the feed section at a temperature that is higher than the initial incoming feed temperature.

$$Y_3 = f_3[(Y_2 + f_{PC3}(\widetilde{P}_3, \widetilde{T}_3, \widetilde{R}_3, \widetilde{S}_3)] \quad (5)$$

Y_3 is the state of the mass leaving the transition section. It is similarly formed from construction and operation variables, and the physical conditions interacting with the mass as was the state of the mass leaving the feed section (Y_2).

The mechanical design is usually such that the mass inside the barrel compresses together by the physical action of the screw. This can be obtained by, for example, increasing the screw root diameter, which is equivalent to decreasing the flight depth, or by decreasing the flight pitch. The mass fills the screw fully in this section. Entrapped air, if present, is squeezed out from the mass by the rising pressure. The mass becomes a true, continuum and the flow continuum mechanics start to function.

In a single-screw machine the mass is being pumped forward due to the action of so-called drag flow, against the pressure from the decreasing space for the mass to employ, and the pressure introduced by the die. A single-screw extruder functions much like a centrifugal pump. The necessary condition for a drag flow to develop is that the mass has a higher adhesion to the barrel wall than it has to the screw element walls. If this is not the case, the mass will just turn around with the screw without being transported anywhere. Frictional surface in single-screw extruders is often increased by rifling the barrel surface. Rifling, however, also allows the mass to flow in a backward direction more freely. It also increases the heat transfer surface.

In many cooking extrusion applications the temperature in the transition section is allowed to rise above the necessary temperature for the initiation of the biopolymer denaturation and thus plastification sets in. This leads to drastic changes in mass physical properties connected with flow and heat transfer. In the initial stages after denaturation the biopolymers form new aggregate structures in the continuum. Thus, viscosity increases rapidly and viscous heat dissipation will also increase.

Two counteracting mechanisms in plastification set in. Mass viscosity will increase due to denaturation and availability of biopolymers for aggregate formation. At the same time the temperature rises due to viscous heat dissipation and often by heat flow from the extruder barrel wall. The newly formed aggregates are also subjected to shear. Viscosity of biopolymer masses is temperature dependent. It would decrease with increasing temperature if no more new aggregates were formed. It is also shear dependent; shear thinning and thixotropic with superimposed structural viscosity. In other words the biopolymer melts are rheologically strongly non-Newtonian. However, at early stages of denaturation and plastification the rise in viscosity is enormous due to the new aggregate formation in the mass continuum.

The viscosity of the biopolymer melt is a strong function of water content; the higher the water content the lower the viscosity. Also functional ingredients can interfere with the viscosity rise, for example, monoglycerides complex the otherwise aggregate-forming starch amylose, which results in a lower viscosity: sugar increases the starch gelatinisation temperature delaying the onset of denaturation by lowering the water 'reactivity', etc. So in practical technological applications complex situations can indeed arise.

Y_3 indicates the state of the mass leaving the transition section. In this heuristic model we take it to be a highly viscous dough, where the denaturation and plastification of biopolymers have started, and are proceeding with the aggregate formation causing a radical rise in the mass viscosity.

$$Y_4 = f_4 [(Y_3 + f_{PE4}(\tilde{P}_4, \tilde{T}_4, \tilde{R}_4, \tilde{S}_4)] \quad (6)$$

Y_4 is the stale of the mass leaving the metering section also called the cooking section. The stage limit can be drawn to the end of the screw flight, or through the void space between the screw end and the die entrance. If the construction of the extruder has a large intermediate space in between the screw end and the die entrance, it can either be handled as a separate stage, or connected to the die or to the metering section. Here we shall use the situation where the metering section is considered to end at the end of the screw flight.

Y_4 is again formed by the entering mass state (aggregate state, temperature, viscosity) and by the mechanical design variables of the screw and the barrel, and by the operation variables, *RPM* of the screw and heat flow indicated by the zone temperature. There is again the interaction of construction and operational variables, and the mass itself in the formation of $\tilde{P}, \tilde{T}, \tilde{R}, \tilde{S}$ profiles.

The basic changes, denaturation and plastification of biopolymers, which were discussed in connection with the transition section, continue in the metering section. The mass temperature in this section is usually the highest in the extruder. Aggregate formation and disruption continues. Viscous heat dissipation and often outside heating raise the mass temperatures into the range 100-240°C in food and feed applications. Shearing can cause conformational changes in biopolymer tertiary structures in high temperature melts. We are now venturing towards the functional properties, maximum peak area, and if heating and shearing are intensive enough, over it.

It was mentioned earlier that physical properties (viscosity, thermal, properties) are a function of mass temperature. The same applies for the reaction and aggregation plastification rates, and to the constants, of the kinetic equations, as limiting factors change. When the mass viscosity reaches its maximum due to the new aggregate formation, there is a radical change in the trend; mass viscosity starts to decrease. We are over the peak. This maximum area, or the conditions connected with it, depends on the raw materials, the mass $\widetilde{P},\widetilde{T},\widetilde{R},\widetilde{S}$ history, moisture content, etc. The following reduction of the melt viscosity is, in some cases, very rapid.

The surpassing of the maximum peak of melt viscosity does not always occur nor is it desired. If the extruder operates under conditions where the shear thinning, due to high shear rates between the flight edges and the barrel surface, and high barrel temperature lead to in proportionally large viscosity differences at the barrel surface and at the screw, slippage can occur which leads to instability of operation. If the mass temperature reaches values which are too high, burning can take place. This will happen at temperatures above 220–250°C, depending on the product, the equipment used and on the operational parameters.

The mass that leaves metering section is at least partly cooked, and normally in a plasticised state. Due to the fact that the pressure inside the extruder is high enough, water is in a liquid state despite the high temperatures. We can say that the mass acts like a polymer melt in its functional properties.

$$Y_5 = f_5\,[(Y_4 - f_{PE5}(\widetilde{P},\widetilde{T},\widetilde{R},\widetilde{S}))] \qquad (7)$$

Y_5 is the state of the mass leaving the extruder die. Die functions include the shaping of the product and the creation of back pressure into the extruder thus affecting the residence time. In the case of fibrous materials such as textured vegetable proteins, the die functions include final orientation of fibre structure. Furthermore, the die induces additional shear into the plasticised mass.

There is always a void space between the screw end and the die, which is filled with plasticised melt. It can be a gap of a few millimetres or a larger reservoir with intermediate holed plates and channels of various constructions. This design detail has a large effect on the shear the mass experiences, and on the residence time in the extruder. In a twin-screw extruder, where no reverse flight elements exist in the screw design, the bulk of the shearing which the mass experiences takes place in the die and in the space between the screw end and the die entrance.

The die is a flow restriction, which creates a back pressure depending on the die design parameters of length, capillary diameter and periphery. If special dies are used, as is the case in protein texturisation, the profile of the die cavity can have 90° channel direction changes. The back pressure which arises is a function of the mass viscosity, and changes during the time the mass resides in the die. Whatever is the pressure needed for the mass to flow through the die, it must be exceeded by the pressure, which is created by the pumping action of the screw or screws. The pressure created by the screws must also overcome the pressure loss in the void space before the die entrance.

The mass passing through the die is, in cooking extrusion applications, still pressurised, and plasticised to the extent of the sum of all the various sections of extruder actions, or the total of action. This mass re-enters the atmosphere through a restriction, the die, with speeds that can be rather remarkable under some conditions. The water in the mass is in a superheated state due to the elevated temperature used in many of the cooking extrusion applications. The water will thus violently boil. It flashes off in an instant, which leads to the expansion of the plasticised mass.

$$Y_6 = f_6(Y_5) \tag{8}$$

Y_6 is the state of the mass after extrusion, which is a function of the state of the mass leaving the die exit. Given appropriate mass composition and appropriate conditions in the extruder, the state of the mass involves enough plasticised material for expansion. If the temperature of the mass is high enough for the water in the mass to boil, violent expansion takes place. Part of the expansion is die swell, which is due to the normal stresses that have been introduced into the mass by its viscoelastic behaviour in the die. Proper explosive expansion needs, however, a plastifying raw material component and superheated water.

In food and feed extrusion cooking various natural and modified starches are needed as plastifying agents if expansion is desired. Proteins do not possess similar expansion ability, they tend to form laminar structure with less expansion. Proteins are also often cooked and extruded with higher water contents than starchy raw materials.

Denaturation of Starches

In order to be able to expand starches must be denatured and the aggregate network worked to a proper state.

If too much heat and shear is introduced into the plasticised starch mass, the state for the maximal expansion is surpassed. This is partly due to the state of the starch, in asmuch as the mass pressure is reduced, when plastification proceeds and the aggregate structures are 'worked over the peak' whereby not enough pressure differential is maintained across the die. Another factor is the lower mass viscosity in a starch mass that has been heated and sheared too much. In this case the stiffening of the plasticised mass happens too slowly in order to hold in the expanding steam bubbles and honeycomb structure. But given enough pressure differential across the die and a proper melt viscosity, starchy masses form a cellular structure when water flashes off at the die exit.

Various factors can interfere with the expansion operation. If there is enough non-plastifying material such as bran in the recipe, and the starch does not fully entrap the inert material in a manner where thick continuous melt layers exist, water is shot off from the structure including many vaporisation nuclei, and no expansion occurs. Monoglycerides do not allow aggregate networks to form, reducing the expansion if present in concentrations above 0.5 per cent. If too much fat is added this also reduces expansion. Fatty acids of proper size complex amylose. Fat also acts as a lubricant in the melt thus reducing heating by viscous dissipation. If too much water is added it also acts as a lubricant, and requires too much thermal energy to vaporise enough of the water at the extruder die exit. This leads to a high moisture content viscoelastic mass that first expands, but due to too high a water content and high temperature after expansion (which leads to too high an elasticity) the initially expanded mass is redrawn to an unexpanded state.

The expanded mass that has been partly cooled upon the explosive expansion by the thermal energy needed to vaporise the water, is often in a viscoplastic state, and can be either cut or stretched and formed into continuous sheets as in the crisp bread production process. In snack food production the expanded mass is cut to pieces of desired lengths.

In some applications no expansion is desired at the die exit. This is achieved by cooking the starchy, or proteinous mass with high moisture contents and cooling the mass below the boiling point of water while the mass still resides in the extruder, or by cooling in special die assemblies before the mass emerges into the atmosphere.

The precooled, plastic mass can be dried and later expanded in an oil bath or microwave oven. Some of these masses have in interesting ground extrudate pasting behaviour. Even though the transparent, plastic like appearance shows that denaturation and plastification have taken place, the ground, water suspended extrudate shows a pasting behaviour much like that of non-denaturated starch. Figure 20.11 shows such an anomaly for barley starch. The WAI value is also low. This anomaly is due to the solubility and diffusional properties of the high water content mass. The plastified aggregate matrix is, in this case, extremely resistant to rehydration. When this mass is dried and later subjected to rapid heating in an oil bath its expansion is due to the plastic like softening properties of the mass. This operation, where the remaining water expands the softened, less viscous mass, has been called alveolation by Menzi.

We have now ventured through a single-screw extruder in a stage-wise manner. A similar systematic, heuristic model can be used to 'think any machine construction through'. There is not enough knowledge of the physical properties of denatured and plasticised masses, material-machine interactions, residence times

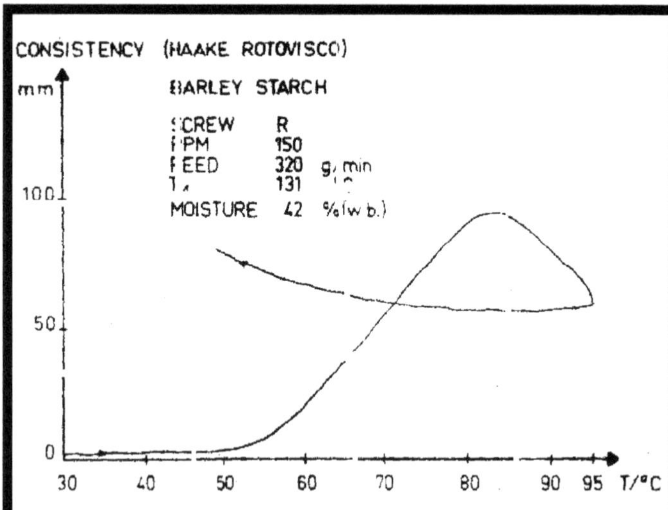

in Pasting
Behaviour of Barley Starch

in various zones of the extruder nor of the mass temperature and pressure profiles to afford exact mathematical modelling for each stage. Mathematical models of extrusion cooking will shortly be discussed below. Before that, however, let us inspect the differences of single- and twin-screw extrusion cookers.

Difference Between Single- and Twin-Screw Extruders

The basic difference between single- and twin-screw extruders is their mass transport mechanisms.

Single-screw extruders are like centrifugal pumps transporting the mass along the barrel by a mechanism of drag flow, which is dependent on the mass adhering more to the barrel wall than to the screw. Twin-screw extruders, on their part, act like positive displacement piston pumps. Janssen and Martelli discuss the action, benefits and problems associated with twin-screw plastics extruders. Clark and van Zuilichem *et al.* refer to the differences of single- and twin-screw extruders. The discussion below, will concentrate mainly on twin-screw, intermeshing flight extruders.

Due to the mass transport by positive displacement, mass transport action, twin-screw extruders introduce less shear into the mass than single-screw extruders. This leads to a smaller viscous energy dissipation in the mass. In some applications, where excess

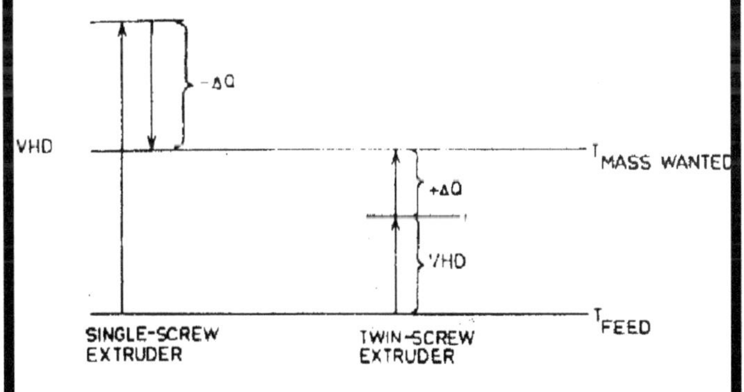

Figure 20.12: Differences in Temperature Formation in a Single- and Twin-Screw Extruder

heat that has been generated through viscous heat dissipation is removed by cooling from a single-screw extruder, some heating is required in a twin-screw extruder.

Twin-screw extruders can have two modes of screw rotation; the screws can rotate in the same direction, in which case they are called co-rotating. They can rotate in opposite directions, in which case they are called counter-rotating. In a co-rotating machine the mass is pumped from one chamber of a screw to another chamber of the other screw in C shaped continuums. In a counter-rotating machine mass has to pass through a clearance between the flight top of one screw and the root of the flight in the other screw. Thus, counter-rotation creates higher mass pressure formation in a shorter distance, but also introduces more shear into the mass and thus also more viscous heat dissipation.

The counter-rotation tends to push the screws apart when the mass is forced through the narrow gap between the nights of one screw and the screw roots of the other screw. Co-rotating machines are self-centering, and thus it is easy to design a self-cleaning screw

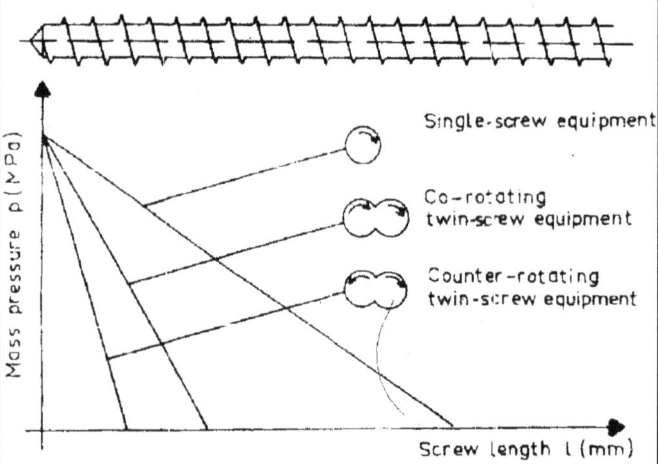

Figure 20.13: Mass Pressure Formation in Extruders for a Single, a Twin-Screw Co-rotating and a Twin-Screw Counter-Rotating Machine

system. The pros and cons for each of the twin-screw designs depends on the intended application.

Twin-screw extruders are at present relative newcomers on the food and feed extruder markets. It seems that machines with straight cylinders will be the standard twin-screw extruders in these industries.

The Screw Design

Figure 20.14 indicates the cross sections and assumed temperature and pressure profiles of a twin-screw.

Twin-screw extruder screws are designed and often manufactured in functional sections. The functions of each section can be analysed in a similar heuristic, sequential manner, which

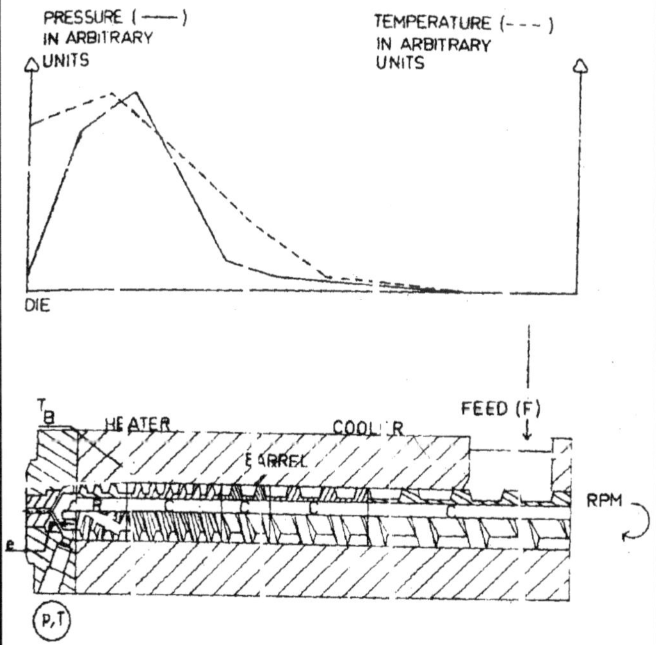

Figure 20.14: The Cross Sections and Assumed Temperature and Pressure Profiles of a Twin-screw

was used above for a single-screw extruder. Figure 20.15 displays a screw design, which is used in a twin-screw, intermeshing flight, co-rotating extruder. It also displays the assumed mass temperature and pressure profiles along the extruder.

If there is no counter pitched screw element in the screw design, the first flow restriction encountered by the mass is the flightless space between the screw and caps and the die entrance, and most importantly the die itself. The machine is filled just in front of the discharge end. In a case where a counter flight screw element is used in the screw design, this results in an effectively filled flow restriction, where the mass has to be jumped through the counter element by preceding compression screw action. Much more intensive shear is introduced, and the residence time in the extruder is greatly increased. This action by the screw can be designed to take place in the hot end of the extruder, or it can be installed in the cooled section to induce effective mixing. Also other types of elements such as non-centred discs for grinding, or for protein dope molecular orientation can be used in the functional screw design.

The functional section design in twin-screw extruders, and the positive displacement material transport mode of these machines, offer great flexibility in the operations design. On the other hand this emphasises the need for systems analysis, and makes exact mathematical extruder modelling difficult.

This design is used in a twin-screw, intermeshing flight, counter-rotating plastics extruder. This design is one of the most elegant constructions available. It has a conical barrel terminal section and increasing root diameter in the screw terminal section. This allows well planned mass pressure, speed and temperature profiles at the final section of the extruder. The residence times for shear and

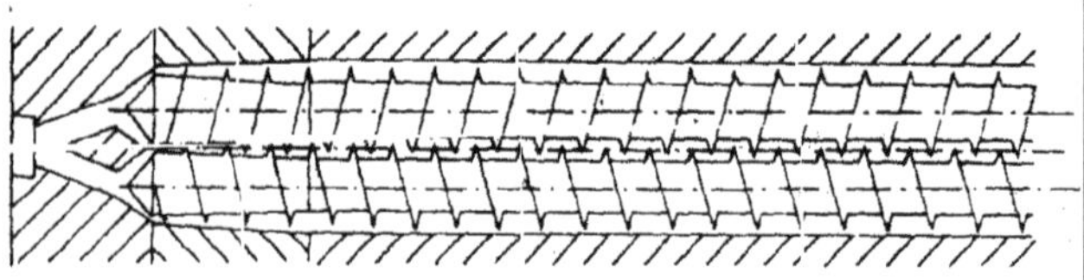

Figure 20.15: Principle Drawing of an ESDE Twin-Screw, Intermeshing Flight, Counter-Rotating Plastics Extruder

temperature sensitive materials such as biopolymers can be made short, and the most wear-plagued parts, the end cone and screws are short.

Twin-screw extruders also have a great flexibility as reactors. This is a result of their mass transport mode and the positive displacement pumping action. They have been used in applications which involve decrystallisation of sugar for low water content boiled sweets, and similar designs have been used as crystallisers at low temperatures. Now applications of twin-screw extruders also include acid hydrolysis of cellulose which has been denatured in earlier sections of the one and the same machine, and direct enzymatic hydrolysis of starch in the extruder. In both cases the biopolymers are denaturated and hydrolysed in a single pass through the reactor.

A twin-screw extruder can also be applied in pelletising of problematic materials which have a high fat content. Here the fat melts enough to be extruded and solidifies upon the exit from the machine. Agglomerator applications are also being worked with.

Twin-screw extruders seem to be more versatile in their material handling ability and application ranges than single-screw extruders. One should, however, remember that they are relative newcomers in the food and feed processing industries. Many new research centres with new ideas and no concepts of 'the impossible' have entered the field with twin-screw machines. Their new applications will, in many cases, be transferred to single-screw machines with innovative designs. Which type of machine is the better for any given job depends on the product, peripheral equipment systems available, technology back-up and economics in each case.

Hierarchical Control System

The conceptual and heuristic models for the basic phenomena and extruder designs are beneficial in a science which is in its infancy.

Dynamic production planning is a matter of operations research and business economics. Steady-state optimisation of processes and direct regulation through controls are the hierarchy levels of engineers and technologists. Figure 20.16 indicates the schematic presentation of the principle of hierarchical control system.

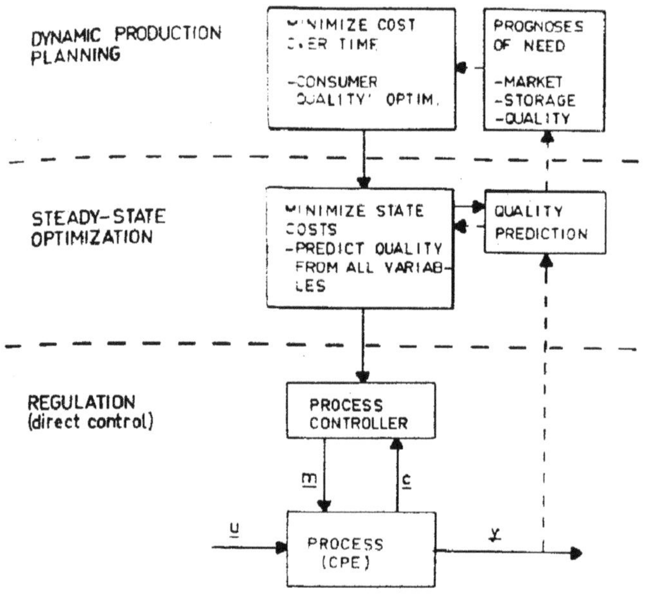

Figure 20.16: The Schematic Presentation of the Principle of Hierarchical Control System

Basis of Modelling can be studied as follows, which dependable variables or variable combinations are the target of modelling?

Which variables are used as the independent variables, or the variables we take into account, when considering how to explain the way in which the dependent variables emerge from the real world? What kind of a model are we going to build? Will it be based on fundamental physical and chemical entities; or will it be an empirical black or grey box model? Branch has discussed this question, which most definitely influences the independent variable selection.

We should remember that any mathematical or heuristic model is but a reflected description of the real world. The purpose of modelling and its usefulness in predicting the real world states, is a measure of the success of modelling.

The products of food and feed industries are intended to be eaten. Thus the final aim of the modelling, in this case, is given in the FPS–the product quality inclusive of functional properties such as sensory entities, safety and nutrition. The regular analytical techniques concentrate on functional properties such as water solubility index, water absorption index, expansion, consistency behaviour during pasting, mechanical strength, etc. or on chemically analysable entities. Here we are discussing a multi-dimensional quality vector, an example of which is given by van Zuilichem *et al.* A lot more work has to be done in defining meaningful quality vectors for extruded food and feed products.

Other dependent variables that need modelling are the variables that give the product and process state anywhere along the processing path. These can be the mass pressure and temperature, viscosity etc. at some point in the extruder. They are necessary as control variables. They should, however, ultimately be tied to the final product quality vectors in order to be useful. To emphasise the point, for a fish feed the decisive ultimate variable is the cost of feed used, or what it costs to gain a unit weight of fish. The pressure at the extruder head is not necessarily an indication of this factor. It can, however, be a necessary indicator for a good product in a proper state at this point.

As no model can be more accurate in its predictive strength than the accuracy of the analytical technique used, and many extruded products are not easy to analyse, there is a need to develop good indicators of product properties. Furthermore, quick, practical analysis for quality control to be used in the plant are worth considering.

Now let us have a look at steady-state modelling.

The steady-state modelling is used to predict chemical reactions, functional properties after processing, and state indicators during processing. There seems to be two apparently different philosophies employed. One is the empirical black box approach; the other one is the basic physical approach. Both have their merits and short-comings. Let us discuss the black box approach first.

The black box approach is based usually on response surface modelling (RSM). In this case the quality indicators are modelled from five independent variables, two originating from the 'recipe' and three processing variables. The second-order polynomial models for quality indicators are obtained through least squares fitting of the analytical data collected from an experimental design. This approach is beneficial when little if anything is known of the real form of the function in the box. One can inspect the effect of recipe and of process variables as they are really manipulated in the world on the end product quality indicators, and on the process state indicators at the extruder head where mass pressure and temperature can be measured.

Figure 20.17 explains the principle of empirical black box response surface modelling, of quality or state indicators from recipe and processing variables.

The equation in the box is a second order polynomial in sum notation ε = the residual after least squares fitting.

These models strictly only apply to the ranges of independent variables used in the experimentation. A problem, which is also encountered when using polynomial models, is the difficulty that arises when physical and chemical interpretations are needed. On the other hand, they can be applied to optimisation by linear programming or simulation.

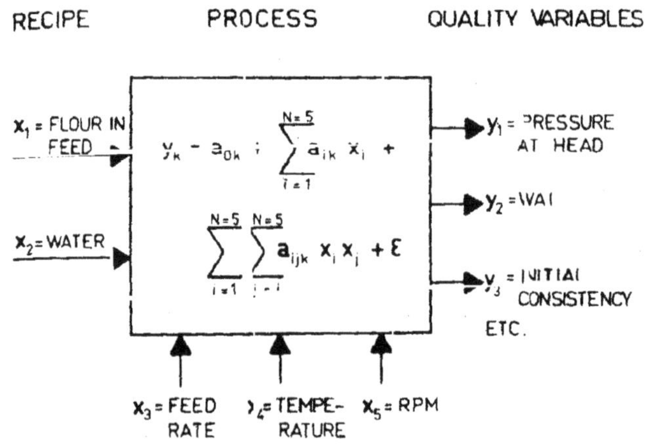

Figure 20.17: The Principle of Empirical Black Box Response Surface Modelling, of Quality or State Indicators from Recipe and Processing Variables

The other approach is fundamental physical modelling. It should ultimately lead to the universal understanding and explanation of the $\widetilde{P}, \widetilde{T}, \widetilde{R}, \widetilde{S}$ profiles through the extruder and of their effects on the product quality indicators. The snag here is that the basic changes in biopolymers and the mass-machine interactions, and the changes in mass physical properties during extrusion are so complex that many simplifying assumptions have to be made. This leads to inaccuracy of the modelling. Besides, many of the 'constants' in the fundamental equations turn out to be functions of temperature or some other process variable.

Tsao and Sizer among others have used first-order reaction kinetics and the Arrhenius equation in the modelling of nutrient retention in single-screw extrusion. This is an indication that reaction kinetics are entering into extrusion studies. More kinetic studies are needed for the extrusion processing conditions in order to take full benefit from the reactor approach to extrusion.

Dynamic models are needed for process control and stabilisation. Hereby the basis is a steady-state operation point and a given set of process state indicators at that point. Let us say that we have mass pressure and temperature as state indicators in an operations point indicated by screw rpm, barrel temperature, and feed rate for a given recipe. If any deviation is observed from the desired state indicators, controls take over and change a process variable so that the desired values of state indicators are reached. This presupposes dynamic models to be known, Olkku *et al.* have performed a study on the dynamics and automation of a twin-screw extrusion cooker.

As biomaterials vary in their composition and behaviour in processing, and the dynamics of most extruders is unknown, much more work in this area is needed before the control applications fully meet the needs of expected automatic control techniques in extrusion cooking.

Extrusion processing carries many number of techniques. Let us try to understand some more details about them.

Extrusion cooking, which can also be called screw reactor processing, is one of the standard techniques employed in food and feed production today and even to a larger extent, in the future. This technique is applied to an increasing number of industrial processes and will be experimented with in new, exotic fields in future. At the same time basic phenomena, their nonlinearity in physical space, and material machine interactions are turned from mystery to manageable practice, even if their will still remain many problems to be solved.

Modelling of extruders is proceeding in its effort, volume and accuracy. Simulation techniques will be increasingly applied in both steady-state mathematical modelling and in dynamic modelling of extrusion processes.

The necessary basis for this depends on the increasing understanding of the basic phenomena under screw reactor application conditions. These phenomena start with biopolymer denaturation at a given temperature, which is reached somewhere in the extruder barrel. This means that the total mass residence time in the whole equipment, which considers that the entire barrel is uniformly equal, is not a phenomenologically true approach. In twin-screw extruders especially the barrel is only sequentially filled by material. So in order to master the extruder, one must consider the effect of various zones on the progress of the basic phenomena. Zones before the onset of denaturation have little if any effect on the plastification and subsequent texturisation. Those zones, which follow the area after the denaturation has started, affect the basic changes, according to their pressure (P), temperature (T), residence time (R) and shearing effect (S) profiles. Here again the effective residence times in each zone have to be considered in order to follow the progress of denaturation and plastification phenomena. Thus, residence time and mass accumulation, profiles are needed for the whole passage through the extruder in order that they can be used in connection with the simulation models.

More detailed reporting of extrusion cooking data has to be emphasised. For example barrel temperatures reported are often but a gross indicator of the actual mass temperatures. Barrel temperatures which are more than 20°C below the desired extrusion temperature have been found when extruding high moisture content 'non-frictional' raw materials. In some cases, where 'highly frictional' materials have been processed with low moisture contents, the observed mass temperatures have been as much as 60°C higher than the set barrel temperatures. This throws some suspicion on the often reported 'extrusion temperature' unspecified data in the literature and on the validity of conclusions drawn on such data. For example one can ask the following question: Is the real effect of water in some reports really due to the changes of water concentration or is it due to resulting changes in extrusion mass temperatures?

The basic and applied research effort in extrusion processing is presently increasing both in volume and in depth. It ranges from low cost extruders for field application to simulation and modelling of the most complicated twin-screw designs. Quality indicators and the basic phenomena are also being worked on. With good reason we can expect this to be beneficial in solving the problems of feeding the expanding world population and in problems connected with exploitation of renewing biopolymer resources.

Plastics extrusion processing uses temperature controls within ± 1°C of the desired set values. In some cases the screw design is altered when small changes are made in recipes. This is not always possible in food and feed processing, where the same machine has to be used often for many products, and recipe changes are a part of the daily praxis. We can, however, benefit from the accuracy of our more advanced colleagues in the plastics field by striving towards more accurate working schemes and modelling.

Part V
Storage of Grains

Chapter 21

Storage and Disinfestation of Grain

Controlled Atmosphere Storage

Two principal types of atmosphere can be used for controlled atmosphere storage and disinfestation of grain.

Low O_2 atmosphere, as used for disinfestion of grain, typically contain less than 1 per cent O_2 with the balance being N_2, CO_2, inert gases, or a mixture of these gases. The most important feature of these CAs is their lack of O_2. They function largely by their anoxic effect and the other gases exert only a minor influence on the action of the atmosphere. Historically, some low O_2 atmospheres produced by metabolic processes, as in "hermetic" storage, or by burning coke in air, were thought to act through their CO_2 content, but this has been shown not to be so. In contrast, insecticidal, high CO_2 atmospheres not only contain a substantial proportion of CO_2, at least 20 per cent and often more than 60 per cent, but also sufficient O_2 to support life if the CO_2 present were replaced by N_2. The change in type occurs at about 3 per cent O_2, with atmospheres rich in CO_2 containing less than this amount of O_2 having the characteristics of low O_2 not high CO_2 atmospheres with respect to insects. The action of CO_2 against molds and on grain appears to be independent of O_2 concentration and thus some atmospheres, *e.g.*, 95 per cent CO_2, 5 per cent (*i.e.*, 1 per cent O_2), can be classed as both low O_2 and high CO_2 with respect to these organisms.

It is important to recognize the distinct types of CA as their characteristic differences have substantial influence on the performance of CA treatments. Sound decisions on the most appropriate atmosphere to use in a particular situation can only be made if these differences are quantified and appreciated.

There is no simple, unique description of the effects of CAs on all insect pests found in stored grains. The level of response varies widely, depending on the target insects (species, developmental stage, and age), the physical environment (temperature and humidity), and on the type and exact composition of the CA used. It will be seen below that neither CA type is more effective under all conditions and thus, the target insect and condition must be considered when choosing the best CA to use in particular circumstances.

It is difficult to make a meaningful comparison of the effects of high CO_2 and low O_2 CAs, partly because of the dearth of data obtained under exactly the same conditions and with the same strains and stages of target insect and partly because of the lack of a well-defined basis from which to start. Some reference level of response to both types of atmosphere is required for a specified organism so that deviations from this reference response can be used to provide comparisons. In the absence of an agreed upon and widely used baseline, the relative response of insects to either 1 per cent O_2 in N_2 or 60 per cent CO_2 in air is taken here as typical for low O_2 and high CO_2 CAs, unless some other standard is applicable for a particular set of data. These two compositions lie around the middle of the possible range of variations of the active component or the two types while still keeping a substantial degree of insecticidal action.

In general, only the lethal effects are important when deciding what type of atmosphere to use and the discussion below is limited to these effects. However, CA systems can have sublethal, usually adverse, effects on insect pests, such as delayed development, impaired metamorphosis, and altered fecundity. Atmospheres with less than 1 per cent O_2 rapidly immobilize, <0.5 min, insects such as *Sitophilus oryzae* and *Tribolium castaneum,* while insects subjected to >75 per cent CO_2 are immobilized in less than 1.0 min. There are insufficient data available to be able to compare sublethal effects of the two types of CA in detail. Generally, the effects are similar in kind. The published information on the subject has been reviewed.

CO_2 can cause significant morality in concentrations about 20 per cent in most susceptible species. However, the speed of response and the minimum effective concentration vary substantially both between species and between developmental stages within a species. Much further work is required before a full understanding of the response range is obtained. Few species have been studied over all their developmental stages and over the range of concentrations and times that may occur during different CO_2 treatment regimes. These may range from high concentrations for a short time for a rapid disinfestation to prolonged exposures at low concentrations for the combined process of slow disinfestation and protection.

The information available at the end of 1985 for establishing carbon dioxide dosage regimes was reviewed by Annis. He used a descriptive method based on pooling of data of wide provenance and was able to approximate the times required to cause high mortality (> 95 per cent). An updated version of this information, edited to include only the species for which comparative data for the two CA types are available.

Firstly and importantly, the times to high mortality at lower concentrations (20 to 40 per cent CO_2) are in many cases not well established. This is usually because experiments were limited to relatively short exposure periods (maximum exposure of approximately 15 d). In several important cases, it is impossible to accurately determine the required exposure period from the published data due to the wide range of reported response to similar combinations of concentration and time. Usually, it is not clear whether this range is truly a reflection of different sensitivity or whether it results from differences in technique only. This means that inadequate information is available at the critical low concentrations (20 to 40 per cent) to define the dosage response surface where the slope becomes very shallow of the concentration axis and therefore, small differences in concentration make for large changes in the time needed to obtain a high mortality.

Figure 21.1 indicates the sensitivity of various stages of *Tribolium castaneum* and *Ephestia cautella* to 63 per cent CO_2 in air and 1 per cent of O_2 in N_2.

Diapausing *Trogoderma granarium* larvae are the most tolerant to high CO_2 atmospheres of any species and stage so far reported. They are tolerant to CO_2 concentrations of 60 per cent or less in air at 25°C and less than 95 per cent mortality has been obtained after 25 d, the longest exposure so far tested. It appears that diapausing *T. variabile* larvae may have a similar response. Other *Trogoderma* spp. are also quite tolerant.

Most species are killed more quickly at the lower O_2 level, but for some this is not so. In particular, *S. oryzae* adults are killed more quickly at 1.0 per cent O_2 than at 0.1 or 2 per cent O_2 under the same conditions. For *S. granarius* adults, at 15 or 20°C, 3.0 per cent O_2 in N_2 kills about as fast as 0.5 per cent O_2 and more rapidly than 1.0 or 2.0 per cent. Against *O. surinamensis* adults (at 15°C), 0.5 per cent O_2 in N_2 was less effective than 1.0, 2.0, or 3.0 per cent O_2.

Only recently systematic investigations of the effect of low O_2 atmospheres (< 4 per cent O_2) containing various proportions of CO_2, but typically with 10 to 20 per cent CO_2, the latter being known

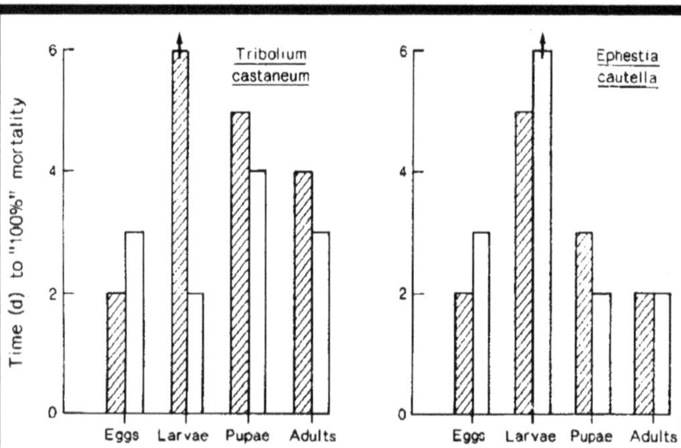

Figure 21.1: The Sensitivity of Various Stages of *Tribolium castaneum* and *Ephestia* to 63 per cent CO_2 in Air and 1 per cent of O_2 in N_2

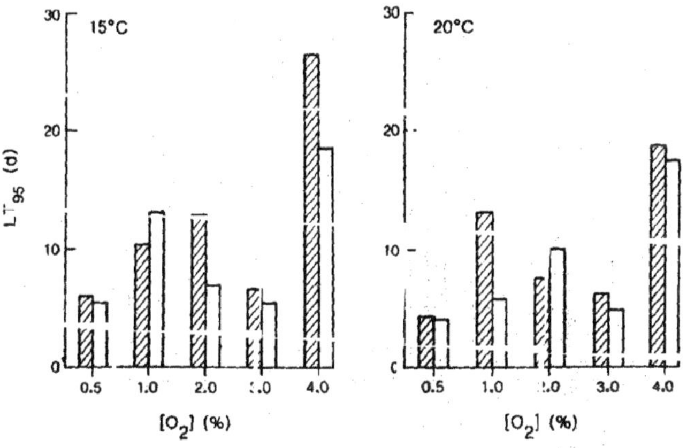

Figure 21.2: Exposure Time Required to Achieve 95 per cent Mortality (LT_{95}) of Various Developmental Stages of *Sitophilus oryzae* Under CO_2 (Left), Helium (Center), or Nitrogen (Right) at 27°C, 60 to 70 per cent

as "burner gas" have been carried out. Some species, *e.g.*, *Tribolium castaneum* as adults, are more susceptible to CO_2 containing low O_2 atmospheres than pure N_2 based atmospheres. The response of others, notably *Sitophilus* spp., appears little affected by the presence of 10 to 20 per cent CO_2. Bailey and Banks concluded that there may be some advantage in practice using a proportion of CO_2 in low atmospheres, although the reduction in exposure time allowable will not be substantial if *Sitophilus* spp. are present. Other authors have claimed significant reduction in exposure times in the presence of > 10 per cent CO_2. It appears that as the O_2 content is increased from 1 per cent, the influence of the proportion of CO_2 becomes more important. This change consistent with an increasing contribution of the high CO_2 type of action on the insect as the low O_2 effect becomes marginal.

Generally, it has been found that the temperature coefficient of the speed of action for the range 10 to 40°C is positive for both types of CA. Exceptionally, Harein and Press found little change in susceptibility of *T. castaneum* larvae at 16, 27, or 39°C to a high CO_2

atmosphere (approximately 48 per cent CO_2 in air) and Zakladnoi found little temperature dependence with the low O_2 type atmosphere, 100 per cent CO_2, at 20, 25, and 35°C, with several species (adults). There is very little sound, systematic data giving the actual trend with temperature, particularly for high CO_2 atmospheres. In this species, the one for which most data are available, the temperature coefficients for both atmospheres are similar though it is generally much more susceptible to high CO_2 CA.

In species such as *S. oryzae* that are relatively susceptible to high CO_2 CA and tolerant to low O_2 CA, the increase in exposure time required at lower temperatures may be much more significant commercially with the latter. Thus, while at 30°C both exposure periods may be acceptable (0.9 and 2.8 d for LD_{95} for high CO_2 and low O_2 CAs, respectively), at low temperatures, *e.g.*, 15°C, the exposure required for low O_2, 12.4 d, may be regarded as excessive, while that for high CO_2, 2.8 d, may not be regarded as excessive. Apparently anomalous results, such as those of Harein and Press, may have led to the belief that the action of high CO_2 atmospheres is less temperature dependent than low O_2 ones. However, a similar effect with insects particularly susceptible to low O_2 CAs might give the opposite impression. From the very limited data available, it appears that the choice of atmosphere is more critical at lower rather than at higher temperatures if the treatment must be successful in minimum time.

Suggested exposure times for complete disinfestation with 0 to 1.2 per cent O_2-containing atmospheres were given by Banks and Annis. The few recent studies with low O_2 atmospheres using tolerant immature stages of *Sitophilus* spp. have tended to reinforce these suggestions. In the absence of the very tolerant *Sitophilus* spp., a reduction of one third in exposure time may be appropriate. In the absence of *Sitophilus* spp., and in the presence of about 10 per cent CO_2 the values can be probably reduced by half without loss of effectiveness. This is consistent with Storey's data that suggest that an exposure of about 12 d at 1 per cent O_2, 10 per cent CO_2, is required at 27°C to eliminate all stages of the most tolerant pests, *S. oryzae* and *Rhyzopertha dominicas.*

The foregoing discussion has largely been concerned with responses to the two types of atmosphere without direct experimental comparison. However, there have been a few studies that do attempt to make a direct contrast. One such study shows a comparison of the response of various stages of two insects to typical low O_2 and high CO_2 atmospheres. It can be seen that neither is generally preferable under this system. For instance, eggs are generally better controlled under high CO_2, while larvae of *Tribolium castaneum*, in particular, are more rapidly killed under low O_2.

It appears that species whose larvae are "internal feeders", *i.e.*, they develop within the grain kernel, are better controlled using high CO_2 atmospheres, while "external feeders", those developing outside the kernel, are more easily controlled with the low O_2 systems. *Trogoderma granarium* and *S. oryzae* clearly represent extreme examples of this generalization. *Trogoderma* larvae, even in diapause, are much more susceptible to low O_2 than to high CO_2 and are "external feeders", while *S. oryzae,* particularly as pupae, is classed as an "internal feeder" and is much more susceptible to high CO_2.

High humidities appear to increase the time taken to kill for both types of CA, but it is not yet clear which is the more sensitive to humidity effects. Under marginal conditions, the effect can be quite pronounced. It has been observed that the lethal action of high CO_2 atmospheres is associated with greater water loss than low O_2 atmospheres, at least from *T. castaneum* adults, but the implication of this effect is not known. The effects of humidity on the action of CAs on insects have been reviewed.

High CO_2 and Low O_2 Atmosphere

The relative effects of high CO_2 and low O_2 CAs may be judged by the compositions required to reduce the linear growth of various mold species by half.

It will be seen that the different species differ markedly in their response and no atmospheric type is always superior. CAs also affect other aspects of mold growth, contributing to the well-recognized overall inhibition of molding by CAs. For instance, at 23°C/85 per cent RH, the lag phase before growth was extended from 4 to 10 d by subjecting *Aspergillus repens* to 15 per cent CO_2 in air and from 4 to 13 d for *A. candidus* under 1 per cent O_2 in N_2. Under the same conditions, 15 per cent CO_2 in air did not affect sporulation and conidia were produced under 1 per cent O_2 in N_2 by these species. Spore germination can be inhibited by CO_2. At favorable temperature and RH, about one tenth of the spores of *Penicillium martensii* germinated under 60 per cent CO_2, 20 per cent O_2, and 20

Table 21.1: Relative Susceptibility of Internal and External Feeders to High CO_2 and Low O_2 Atmosphere

Species	*Stage*	*Type of Feeder*	*Atmospheres Compared*		*Conditions*	
			Quicker Acting	*Slower Acting*	*Temperature (°C)*	*RH (%)*
Rhyzopertha dominica	E,L,P,A	Internal	60% CO_2 in air	99% CO_2, 0.2% O_2	27	43
Sitophilus oryzae	E,L.P,A	Internal	60% CO_2 in air	99% N_2, 1.0% O_2	20–30	–
Sitophilus zeamais	E,L,P	Internal	60% CO_2 in air	99% CO_2, 0.2% O_2	27	43
Ephestia cautella	E,L	External	99% N2, 1.0% O_2	63% CO_2 in air	27	66
Ephestia cautella	P	External	63% CO_2 in air	99% N_2, 1.0% O_2	27	66
Tribolium castaneum	E	External	63% CO_2 in air	99% N_2, 1.0% O_2	27	66
Tribolium castaneum	L,P,A	External	99% N_2, 1.0% O_2	63% CO_2 in air	27	66
Trogoderma glabrum	L,L(d), P	External	99% CO_2, 0.2% O_2	61% CO_2 in air	32–36	50
Trogoderma variabile	L,P	External	99% CO_2, 0.2% O_2	61% CO_2 in air	36	50
Oryzaephilus surinamensis	L,P,A	External	98% CO_2, 0.4% O_2	62% CO_2 in air	16–32	50

per cent N_2 compared with those in air after 16 h. Presumably, the effect would be more pronounced under limiting RH conditions.

Both high CO_2 and low O_2 atmospheres have been found to be beneficial as storage atmospheres for grains and other durable foodstuffs stored under conditions where molds are active and are the predominant cause of degradation. This is typical for storage humidities of 70 to 85 per cent. Below 70 per cent, molds are dormant and there is usually no advantage in using CAs rather than air storage as far as microbial attack is concerned.

Until recently there was very little reliable comparative laboratory data for mold proliferation under various combinations of N_2, CO_2 and O_2 at <85 per cent RH, with most workers preferring to work at high RH, where mold growth is rapid, but differences are often insignificant. Comparison of observations by different workers is hampered by the range of, strains and species of mold involved and problems of taxonomy and methodology, particularly quantitation of effects. Furthermore, RH, temperature, and atmospheric composition all interact in their influence, particularly under conditions where one or more factors are becoming limiting, adding further confusion. However, generally it is known that the O_2 content of atmosphere must be low, usually < 1.0 per cent and often <0.20 per cent, before growth of some important species of storage fungi is strongly inhibited.

Both high CO_2 and low O_2 CAs affect toxin formation by molds. Generally, in both laboratory and field studies, a reduction in growth of toxogenic molds has been found with exposure to CA compared with air, with the inhibition of molding by the CA usually leading to a proportionally greater reduction in toxin production. Neither type of atmosphere appears to be particularly advantageous in its effect on toxin production.

At equilibrium RH of >85 per cent, if the storage is well sealed and air ingress is thus restricted, the natural respiration of the grain and associated microorganisms quickly reduces the O_2 level in the storage atmosphere to <0.2 per cent. Under such conditions, there appears to be no advantage in creating CAs of either type in the store as molding is already inhibited. At yet higher storage RH, >90 per cent, lactic (anaerobic) fermentation takes place. The *Lactobacillus* spp. responsible are not inhibited by either CA type and again no advantage is gained from their use. However, when there is sufficient air ingress to maintain mold growth, there may be a case for applying either type of CA. However, unless strictly anaerobic conditions are created, yeasts continue to develop with low O_2 CA. The use of high CO_2 though clearly inhibitor to moling, has been found actually to stimulate yeast growth, though the effect may have been indirect by removing competition. Further research is necessary to clarify the action of high CO_2 atmospheres on microorganisms in grain under these conditions.

Generally, storage in either high CO_2 or low O_2 atmospheres produces only slight differences in quality and viability of grain compared with storage in air under the same conditions. At moisture contents (MCs) where molds and other microorganisms are active, the effect of the storage atmosphere on grain itself may be confused with an indirect effect through action on the microorganisms. A further source of confusion arises from practical difficulties in controlling humidity and temperature rigorously under conditions of rapid metabolism. Accumulation of CO_2 in a storage can often accompany development of poor storage conditions resulting from the heat and moisture developed by metabolism. The loss of quality may then be attributed to the CO_2 rather than the general poor storage environment. Statements in the literature that CO_2 is detrimental to grain quality or viability in storage probably can be explained by the influence of other uncontrolled factors, notably humidity and temperature. For instance, CO_2 is said to reduce germination of barley, while Shejbal asserts that under 60 per cent CO_2 negative effects were observed on quality of several grains within weeks or months of storage. None of these have been supported with specific data.

The field and laboratory data on effects of CAs on grain quality have been reviewed. There was a lack of good comparative data for the two types of atmosphere at the time, but this is slowly being remedied. However, it was known then that the effects of enhanced CO_2 were independent of those of lowered O_2, at least for storage of high-moisture grain, though whether the effect is directly on the grain itself or on associated detrimental microorganisms has not been determined. The effect of high CO_2 concentrations in air, *i.e.*, with both raised CO_2 and lowered O_2, was not determined.

Recently, laboratory studies have demonstrated that, under some conditions, the presence of CO_2 (*e.g.*, 60 per cent) in a storage atmosphere with a certain O_2 concentration may sometimes cause a slightly reduced seed longevity compared with N_2-based atmospheres. This is so for wheat held at 60°C. At this temperature, molds are inactive and the effect observed can be expected to be that on the grain itself. CO_2 has a slight detrimental effect at 60 per cent RH on retention of germination of one but not another rice variety and of maize, but not apparently on barley. The detrimental effect of CO_2, at least on rice, maize, and wheat, becomes more pronounced at higher temperatures (*e.g.*, 47 and 60°C) and from the trends observed, may not be present at all below 30°C.

Figure 21.3 indicates the nature of viability of wheat.

Since the viability of all grains at <16 per cent MC so far studied is better retained in N_2-based low O_2 atmospheres than in air, but the effect of high CO_2 atmospheres varies with grain and temperature, and may sometimes be detrimental compared with air, the use of CO_2-free low O_2 atmospheres is to be preferred if preservation of

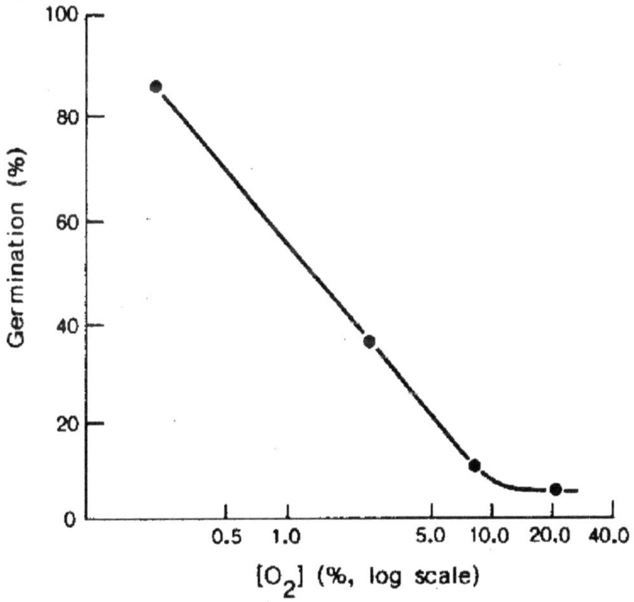

Figure 21.3: Viability of Wheat, 18 per cent MC, Stored Under Various O_2 Concentration in Nitrogen for 16 d at 30°C

germination is of primary importance, unless there are specific data on the effect of CO_2 under the particular conditions and with the grain concerned. Note that very anoxic atmospheres (<0.01 per cent O_2) do apparently kill actively respiring wheat grains (>16 per cent MC) and 0.2 per cent O_2 in the storage atmosphere can be slightly less effective than 2.0 per cent O_2 in preserving germination of maize or barley. However, atmospheres of 0.2 per cent O_2 or less are not usually achievable in practice because of air ingress by leakage and permeation.

Control of Gas Leakage

CO_2-rich atmospheres are substantially more dense than air under the same conditions of temperature and pressure. For instance, a typically high CO_2 CA, 60 per cent CO_2 in air, is 1.3 times as dense as air. When such an atmosphere is made in a storage, the density results in a hydrostatic pressure difference across the enclosure walls proportional to the height of the storage and concentration of CO_2 in the CA. This phenomenon, referred to as the "chimney" or "stack" effect, also results when there are differences in density caused by different internal and external temperatures or differences in both comparison and temperature. The hydrostatic pressure of the column of CO_2-rich gas causes the CA gases to be lost through leaks at the base of a storage and air to be drawn in through leaks at the top, creating a region of diluted air and possible ineffective CA. This "falling out" of high CO_2 atmospheres has been observed in tall silo bins. Though tall structures are particularly prone to "falling out", there is evidence that the stack effect can be the dominant cause of loss of high CO_2 atmospheres even in squat structures, such as sheeted bag stacks.

The most obvious difference, the density of the added gases and target atmospheres relative to air, results in several major practical differences. The density of high CO_2 atmospheres is substantially greater than air, while those for low O_2 atmospheres are similar or slightly less than air. Primarily, the relative density influences the leakage rate from the enclosure under CA and the efficiency of creating the CA. These factors in turn affect the maintenance rate required for a given composition, costs of the CA, and safety. In conjunction with considerations of how the gases affect the insect pests and other organisms, the physical properties define the composition of input gas acceptable and how best to create and maintain a particular CA in a specific situation. Other differences between high CO_2 and low O_2 atmospheres arise from the chemical reactivity of CO_2, the relative ease with which CO_2 can be generated or O_2 can be removed from air, sorption behavior, current sources, and potential sources of input-gases to create CAs.

Table 21.2: Important Physical Properties of Source Gases for Controlled Atmosphere at 25°C, 101 kPa, Unless Otherwise Stated

	CO_2	N_2	Burner Gas
Molecular weight	44.0	28.0	30.1 (equivalent)
Gas, relative density (air = 1)	1.53	0.97	1.04
Density (kg m^{-3})			
Solid	1560 (–78°C)	–	–
Liquid	–	806 (–196°C)	–
Gas	1.811	1.146	1.23
Specific volume (m^3 t^{-1})	552	872	813
Boiling or sublimation point (°C)	– 78.5	– 196	–
Vapor pressure (MPa)	6.4	–	–
Specific heat (C_p) (KJ kg^{-1} °C^{-1})	0.858	1.037	1.00
Latent heat of vaporization (KJ kg^{-1})	233	199	–
Solubility in water (vol gas/ vol water)	0.76	0.02	–

Low O_2 atmospheres are similar in density to air under the same conditions, so any chimney effect operating with low O_2 CA is largely associated with temperature-induced density differences, not those resulting from different compositions. The total forces concerned are much less than those induced by high CO_2 atmospheres and consequently, the potential for gas loss is less, too. Since the other forces causing gas leakage from storages, wind effects and temperature and barometric pressure variation, are independent of gas composition, total rates of leakage of high CO_2 atmospheres can be expected to be greater, sometimes much greater, than low O_2 atmospheres under otherwise identical conditions (Figure 21.3). Unfortunately, it is not possible to test these theoretically derived conclusions as there have been no full-scale comparative trials using the two types of CA.

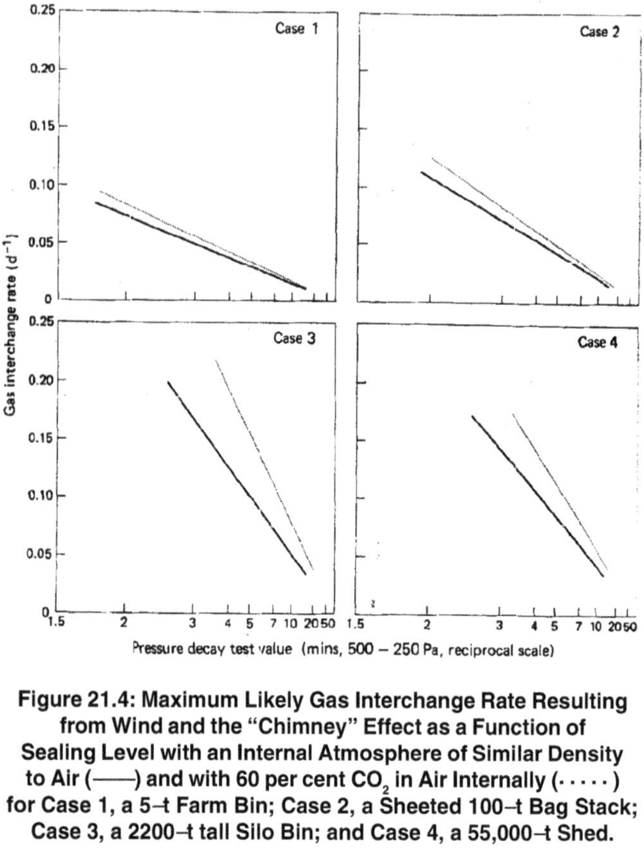

Figure 21.4: Maximum Likely Gas Interchange Rate Resulting from Wind and the "Chimney" Effect as a Function of Sealing Level with an Internal Atmosphere of Similar Density to Air (——) and with 60 per cent CO_2 in Air Internally (· · · · ·) for Case 1, a 5–t Farm Bin; Case 2, a Sheeted 100–t Bag Stack; Case 3, a 2200–t tall Silo Bin; and Case 4, a 55,000–t Shed.

With storage enclosures built of permeable material, particularly plastic membranes, gases may be lost directly through intact, leak-free walls. The rates of permeation vary with the material and are specific to a particular gas. CO_2 tends to permeate more rapidly than O_2 or N_2. This differential permeability and the high CO_2 concentration difference across the membrane can lead to a lowered pressure in very well-sealed systems when high CO_2 atmospheres are used. For instance, when grain is packed under CO_2 in plastic laminates, a substantial decrease in gas pressure occurs. Typically, for wheat at 11 per cent MC, 20°C, about 35 kPa below atmospheric develops within 24 h. This results from a combination of sorption, the major effect, and differential permeability. The sorption of CO_2 by grains can, in theory, produce dangerously reduced pressures in very well-sealed storages leading possibly to structural failure. Reduced pressures attributable to sorption of CO_2 are often observed during treatment of large bag stacks of rice sealed in PVC enclosures. With CO_2-free low O_2 atmospheres, where there is little sorption and only small concentration differences across the membrane, the pressure changes in the enclosure can be expected to be slight.

The degree of sealing of a storage that enables CO_2 to be used in a "one-shot" system is the same as the optimum for use of low O_2 atmospheres where the, unit cost of gas for purging and maintenance is high. This level of gas tightness can be calculated theoretically and has been found to be a good guide in practice.

Creation of CAs

The simplest form of creation of CAs involves the addition of gases to the base of the grain bulk. CAs are usually produced initially by purging the air from the storage enclosure with either CO_2 or a low O_2 gas, N_2, or N_2/CO_2 mixture. These give the high CO_2 and low O_2 CAs, respectively. Exceptionally CO_2 can be used to give the low O_2 type of CA.

Most CAs for grain storage are created at present by adding gases that have been produced off-site. Nitrogen can be supplied as a compressed gas in cylinders or, for larger applications, as a cryogenic liquid in insulated tankers. CO_2 can also be supplied in these ways and additionally as "dry ice". The latter form is particularly convenient as it requires only insulation for storage and handling, avoiding the heavy metal cylinders or specialized transport vessels and potentially dangerous high pressures usually associated with compressed gases and cryogenic liquids. Occasionally, storages to be treated with CA may be close enough to an industrial source of CO_2 or N_2 for direct supply as a gas to be economically feasible. For instance, CO_2 gas is supplied by pipeline to the large grain storage at Kwinana, Western Australia, from the adjacent nickel refinery" where CO_2 is produced as a by product of ammonia manufacture.

Supply of gases for CA creation from off-site sources may often be the most economic system. This will be particularly so for single applications of CA where substantial investment in capital items cannot be justified. However, it would be logical to generate gases for CA on-site, if possible, utilizing air as the source of N_2 or easily transported materials such as propane, charcoal, or agricultural waste as sources of CO_2. This approach potentially avoids the large costs associated with the transport, handling, and supply of pure gases, particularly to distant sites, and may make supply of CA gases independent of organizations outside the commodity storage system using the CA.

On-site generation systems for CA for disinfestation of grain and similar commodities are still under development. Potential methods have been reviewed in detail by Banks and Zanon. Systems generating low O_2 atmospheres by burning propane or natural gas in air currently appear to be the most promising and have been tested experimentally at full scale by a number of researchers. Other systems for generating N_2 from air, notably pressure-swing absorption (PSA) and membrane separation systems, have recently been refined to the level where they can produce from air gas with O_2 content low enough (<0.5 per cent) for use as a source of low O_2 CA, but their high capital cost at present inhibits their use with stored grain. There are as yet no details published of full-scale trials using these systems for disinfestation or preservation of grain. On-site generation of high CO_2 atmospheres for grain storage has not yet been attempted though there are several systems, including the PSA and membrane separation systems, that can enrich gas streams containing about 10 per cent CO_2 to >60 per cent CO_2 and are thus potentially suitable. Such gas streams can be produced by burning carbonaceous material, *e.g.*, straw, coke, and hydrocarbons in air.

The difference in physical properties of the gases producing low O_2 and high CO_2 atmospheres results in several technical differences affecting their efficient application. The low O_2 atmospheres based either on N_2 or use of hydrocarbon burners are always created by adding the atmosphere source as a gas. High CO_2 atmospheres may be created by adding CO_2 in gaseous form, but the CO_2 can also be added as a solid dry ice or as a liquid under pressure. In the latter technique, known as "snow shooting", liquid CO_2 is released into an unobstructed space, such as the head space of a bin, through some form of nozzle. The liquid turns into a mixture of gas and dry ice, as dry ice "snow", as it emerges from the nozzle and pressure is released. The various ways of adding CO_2, compared with N_2-rich gases, provide a flexibility that can be used to advantage in some situations.

The added gases displace the storage atmosphere, which is typically vented through an opening in the roof. This process can be used for creation of both types of CA. Procedures involving addition of CO_2 into the grain stream during filling or into the head space can use substantially more gas than when gas is added to the base of the store. However, this lower efficiency may be acceptable where the latter process is inconvenient, such as with freight containers, or where there is no introduction point already fitted to a store to be treated.

Substantial differences in purging efficiencies can be expected when N_2 or burner gases are used to create CAs compared with CO_2 when applied to the base of a store at the same rates and with the same distribution ductwork. These differences arise from the comparatively large difference in relative density of CO_2 to air compared with that of low O_2 atmospheres. The large difference limits the degree of dispersion of the CO_2 purging front as it moves through the grain bulk and also causes the added CO_2 to spread horizontally through the bulk from the point of injection. The same behavior can be expected with N_2/CO_2 mixtures produced from burners, but to a much lesser extent since these are slightly more dense than air. However, if N_2 is used and it is applied to the base of the store, there will be a tendency, for the gas to rise rather than spread horizontally from the point of application since it is appreciably lighter than normal storage atmospheres. In contrast CO_2 spreads rapidly over the floor before purging the higher regions. Figure 21.5 shows the quantity of N_2 or CO_2 required to purge points at the base of a squat cylindrical bin. The similar values for CO_2 show an even and rapid purging, while the spread of values for N_2, with large values at the bin center, demonstrate a poor distribution over the floor. There may be a case for provision of distribution ductwork when fitting a large, squat storage for low O_2 atmospheres to improve purging efficiency but this is usually unnecessary for high CO_2 atmospheres if the CO_2 can be added at the base of the enclosure. Unfortunately, there have been no comparative trials carried out on the various gases used for CA added under identical

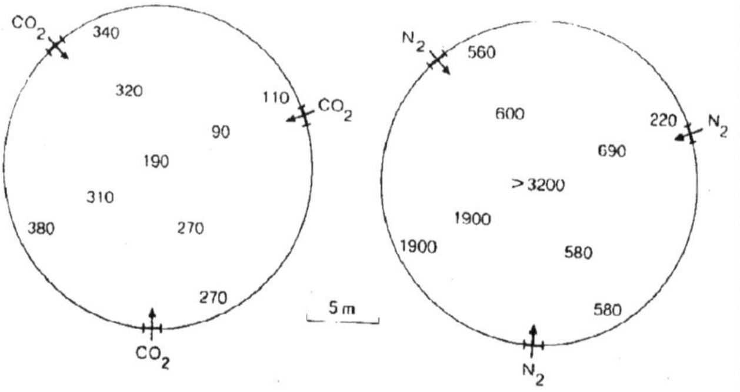

Figure 21.5: Volumes of CO_2, (Left) or N_2, (Right) Required to Reduce the 0_2 Concentration at Floor Level to 2 per cent from Normal Air, Introduced at 3.0 and 1.7 m' min^{-1}, Respectively, at Three Points in the Wall at the Base of a Well-Sealed Bin, Diameter of 21.6 m and Containing 6800 t of Wheat.

conditions of flow rate, temperature, storage configuration, etc. so that a reliable, quantitative measure of the differences in purging efficiencies has not been obtained.

The density difference between CO_2 and air can also be expected to maintain a well-defined interface between the added gas and that being displaced when CO_2 enters an unstructured space such as a head space during purging, whereas with N_2 and, to a lesser extent, bummer gas, free mixing can be expected. Free mixing requires much more gas than plug flow to achieve the same effective change in concentration on purging. To create a 1 per cent O_2 atmosphere from air using pure N_2 theoretically requires addition of only one volume by plug flow, but three volumes by free mixing. This difference together with differences in purging results in lower efficiencies of purging when using N_2 or burner gas rather than CO_2 for CA creation.

Purging of large horizontal storages and bunkers with either N_2 or burner gas has not yet been attempted, but low efficiencies may be expected unless distribution ductwork is used. In contrast, CO_2 added as a gas at a single point is surprisingly efficient at displacing the normal atmosphere even in large structures many

Table 21.3: Efficiency of Purging and Gas Usage for Creation of High CO_2 and Low O_2 Controlled Atmospheres

Source Gas	Target Composition	Efficiency (%) Average (Range)		Number of Trials	Equivalent Usage for Filling Ratio = 0.95		Expected Usage at Average Efficiency Filling Ratio = 0.6 (t/1000 t)
		E_1	E_3		m^3/1000 t	t/1000 t	
CO_2	70% CO_2 in air	74 (70–79)	–	3	0.50 (0.53–0.47)	0.91 (0.96–0.85)	1.4
		74 (62-94)	–	26	0.50 (0.60–0.39)	0.91 (1.09–0.71)	1.4
		44 (38–50)	–	3	0.83 (0.96–0.73)	1.50 (1.74–1.32)	2.3
N_2	1% in air	–	82 (73–92)	8	0.79 (0.89–0.71)	0.91 (1.02–0.81)	2.6
		–	65 (55–79)	9	1.00 (1.18–0.82)	1.15 (1.35–0.94)	3.3
Burner gas	1% in N_2/CO_2 mixture	–	81(74–89)	3	0.80 (0.88–0.73)	–	–

times as long as high. For instance, efficiencies (E_1) of 73 and 43 per cent were found in purging of a 16,000 t shed (height to length, 1.9) and a 10,000 t bunker (height to length, 1 : 13), respectively.

Effective concentration required for high CO_2 CAs results in a requirement of about 1.8 times as much N_2 as CO_2 by volume, in order to create equivalent CA atmospheres in a well-filled and sealed cylindrical bin. The requirement for burner gas appears to be about 1.6 times by volume of that for CO_2. Extrapolation from the efficiency figures for bins on the basis of E_3 suggests requirements for N_2 may be 3.3 times by volume, 2.1 times by weight, those of CO_2 in sheds or partly filled bins.

Usage of CO_2 in One Shot System

The dry grain can be disinfested by applying one shot system. This uses a single application of CO_2 with no subsequent introduction of gas needed to maintain the CO_2 concentration and relies on natural leakage being low enough to maintain adequate concentration levels for the required exposure time. In contrast, at least with dry grain, disinfestation will not normally be achieved using a "one-shot" system with low O_2 atmospheres as the gain of O_2 will be too rapid for the O_2 content to stay low enough for sufficient time. The ability to use a "one-shot" system with CO_2 is a major advantage in practice.

In small, well-sealed and squat storages of less than about 200-t capacity, CO_2 can be used to create high CO_2 atmospheres using the "one-shot" system and no further action is required to achieve disinfestation. Disinfestation with CO_2 of 200 t bag stacks of rice sealed in PVC enclosures is an example of this process now in commercial use. However, if the "one-shot" system is to be used in well-sealed large storages and particularly in tall silo bins, it is necessary to provide some recirculation of the internal gases during the holding period. In tall structures, containing CO_2, the density of the atmosphere tends to cause leakage at the base of the structure with air being drawn in at the top creating a zone of reduced CO_2 content. Often, natural convection currents are insufficient to mix the atmosphere within the structure adequately; to counteract this tendency, an artificial means must be employed to ensure adequate CO_2 concentrations for insect control in the top of the structure.

Recirculation is not required with low O_2 atmospheres. Their densities are close to that of air and composition-related losses from the stack effect are likely to be small and insufficient to cause zones of excessive O_2 content. However, the unavoidable losses expected even in a well-sealed store are such that a low O_2 atmosphere is quickly diluted to ineffective levels. For a typical loss rate constant for a well-sealed enclosure of 0.04 d^{-1}, an atmosphere of 0.2 per cent would be diluted to > 1.0 per cent in <0.8 d if no low O_2 gas is added. This sensitivity of very low O_2 atmospheres to leakage necessitates some form of continuous or frequent additions of gas of very low O_2 content throughout the exposure period to maintain the atmospheric composition. This usually takes the form of a slow addition of N_2, either continuously or on demand from a pressure sensor. Low O_2 atmospheres can be maintained using further additions of burner gases or by recirculation through a catalyst that combines the O_2 from the storage atmosphere with propane. Many of the N_2 generating systems described by Banks have the potential to be adapted to maintain atmospheres.

Both high CO_2 and low O_2 atmospheres require maintenance when long or indefinite exposures of dry grain under CA are needed. Under such conditions, there is no significant O_2 uptake by respiration. For instance, in one system for disinfesting stored wheat in western Australia, a 40 per cent CO_2 atmosphere is maintained over a period of a few weeks using continuous addition of CO_2. Many authors have described holding dry stored products under low O_2 atmospheres for long periods using occasional or continuous input of N_2 from a liquid source or of burner gas to maintain the appropriate atmospheric composition.

There are no published figures on the maintenance rates for strictly comparable full scale situations for high CO_2 and low O_2 atmospheres. The published maintenance rates for low O_2 atmospheres have generally been for well-sealed structures using N_2 as a source of the CA, while the rates for CO_2 have been for much more leaky structures. Although the leakage rate may be greater with CO_2 because of the chimney effect, it can be shown that theoretically about 13 times as much N_2 is needed to counteract a given level of leakage if there is rapid free mixing within the enclosure. In practice, the maintenance gas works partially by diluting the incoming air and partially by displacing it. Since the latter process requires the same quantity of gas irrespective of type, the discrepancy between the requirements for low O_2 and high CO_2 are likely to be much less in practice than in theory, but will vary with the situation and type of enclosure. On balance, it is likely that more N_2 will be required to maintain a low O_2 atmosphere than CO_2 to maintain a high CO_2 one. For costing purposes, we estimate that CO_2 requirements will be between one half and one quarter of the rates for N_2, in the absence of any reliable field data on the subject.

In many applications of CA, the composition of the atmosphere is allowed to return to normal by natural dilution and loss over a period of weeks or months. There may, however, be occasions when it will be necessary to remove the CA rapidly so that the grain may be moved or personnel may enter a store to inspect or handle the grain. Under such conditions, it will usually be possible to remove a low-O_2 atmosphere rapidly so that the O_2 level is above the accepted level for workspaces, 18 per cent, simply by blowing air in rapidly. Under similar conditions it may take much longer to vent a CO_2 atmosphere to give a concentration below the TLV for workspaces for CO_2, 0.5 per cent, because some CO_2 is retained within the grain and is only slowly released. The analogous situation for low O_2 atmospheres, sorption of O_2 to below the TLV, after venting with air, is most improbable.

Both high CO_2 and low O_2 atmospheres can be lethal to humans and can present safety hazards when used poorly. The hazards and physiological effects of the two types of humans are quite distinct. Persons succumbing to either type of atmosphere recover completely if removed quickly from their influence and given resuscitation. The most notable distinction is that the effects of high CO_2 are very unpleasant and persons do not willingly stay exposed even to moderate concentrations of CO_2 (*e.g.*, 5 per cent). On the other hand, CO_2-free, O_2-deficient atmospheres (*e.g.*, 14 per cent O_2 in N_2) have pleasurable effects and unconsciousness can occur without warning. In low O_2 burner gas-based atmospheres, the CO_2 content is inadequate to give a reliable warning of a hazard that may present.

The well-known tendency of high concentrations of CO_2 to accumulate in low-lying and enclosed regions is a potential cause of safety hazards when using high CO_2 atmospheres, particularly in leaky enclosures. Because their density is similar to that of air, low O_2 atmospheres are unlikely to present a similar type of hazard.

Both kinds of atmosphere require free ventilation around the enclosures in which they are used to ensure that there is no external hazard to personnel.

Now let us try to understanding about some of the differences between high CO_2 and low O_2.

Table 21.4

High CO_2	*Low O_2*
Most effective against insects developing within the grain kernel	Most effective against insects developing externally to the grain kernel
Can tolerate substantial dilution before being rendered ineffective against molds and insects	Easily diluted to levels ineffective against molds and insects
Adverse effects on viability and quality sometimes observed	No adverse effects on grain quality
Sorbed on grains and new concrete	Inert to grains and structures
Density greater than air	Density similar to air
Loss rate accentuated in tall structures; requires recirculation	Composition does not increase loss rate from tall structures; does not require recirculation
Does not need ductwork to achieve high purging efficiency	Requires ductwork for efficient purging of large, squat structures
Can be used as a "one-shot" treatment in sealed enclosures	Requires maintenance after purging
Requires gas supply from off-site sources	Requires gases easily generated on site
Unsafe concentrations easily detected by humans.	No warning with concentrations unsafe to humans.

There are many different features of high CO_2 and low O_2 atmospheres that influence the conditions to which they are, particularly suited. In some situations, cost or availability will be an overriding consideration, but in others, a choice can be made as to which type of atmosphere is more appropriate.

Despite much research effort, it is still not possible to provide soundly based distinctions between high CO_2 and low O_2 atmospheres in many of the relevant areas of interest. It has only been within the last 15 years that there has been a recognition that the two types of CA should be compared and that there is often a choice with one atmosphere, distinctly more appropriate in certain circumstances. A consequence of this lack of recognition is that there are no published comparisons of high CO_2 and low O_2 atmospheres under field conditions and the comparative behavior of the atmospheres during both purging and maintenance must thus be inferred from individual studies in different situations. Clearly, a set of well-monitored trials with both CA types under comparable conditions would be most useful, serving to test some of the assertions above.

Figure 21.6 indicates how to choose the most appropriate type of controlled atmosphere.

Control of Yeasts and Molds

A large majority of yeasts and molds require dissolved or gaseous oxygen in order to develop, and strict anaerobic fungi are unknown. On the other hand, microaerophilic and facultative anaerobic behaviors have been definitively demonstrated for several species such as *Fusarium* genera and for different mucorales. However, such an ability for anaerobic growth in these species comes at the expense of greater heterotrophy. In other words, such behaviors are observed in experimental conditions only when the medium contains certain growth factors such as biotine, thiamine, or nicotinic acid which have become essential for such various species as *Mucor rouxii, Aspergillus niger, Penicillium* spp., or *Fusarium* spp., for example.

According to Gunner and Alexander, inorganic ions such as MNO_2, NO_3, and FE^{3+} can, to a certain degree, replace molecular oxygen as terminal acceptor so that a reduction in requirements for these essential biotic factors is possible. In the same way, the biosynthesis of ergosterol, an essential component of the fungal membranes, is blocked by the absence of oxygen, and it becomes necessary to supply *Saccharomyces cerevisiae* with this metabolite in order to obtain growth in anoxia.

If the great majority of fungal species which colonize grains at the moment of harvest show an absolute need for oxygen, it is equally clear that these needs are often rather low and in some cases extremely low. Most laboratory studies show that partial oxygen pressures of about 3 to 5 per cent are generally not limiting. Below

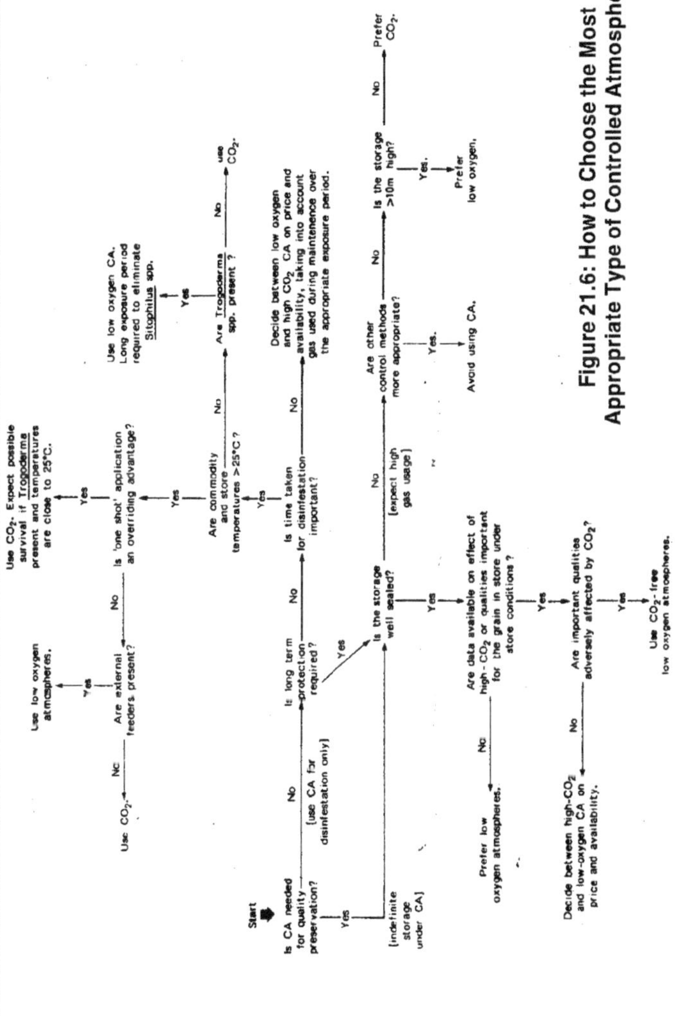

Figure 21.6: How to Choose the Most Appropriate Type of Controlled Atmosphere

1 per cent, the sporulation of most species is greatly decreased, but the most tolerant genera and species seem still capable of growth at less than 0.01 per cent oxygen.

No gaseous composition of oxygen, carbon dioxide, and nitrogen is capable of blocking or even significantly slowing down the fermentative processes of the anaerobic or facultative anaerobic bacteria. The retardations in bacterial growth which have been reported to exist in other fields of food protection seem negligible in the case of grain storage at high humidity. Recent experiments conducted on rice stored under carefully controlled conditions have shown that artificial atmosphere of pure carbon dioxide and strict anoxia, obtained by the natural consumption of intergranular oxygen, allowed the same multiplication of lactic bacteria.

Lack of current available data prevents a general understanding of the effects on molds and yeast growth of a reduction in partial oxygen pressure combined with a lowering of water activity through drying. In their experiments on pure species representative of field mycoflora and storage fungi, Magan and Lacey clearly showed a synergy between the simultaneous reduction of available oxygen and the a_w on mold growth. For *P. roqueforti*, considered particularly tolerant to anoxia, a reduction in partial oxygen pressure to less than 0.14 per cent extends the lag phase up to 4 d at a_w of 0.98 (the highest a_w possible in laboratory c conditions) and to nearly 20 d at a_w of 0.821. Similar effects are observed when the temperature is lowered. Linear growth of *F. culmorum* is in this case reduced by half at 23°C if the residual oxygen is about 0.1 per cent, whereas at 14°C, the same result is obtained with partial oxygen pressure at about 10 per cent.

However, interpretation of these laboratory data is difficult to apply directly to the very complex systems represented by damp grains in airtight conditions or under MAs. They do not, for instance, allow prediction of the effect of a reduction of available oxygen on mixed microbial populations when several species are in competition. In particular, when fermentation occurs on very damp grains, the organic acidity of the grains increases progressively and thus, retards the development of the molds.

In hermetically sealed silos, the damp grains and their microflora always generate more or less heavy concentrations of carbon dioxide, the effect of which on the ecosystem is often difficult to interpret. Carbon dioxide has long been considered an important inhibitor of microorganisms in controlled atmosphere (CA) storage. Although numerous works agree as to an inhibitory effect brought about by atmospheres containing very high concentrations of CO_2, many hygrophilic fungal species are clearly stimulated by partial carbon dioxide pressures of about 10 to 15 per cent and at least some of them are capable of using the CO_2 as a source of carbon, to synthesize tricarboxylic acids and associated amino acids.

In oxygen-free conditions and under strong partial CO_2 pressure, the conidia of molds are progressively destroyed and the rate of destruction increases as the temperature rises. This toxic effect of strong CO_2 concentrations does not seem to be clearly explained and could only be the result of the progressive acidification of the environment. A great practical importance may be found in this slow process of autosterilization since it results in the progressive disappearance of most of the fungal components of the microflora of very damp grains, to the benefit of the yeasts and some very resistant species of fungi.

Seed Viability

The germinative capacity of grains, considered as a sensitive criterion of conservation quality, seems especially difficult to maintain at a high level when water activity goes above values of about 0.85. Indeed, as Christensen and Kaufmann have pointed out, as soon as the moisture of the grain rises to a point where storage molds may grow, the viability of the grain decreases. The faster the molds invade the germ, the faster the germinative capacity diminishes. Experimenting on peas, Fields and King confirm the essential part that molds, play in the loss of viability during storage. Consequently, an oxygen-free atmosphere which prevents the increase of these microorganisms should equally well enable the germinative capacity to be maintained even at high humidity. The protective effects of oxygen-poor atmospheres on seed viability during storage have been confirmed by various other studies.

Wheat stored under nitrogen atmosphere ($O_2 \leq 0.04$ per cent) at 18 per cent MC maintains a high rate of viability (97 per cent) for 2 months, whereas for the same grain in open-air storage, the viability rate falls to only 18 per cent during the same period. However, it seems that at such MC and even when fungal growth is completely inhibited by anoxia, viability cannot be preserved for extended periods. Wheat stored at 17.4 per cent MC under technical nitrogen atmosphere ($O_2 \leq 0.2$ per cent) at mean ambient temperature maintains its viability better than the same grain in the presence of oxygen. This viability decreases from 94 to 70 per cent in 10 weeks and falls to 0 after 32 weeks because of very slow mold development. However, without any doubt, for dry grain (12 to 14.5 per cent MC) the viability and germinative energy are perfectly well maintained for up to several years and possibly improved under very low residual pressure ($O_2 \leq 0.3$ per cent). The presence of CO_2 in more or less high concentrations seems to have only a slight influence on dry or moderately damp grains. The behavior of dry grains (11 to 12 per cent MC) is worth noting here. In the absence of any activity on the part of the microorganisms due to low hydration, Roberts and Abdallah suggest, in the case of dry barley, that the loss of viability in the long term is in fact linked to the presence of intergranular oxygen and that the germinative capacity remains significantly higher under nitrogen atmosphere than in air.

According to Shejbal, wheat stored under technical nitrogen atmosphere ($O_2 \leq 0.2$ per cent) and subjected to a slight constant overpressure of this gas maintains a higher viability than under pure nitrogen ($O_2 \leq 0.01$ per cent). This suggests a process of germ asphyxiation when oxygen is completely lacking and/or when cellular damage is caused by metabolic activity on the part of the embryo itself when the humidity increases. It is not possible from the available published data to be more specific about whether the strictly oxygen-free atmospheres could, by themselves, have a negative effect on the germinative capacity. Although it has long been thought that a certain amount of respiratory activity was necessary for the embryo to maintain its viability, several physiologists believe, on the contrary, that not only is oxygen unnecessary, but that it might in fact be harmful.

The oxygen reaches the embryo through dissolution in the water which is absorbed by the peripheral tissues of the caryopses or kernels. These envelopes, in the case of a great number of seeds and especially cereals, contain easily oxidizable phenolic compounds which act as a trap for oxygen and prevent its access to the embryo. It would seem, therefore, that the grain ensures its longevity by a limitation on residual metabolism which is dependant on the degree of hydration and on the available oxygen. In the absence of any microbial action and depending on the residual hydration, eventual fragmentation of the DNA of the embryonic cells will occur, leading more or less rapidly to the death of the embryo. Recent experiments on wheat at 15 to 33 per cent MC would appear to show, on the contrary, that the oxygen in the intergranular atmosphere maintains a certain metabolic activity at the embryo level, thus providing, to a certain extent, for repairs to cellular damage undergone by the cells because of heightened hydration. Decarboxylasic activity was also studied in rehydrated wheat germ, and could, apart from any microbial intervention, lead to damage to the embryo (loss of germinative energy, abnormal germination).

Taking everything into consideration, the possibility of maintaining the viability of damp grains for long periods appears problematic, regardless of the conditions under which they are stored. It is possible to conclude that from this point of view very oxygen-poor atmospheres are always beneficial to damp grains since they prevent the development of molds which, in open air, would rapidly kill the germs. On the other hand, it is apparent that these storage conditions cannot rival the benefits of drying techniques and cannot be used, for example, for seed production of long-term storage of barley for malting. However, according to Peterson *et al.*, in an emergency situation where drying is not immediately possible, airtight atmospheres may preserve the germinative capacities of the grain very effectively for short periods (1 to 2 weeks for grains at 18 per cent MC). As a general rule, this would permit the use of controlled atmosphere prestorage in acceptable conditions while waiting for drying.

It has been known for a long time that the storage of damp grains in noncontrolled atmospheres rapidly leads to a modification of glucidic fraction under the action of the α- and β-amylases which convert starch into dextrine and maltose. The maltose thus formed is consumed simultaneously by the microbial respiration as soon as moisture content (MC) goes over 15 per cent and oxygen is available in sufficient quantities. The general evolution of the glucidic fraction consequently leads to a loss of dry matter which is progressively transformed into carbon dioxide, water and heat by the microorganisms. The increase in reducing sugars (maltose) due to the action of the amylases of the grain itself or those of the micro-

.organisms is followed immediately by a decrease in nonreducing sugars due to microbial respiration. The works of Glass *et al.* on wheat stored at MCs ranging from 15 to 18 per cent in air or under nitrogen atmosphere ($O_2 < 0.4$ per cent) for 4 months at 30°C show a highly significant accumulation of reducing sugars in the grains kept free of microorganisms. Other results on wheat stored for 2 months at 30°C and at 20 per cent MC under nitrogen atmosphere or under CO_2 confirm this accumulation of reducing and nonreducing sugars when respiration is inhibited.

According to Iwasaki and Tani, the α- and β- amylasic activities in paddy stored for a year at 15 per cent MC are not significantly influenced by the oxygen concentration in the intergranular atmosphere, thus explaining the accumulation of nonreducing sugars where the moisture content of the grain was too low for fermentative microorganisms such as yeasts or lactic bacteria to use them. Consequently, a condition of anoxia for intermediate moisture grains generally leads to a significant reduction of loss of dry matter. This advantage, however, may be offset by the simultaneous appearance of free sugars which favor the fermentation process or which might enter into non-enzymatic browning reactions. These latter aspects have not yet been studied to the best of our knowledge.

The lipidic fraction of grains and seeds is the one which breaks down most rapidly after the harvest, during open air storage. Two principal mechanisms may operate here. The first, direct oxidation from the oxygen contained in air, leads to the formation of typically rancid odors and off-flavors. This direct oxidation is always the main problem in the storage of oilseeds, particularly those seeds which have suffered mechanical damage during the harvest and whose lipids are no longer protected by the natural antioxidant in the seed envelopes. The second mechanism, caused primarily by molds, is the hydrolysis of fats, especially triglycerides. According to Daftary and Pomeranz, on poorly stored wheat in open air, an increase from 10^3 to 2.10^6 molds per gram is accompanied by a decrease in total lipid content of about 40 per cent with considerable losses in triglycerides as well as in phospholipids and glycolipids.

The high lipolitic activity of the xerotolerant molds and the favorable effect of the lipids on the growth of these microorganisms at low thermodynamic water activity are well known. At least during the first stages of alteration, the result of the fungal lipase activity is a considerable increase in the contents of free fatty acids, especially linoleic and linolenic acids in the case of cereals. Measurement of this increase in "fat acidity', in the grains and their products is indeed often used as a sensitive index of grain quality during storage. The elimination of intergranular oxygen should provide good protection for the lipids during storage of grains with intermediate MC by establishing a barrier to both direct oxidation of lipids and their hydrolysis by microbial activity.

This effect is largely confirmed by numerous laboratory and scale experiments on cereal grains as well as on oil seeds. In this way, Poisson *et al.* observed that fat acidity stabilized on mechanically damaged sunflower seeds which had been stored at a_w of 0.75 in completely sealed containers at mean ambient temperature, whereas in open air storage, the level of fat acidity increased about 15-fold in 3 months. On wet rapeseed cakes (15 per cent MC) stored in a well-confined atmosphere, the alteration of the residual lipidic fraction was significantly slowed, but nevertheless doubled in 4 months. The same slow acidification was observed in wheat and maize stored for several months at about 18 per cent MC under technical nitrogen ($O_2 < 0.5$ per cent).

In imperfect anoxic conditions, it is always possible to invoke the processes of autoxidation and/or hydrolysis by grain enzymes of slow-developing microorganisms. Another possible hypothesis is that in which microbial lipases were formed (under the influence of field mycoflora) prior to the appearance of a protective anoxia and maintained acidifying activity even in the absence of any new microbial development.

The quality of the seeds at the beginning of storage is, of course, of first importance in this regard. Oil seeds which were already damaged (*i.e.*, with high initial fat acidity values) prior to their storage in neutral atmosphere, underwent deterioration of lipids which could not subsequently be controlled by anoxia. This may support the hypothesis of continuation of a slow microbial lipase activity during airtight storage. At the present time, no data exist on the exact life span of these enzymes in anoxia on damp grains, nor is there any existing work to be found which might cast light on the influence of the gaseous O_2 and CO_2 environment on lipase activity.

It is nonetheless possible to imagine that this residual acidification would occur with much less intensity in an anaerobic atmosphere than in free air storage. In fact, in the latter situation, the free fatty acids measured at any given moment would only represent the difference between the fatty acids produced by lipase activity and the consumption of fatty acids by growing microorganisms. In anoxia, the free fatty acids are no longer metabolized by molds, and, for the same level of free fatty acids, the lipidic fraction which is broken down is probably lower.

Only a few data exist on biochemical modification for the proteins, peptides, and amino acids of damp grains stored in MAs. As Pomeranz points out, even in open air storage conditions, microorganisms must attain rather advanced levels of development for important modification to be observed in the total nitrogen or protein contents of grain. Alexeeva *et al.* have observed a great stability in the proteins of maize stored for 2 months at about 20 per cent MC and temperatures ranging between 15 and 20°C under anoxia ($O_2 < 1$ per cent). Analogous observations have been made for maize stored at 17.4 per cent MC under nitrogen atmosphere for 8 months.

Cereals are generally considered to be important sources of vitamins such as thiamine, inositol, riboflavin, and vitamin E. According to Young *et al.*, the destruction of vitamin E in damp cereals treated with propionic acid is due more to the dampness of the grain than to the acid treatment. Recent work by Rice *et al.* on damp barley (24 per cent MC) stored in anaerobic jars at mean ambient temperature of 12 to 15°C suggests, on the contrary, that the degree of dampness of the grains was only a secondary factor in the stability of the tocopherols which remained perfectly conserved in strict anoxia. To some extent, these results have been confirmed by other authors and thus favor storage under MA since other recommended treatments for the microbiological stabilization of damp grains, such as propionic acid, may have the opposite effect from that intended and probably lead to a destruction of the vitamin E.

Technological Qualities of Dry Grains

The essential technological qualities of dry grains are conserved under MAs as in open-air storage. In wheat with 10 to 20 per cent MC, the gluten content remains unchanged during storage under nitrogen, but its quality and thus its value to bakers cannot be maintained longer than 2 or 3 months. In general it is considered that milling properties of intermediate moisture wheat do not appreciably change during MA storage, but after 2 or 3 months, the loaf volume obtained decreases significantly. For short periods of storage, only the rheological properties of the dough, as established by alveograph or farinograph, are modified and the gluten becomes "short". Thus, in fact, the functional properties of the proteins are modified during grain storage in humid conditions under MAs apart from any microbial action.

Very little definite information appears to be available concerning protein modification in grain stored in humid conditions. Modification of aggregative properties leading to a decrease of the water soluble fraction are known for dry grains, however, and may result in slightly lowered digestibility.

On damp maize (21 to 22 per cent MC) stored at 18°C in gastight mini silos, Richard-Molard *et al.* have shown a slow modification of the functional properties of proteins without any significant variation in the total content after 1 year of storage. These modifications were made evident by the classic turbidity test commonly used in the starch industry to show up possible alterations in proteins due to drying at too high temperatures. In the case of the storage of damp grains under airtight conditions, no thermal degradations can occur and therefore, structural modifications should perhaps be proposed to explain at least the partial insolubility of the proteins. It is also likely that the increase in acidity, caused by low fermentative activity on damp grains, by residual lipasic activity, or by carbon dioxide which is always produced or added to in these conditions, has a negative effect on the functional properties of proteins. Analogous observations have been reported for other grains, such as soybean, but the question does not seem to have been studied in much detail as yet and further research in this field would be necessary.

The persistence of some enzymatic activities under MAs implies various consequences for the technological, organoleptic, or nutritional qualities of the products. For example, the α-amylase activity of freshly harvested rice is responsible, to a certain extent, for "stickiness" in cooking. So one can assume that storage of wet rice under MA will reinforce this characteristic which may be considered detrimental or advantageous to the quality of the rice depending on the country and the alimentary customs of its consumers.

In most of the experiments on storage under MAs, modifications in the organoleptic properties of the stored grains and their products can be observed from the moment when the moisture content is higher than 16 per cent, that is to say, when a_w is above 0.80. For wheat at about 17 per cent MC stored in confined atmospheres at 22°C (residual oxygen level unspecified), Guilbot and Poisson have shown the appearance of unusual odors after about 1 month and after only 10 d if the MC reached 21 per cent. According to Shejbal *et al.* rice stored in anoxia at over 16.5 per cent MC rapidly takes on a strange off-odor resulting from "anaerobic respiration" which makes it unfit for human consumption.

The possible causes of the appearance of stale, sour, fermented, or rancid odors in stored grains are obviously numerous, but in most laboratory or scale experiments, the precise origin of the responsible volatile components has not been sought. It is therefore extremely difficult, as Banks points out, to know to what degree these processes condemn the very principle of MAs for damp products or, on the contrary, to what extent they must lead to a search for a new means of perfecting the silos in order to limit as much as possible such organoleptic modification.

Since oxidation and respiration mechanisms are inhibited in anoxia, the volatile components responsible for the rancid or moldy odors do not appear, and the fermented or sweet sour smells are the most often described. When conservation does not exceed a few weeks or months, off-odors can often be eliminated by airing, drying, or cooking the products, but for damper grains or those which have been stored for a longer period, the modifications appear to be irreversible, thus rendering the products unfit for classical industrial uses.

Although they have never been completely explained, these processes are obviously of great importance. The question arises of whether the odors are produced by the proper enzymes of the grain itself, which can remain active for a long time in anaerobiosis, or if insufficiently controlled microbiological processes occurring during storage are responsible. According to Taylor, rice in particular is subject to anaerobic respiration, and this idea is examined by other authors. It is equally possible, however, that yeasts can have a very slow fermentative activity which does not correlate with a measurable growth when a_w is particularly low. Recent laboratory experiments on rice at a_w of 0.87 (18 per cent MC) stored at 25°C have shown that yeasts such as *Hyphopichia burtonii*, frequently found on damp cereals stored in airtight conditions, are capable of growth at oxygen levels as low as 2 to 4 µg/g of grain per day, even in the presence of 95 per cent carbon dioxide. Undoubtedly these microorganisms are able to produce the volatile metabolite responsible for the off-odors, even in conditions of such low residual oxygen that it cannot be measured by ordinary means.

A particularly interesting study on this question was carried out by Anthony Raj and collaborators on damp rice (30 per cent MC) stored in different sorts of airtight packaging. Within a few days, the control grains developed peculiar odors, whereas the grains which had previously been subjected to superficial disinfection by immersion in an antibiotic solution kept their organoleptic characteristics intact. It would be interesting and valuable to compare results obtained in completely airtight systems, where volatile products accumulate and impregnate grains made more permeable by the humidity, against those observable on grains stored under nitrogen, *e.g.*, with a permanent flow of gas which would eliminate those volatile components just as they are being formed.

Multiplication of Lactic Bacteria

The multiplication of lactic bacteria was shown to be slow but significant on maize at 21 to 22 per cent MC at 18°C in confined atmosphere, but could not be observed on rice at 18 per cent MC stored in comparable conditions. The limit a_w for the growth of lactic bacteria is most certainly in the neighborhood of a_w 0.90[9] or even a bit lower in ideal laboratory conditions, but it is necessary to be more specific about these data in the case of damp grains under MAs.

Although detailed results are still unavailable, it is highly probable that the species which make up the natural lactic flora of damp grains are close to those known as determinants in green silages. *Leuconostoc, Streptococcus, Pedicoccus,* and *Lactobacillus* are the genera most frequently cited. As mentioned above, laboratory tests have shown that the MAs usually found in grain storage, including strict anoxia or pure carbon dioxide, cannot effectively reduce or prevent the development of lactobacilli (*Lactobacillus brevis,*

L. plantarum). Consequently, it is clear that without any complementary bactericidal treatment, grains stored under MAs at high MC will inevitably undergo lactic fermentation and yeast development and therefore, changes in their organoleptic and technological characteristics.

According to McDonald, since the lactic bacteria found in such ensiled products cannot hydrolyze starch, their development is limited by the low initial level of fermentable sugars present at harvesting. On the other hand, the persistence of amylasic activities in damp grains frees sugars which can be assimilated by lactic bacteria and yeasts and thereby favors the fermentative and acidification processes of the grains. As in the case of green silages, the lactic and, to a lesser extent, the acetic acid thus produced protect the grain from the development of undesirable bacteria such as *Clostridia*, thereby allowing the cereals thus fermented to be used especially as animal feed. It is, however, unclear whether the fermentative processes involved, either by yeasts or lactic bacteria, develop in the same way on grains at 20 to 22 per cent MC or on very damp grains.

According to Hyde and Hakkarainen *et al.* acidifying lactic fermentation would reach a significant level for grains only above 25 per cent humidity, *i.e.*, at a_w of about 0.95, which would effectively correspond to a hydration level no longer limiting to lactic bacteria in general. It is therefore likely that for the dryer grains, between about 19 and 23 per cent MC, in the practical conditions of anoxia most frequently encountered, the dominant process would be the fermentative activity of yeasts, especially alcoholic fermentation. Concordantly, Meiering *et al.* observe only a low acidification of grain at 22.5 per cent MC, whereas its pH rises radically at 28.5 per cent MC. In maize at 21.5 per cent MC, stored in experimental airtight silos where yeasts (*Candida* sp.) gradually became predominant, a low but regular increase in acidity was also noticed during a 1-year storage period. According to Burmeister and Hartmann, the genera and species which develop on damp maize are principally *Hansenula anomala* (Hansen) Sydow and *Candida krusei* (Cast.) Berkout. These yeasts are capable of assimilating lactic acid.

Numerous authors make note of the genus *Candida* found on most of the damp grains and seed grains stored in airtight silos. On humid rice, Teunisson has isolated, in particular, *Endomycopsis chodati* (Nechitch) Wickerham and Burton, also known under the name of *Hyphopichia burtonii* (Boidin) Von Arx and Van der Walt. The imperfect stage, *C. chodati* or *C. variabilis*, is apparently particularly well adapted to the grains stored under very low residual oxygen concentration at humidities of around 18 to 23 per cent. Nevertheless, this very particular microflora of damp grains in airtight silos is less well known than the molds on stored grains in open air atmospheres. Further studies must be undertaken in order to determine their possible pathological effects on the people handling these grains and on the animals which eat them.

The fermentative processes do not seem to introduce major biochemical modifications into the composition of such grains. When MC is up to approximately 25 per cent, dry matter loss is less than 1 per cent, whereas on very damp grains (35 per cent MC or more), the "cost" of fermentations can reach 4 per cent or even more as dry matter losses. As long as anaerobiosis is effective, the modification of the lipidic fraction, estimated by the increase in fatty acids, remains moderate on maize at 28 per cent MC after 2 months of conservation. On the other hand, the level of vitamin E cannot apparently be maintained on barley at 28 per cent MC stored for 10 months in airtight conditions.

As described above for grain at intermediate humidities, a progressive decrease in soluble protein content can be observed. For example, on soybean at 30 to 33 per cent MC, the loss becomes greater as the mean storage temperature is raised. Thus, at 30° C this fraction is diminished to about 30 per cent of its original value, whereas at 5°C, it is practically unchanged. In maize hermetically ensiled at very high MC, microbial proteolysis leads to an important increase of non-protein nitrogen content, whereas a radical decline in essential amino acids such as lysine and arginine can be observed in the soluble nitrogen fraction.

Nevertheless, the consequences in the nutritive value for animals seem relatively negligible; authors admit that the feed efficiency of maize and also of damp barley is very close to that obtained for dry grain, with regard to growth rates as well as for carcass quality. In the case of ensiled wet maize, Aumaitre reports results slightly lower than those obtained with dry grain for pig feed. On chicken feed, however, rather better results were obtained with grains stored for 6 months at 21 to 22 per cent MC. The difficulties in experimentation in the area of livestock feed and the problems of determining with precision the condition of damp grains in so-called airtight silos are well known. The experiments of Livingstone *et al*. are interesting in this regard as they lead to the conclusion that damp grain stored in hermetically sealed polyethylene bags gives less effective storage results than drying the same grain to 14 per cent MC at harvest time. In fact, it appears that during these experiments, the storage construction material did not prevent a certain development of molds. Thus, the importance of strict, well-controlled anoxia can be seen. The general conclusion seems to be that from a nutritional point of view, the damp grains are at least equal to dry grains with regard to the degree that they remain free of molds.

Storage of Wet Grains

The first reported experiments in large-scale damp grain storage were conducted in France at the end of the 19th century. According to the criteria for quality at the time, wheat was successfully stored by Doyere in semiunderground silos at up to 17 per cent MC for several months. Similarly, Muntz stored maize and barley at MC of less than 17 per cent in silos with walls of masonry lined with bolted metal plates. It had already been noted that, at higher MC, some fermentation occurred, resulting in the production of carbon dioxide, but that the grain was nonetheless "acceptable to the horses". Just prior to the World War II, Blanc succeeded in storing wheat for long periods at 16 per cent MC in above-ground welded metal silos without excessive deterioration of its baking qualities.

Beginning in the 1950s, short-term storage of wheat an intermediate humidities (16 to 19 per cent MC) began to be used in the French countryside in airtight metallic grain elevators. These silos, generally made of welded metal and with a capacity of between 100 and 200 tons, were erected in groups of several dozen. Detailed information on the evolution of grain quality and the effectiveness of the water and air tightness is nevertheless not available. Despite the appearance of off-odors in the dampest grains (or those stored for the longest time), these products could be mixed with dry grains, thus giving an acceptable medium baking quality. It is, of course, impossible to know if this grain would pass the standards of today.

Because of climatic conditions during the harvest in Great Britain and North America, as well as the evolution in elevage techniques, the grains used in these countries for animal feed are maize and barley, and research has concentrated on means of storing them in a damp state. Foster *et al.*, in the U.S. stored maize at 18 to 27 per cent MC in silos, holding up to about 1700 bushels, made of vitreous enameled steel sheets ("glass-lined" steel) which were assembled by bolting over sealing mastic and said to be gas tight.

In order to minimize pressure variations on the walls, these silos are equipped with a relief valve preadjusted to levels established by the silo manufacturer which are intended to prevent pressure variations which could jeopardize the gastight seal. However, depending on the case, such a valve might allow the introduction of air or the release of carbon dioxide. Grain was stored perfectly at 18 per cent MC for the six autumn and winter months following the harvest. After a year, however, it showed slight darkening and gave off a sour fermented odor.

Above 23 per cent MC, these modifications became apparent after only 1 month of storage. The grain became difficult to handle with a classical portable grain "auger" and a "white powdery mold" was observed, in a very short time, to cover the grains removed from the silo. These white powdery molds are undoubtedly very close to the yeasts described at the same period by Teunisson on very damp (between 21.5 and 32.4 per cent MC) stored rice which only develop very slowly on drier grains (18 to 20 per cent MC) stored for 7 months "in a glass-lined bin" containing 260 bbl. It should be noted that these yeasts were neither homogeneous nor equally distributed in the mass of damp rice. As a matter of fact, it is at the top of the silo that the microbiological evolution is the most intense probably as a result of easy oxygen penetration via the devices mentioned above (breather bag, valves, etc.). It is also on the upper layers of grain that a rise in MC may be observed, due to the effects of water migration and condensation. The milling qualities of these grains after damp storage were quite acceptable, but the finished products had to be rejected because of unusual color and off-odor.

The experiments carried out in Great Britain on wheat, and especially those conducted on barley, have confirmed the limitations of storage procedures in confined atmospheres for wheat intended for bread-making. On the other hand, the advantages for storage of cereals destined for animal feed were at the same time firmly established. Thus, about 10 tons of wheat stored at 18 to 19 per cent

MC in highly impermeable and airtight welded metal silos took on a sour-sweet odor which could not be dispelled by ventilation or hot-air drying. Technological qualities remain acceptable for 4 to 5 months, but diminish rapidly thereafter. In general, it may be concluded from these silo plant scale experiments that even using state-of-the-art gas tight silos, storage in confined atmosphere is of little use for grain destined for human consumption, except, of course, in emergency situations when prestorage before drying cannot be avoided. An especially interesting observation has been made by these researchers and has been subsequently confirmed by others. If damp grain is stored for a period long enough, for 4 or 5 months at about 18 to 23 per cent MC, the molds are not able to develop rapidly when the grain is taken out into the open air. If this kind of grain is not dried or used immediately after opening, it is rapidly invaded by storage yeasts, but there appears to be no serious risk of mycotoxin formations. The application to farm animal feed is indisputable, and in all the tests on storage of damp grain for animals, the zootechnical results were considered satisfactory even when the silos were opened intermittently to take out grain. Similarly, tests of storage of damp grains hermetically sealed in heavy gauge polyethylene sacks gave very positive results.

It has thus been definitively established that the sine qua non for success in damp grain storage under controlled atmospheres is the establishment of an atmosphere as oxygen-poor as possible. It is now necessary to examine how this condition was fulfilled in the experiments carried out up to the present time.

The Evolution of Gas Concentrations

The influence of temperature is of first importance in the evolution of gas concentrations.

Figure 21.7 indicates the intergranular oxygen consumption, by grain/microorganism ecosystem.

The speed of intergranular oxygen consumption depends very closely on four major factors: the level of water activity, the temperature of the system, the microbiological condition of the grains, and the air tightness of the silo in question. These factors are highly interdependent. Each can influence decisively the microbiological evolution on grains and thus the biological demand for O_2 of the ecosystem at any moment.

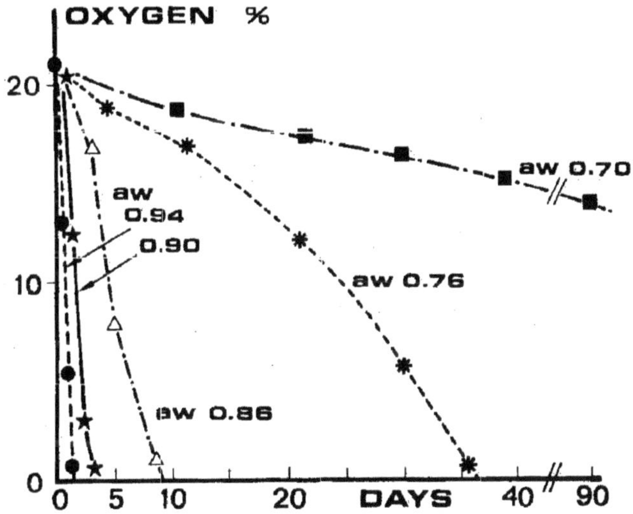

Figure 21.7: The Intergranular Oxygen Consumption, by Grain/Microorganism Ecosystem

It is very easy to imagine the number of very different situations which might be encountered in practice. One might wish to conserve wheat with intermediate MC (17 to 18 per cent) in a temperate zone where the grain would become cold at temperatures below 10°C during the winter months, or, on the contrary, to store very damp rice (22 to 24 per cent MC) harvested in a tropical climate where the average storage temperature would be 30°C. Undoubtedly, this diversity of possible situations, made even more variable by the differences, in the length of storage desired, has prevented the establishment of simple tables which, for a given MC and a known average temperature, would give the aptitude of a certain quantity of grain to maintain a protective anoxia.

When dealing with structures of a large size, it is extremely difficult to obtain and maintain complete anoxia in practice. The best results for capacities up to 200 or 300 tons have been obtained using silos made of bolted metal plates sealed with mastic. To avoid corrosion by acid byproducts of fermentation, galvanized metal or, more frequently, metal treated with epoxy resins or with a vitreous-enamel finish is used. Silos of this type are today in common use in Europe and North America for storing very damp grains. They are, however, never completely airtight as they must be equipped with pressure valves and grain-unloading systems. In France, these silos are used particularly for storing very damp maize (MC $\geq$ 35 per cent) and are often equipped with special sweep-arm systems for grain removal when there is a danger of "bridging".

Welded sheet-metal silos could prove to be even more effective, especially when furnished with seam quality control techniques such as X-ray scanners. However, until now, the cost of this technique has been prohibitive and it is only used in experiments involving a few cubic meters of damp grain. It must be pointed out that, if a silo, where an applied excess pressure of 500 Pa in full storage drops to 250 Pa within 5 min, can be regarded as satisfactory for airtight storage and fumigation of dry grains, its gas tightness is certainly inadequate for damp grain storage. For this purpose, molded fiberglass silos would undoubtedly prove very useful, but, to the author's knowledge, there are no reports published on the matter of very damp grains stored in large quantities in such structures.

The use of flexible bag silos made of plastic film (butyl rubber, polyvinylchloride, polyethylene, or more complex films) would appear to be restricted to smaller capacity bins (10 to 50 tons) and to require a frame of metal or masonry to avoid simply "piled-up" storage conditions. Good results have been obtained in this type of silo for grains having different MC, as has been the case with storage in sacks. However, although this solution is not initially expensive, the plastic films do not last very long. In addition, they are quite fragile and risk being accidentally pierced during loading or by birds, rodents, or hunters' bullets during storage. This disadvantage could be overcome by the inclusion of the flexible bag silo in a sheet metal container rather than in a cylinder of spot-welded mesh.

Another important consideration in the use of this type of silo is the permeability of the plastic used. As in the case of welded or bolted silos, the biological demand for oxygen increases very rapidly with the MC of the grain and the partial oxygen pressure in the silo is always very low (less than 0.5 per cent). This measurement, however, does not represent the amount of oxygen which has actually penetrated the silo due to very slight leakage and the porosity or the permeability to oxygen of the materials used. In recent experiments on maize stored at intermediate humidity (20 to 22 per cent MC), the use of silos with a capacity of 8 m^3, made of a complex metallized plastic film resulted in satisfactory conservation for a whole year. Film permeability to oxygen was about 43 $cm^3/m^2/d$ at 20°C. Very slow development of the storage yeasts *H. burtonii, Candida* spp., and *Aureobasidium pullulans* was observed in the grain mass of the silo.

These microorganisms at a thickness of several centimeters could be seen on the grains located in close proximity to the plastic film all around the silo, *i.e.*, in the zone where a very low amount of oxygen was permanently available. From the first days of the test, measurements carried out using a magnetodynamic O_2 meter, and taking into account the concentration of CO_2, showed a result of residual O_2 below 0.1 per cent. For very damp grains due to the high microbiological demand for O_2, it is nearly impossible to measure precisely the low quantities of oxygen which effectively penetrate into the interior of the silo. Taking into consideration the different biological demands for oxygen involved, it is clear that measuring a residual oxygen pressure of 0.5 or 1 per cent in a system at a_w of 0.95 has a completely different significance, compared to the same measurement on grain with a_w below 0.80. In the second situation, the ecosystem has extremely low oxygen requirements and the silo is certainly very airtight. In the first case with very damp grain, at least during the first months of storage, the ability of the ecosystem to utilize oxygen remains very high and the airtightness of the silo is therefore questionable and could result in significant losses in dry matter, possible development of microaerophilic molds, discoloration of the grains, etc. For this reason, results are often difficult to compare between different experiments and it is necessary to be circumspect when talking about "airtight" silos.

The respiratory processes which are, in theory, based on the exchange of 1 mol of oxygen for 1 mol of carbon dioxide cannot be the cause of important variations of the internal pressure in hermetic silos. On the other hand, the processes of fermentation, particularly alcoholic fermentation, produce excess carbon dioxide, and for grain at 22 to 24 per cent MC, the carbon dioxide concentrations in the intergranular atmosphere can rise to over 90 per cent, thereby leading to a considerable increase in total interior pressure. According to

Hyde, such increase could temporarily reach several atmospheres, with grain at above 23 per cent MC. Nevertheless, according to the author's experience, increases in internal pressure in silos of various types filled with wet maize (20 to 30 per cent MC) were observed only during the few days following ensilage and at mean ambient temperature they never exceeded a few fractions of an atmosphere.

One phenomenon which stands in opposition to excess pressure of CO_2 is the absorption of a considerable part of the carbon dioxide produced by the grains themselves. This phenomenon was studied in detail by Yamamoto and Mitsuda, who stated that a grain can absorb or desorb from 0.3 to 0.5 ml of CO_2 per gram, depending on its humidity. This process explains as well the collapse of the plastic films onto the grains, which always occurs at the outset of storage of grains under artificial atmospheres of CO_2.

Daily or seasonal temperature fluctuations naturally bring about variations of pressure within the silos. Even though the silos are equipped with more or less elaborate regulating equipment such as breather bags or valves, it is always useful to minimize thermal influences by storing in sheltered silos or by applying special paints or by using solar reflective material. The ideal conditions would probably consist of a double-walled silo designed to reduce pressure and thermal variations at the same time, but the cost of such an installation is still very high. Another beneficial result of the use of such silos would be the reduction of moisture migration and water condensation which cause increased local dampness detrimental to the stored grains. The phenomena of heat transfer and moisture migration in airtight silos described by some authors, but not by others, remain poorly defined and consequently difficult to manage and overcome. In the author's experience, the choice of geometrical form as well as the airtightness of the silo has a role to play. In an experiment lasting several months on a pyramid-shaped plastic silo having a capacity of about 10 m^3 of maize at 21 to 22 per cent MC, considerable moisture migration and recondensation occurred, as a result of rodents piercing the impermeable envelope, thus inducing drying in some parts of the bulk and moisture migration in others. The upper part of the grain mass, which was subject to condensation due to the particular shape of the silo, showed a MC of above 40 per cent, whereas the southern side exposed to the ray of the sun dried out and showed less than 15 per cent MC.

Modifications of Gas Concentrations

When the MC of the stored grains and seed is too low for rapid auto consumption of intergranular oxygen by presumably nonpathogenic and nontoxigenic microorganisms, it becomes not only worthwhile but possibly indispensable to create and maintain an artificial anoxic atmosphere. The use, on an industrial scale, of an atmosphere rich in carbon dioxide for controlling insects on dry grains is today more and more widespread. Large-scale experiments utilizing CO_2 on damp grains are, on the contrary, numerous and aside from the beneficial effect of chilling the grain when it is flushed with compressed gas, no conclusive results seem to have been obtained in this field. The same is true for the use of propane gas, which always contains 1 to 2 per cent residual oxygen, not to mention the impurities resulting from imperfect combustion which can accumulate on the grains.

The best large-scale results in the use of gases for storage are those obtained in Italy, using nitrogen, for wheat, barley for malting, and oil seeds. The silos used were constructed of fiberglass, metal, welded sheet metal, or sheet metal crimped using the Lipp process, leading to a helicoidal construction, with welded bottom and roof.

In general the following procedures are observed once the silo is filled: purging of the intergranular atmosphere with pure or technical nitrogen followed by light flushing of the silo during the whole period of storage. Maintaining a slight excess of pressure assures anaerobiosis even in the case of possible air leakage into the silo. Expert technicians should, of course, be responsible for calculating and installing the gas distribution systems in accordance with the characteristics of the storage units, taking care to optimize their use in each given geographical or climatic situation. The nitrogen used for the initial filling and the continuous flushing may be supplied in the form of pure liquid gas, but may also be produced on site by cryogenic separation of the air or by separation through molecular sieves. The choice must obviously depend not only on local economic and technical conditions, but also on the degree of purity of nitrogen to be used. Technical, nitrogen may contain up to 0.2 per cent oxygen, part of which becomes introduced continuously during flushing. Although the development of fungi will be significantly inhibited compared to open-air storage, it

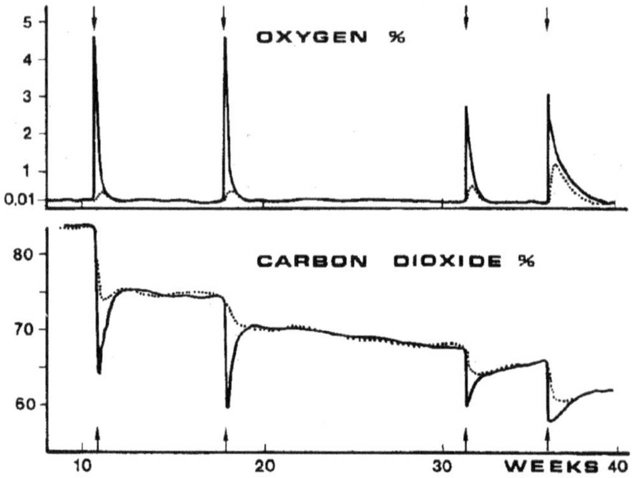

Figure 21.8: Modifications of Gas Concentrations at the Top (-) and the Bottom (—) of a Hermetic 10-ton bin of Wheat (23 per cent MC) after Successive Openings (35 to 40 min each)

will not be completely inhibited in these conditions. This is all the, more true with damp grains (above 19 to 20 per cent MC), where the development of yeast is unavoidable. Pure nitrogen ($O_2 \leq 0.01$ per cent), however, can block the microbial activities on grains having intermediate MC, but the cost is higher.

The most remarkable aspect of this critical review of present possibilities and applications of storage of damp grain or grain at Intermediate MC under inert atmospheres is the fact that so little work has yet been devoted to these questions, compared to other areas of application of controlled atmospheres. The limits of these procedures appear, nonetheless, to be nearly completely defined and it is possible to summarize them as follows: complete inhibition of xerotolerant and hygrophilic molds may be obtained, provided that as complete anoxia as possible (partial residual oxygen pressure not higher than 0.05 per cent) is secured. In this way, the problem of the contamination of the grains by mycotoxins present on the grains after the harvest can be overcome. For very damp ensiled products intended for animal feed, these methods are equal to conservation by chemical agents such as ammonia, SO_2, or propionic acid.

For grains of intermediate MC, even imperfect anoxia ($O_2 < 1$ per cent) allows significant slowing of the process of alteration by microbial action in comparison with open-air storage, but cannot rival classic drying techniques for long-term storage. Numerous technological, qualities, such as viability, functional properties of the proteins, etc., are rapidly altered by high levels of hydration, probably even in total absence of microbial activity. The use of anaerobic atmospheres can only be justified, therefore, for short-term storage and especially in emergency situations or for prestorage while waiting for the drying processes to become available. For these grains at intermediate MC ($a_w = 0.90$), the biological demand for oxygen, essentially due to molds, is relatively very low. The question, therefore, is raised as to the value of insisting on an artificial anoxic atmosphere through nitrogen flushing as soon as the silo is sealed or, on the other hand, of waiting for auto consumption of the intergranular atmosphere. With strictly airtight impermeable systems, it is almost certain that the auto consumption of the intergranular oxygen would have very few measurable consequences on the product, no matter how long it takes to reach a state of complete anoxia. In practice, silos never offer much complete gas tightness so that artificial anoxia obtained by a permanent slow flow of nitrogen may be the best technical solution to protect products of high value. This could be especially true for oil seed which are particularly sensitive to oxidation.

As a matter of fact, today's most widespread application of MAs for wet grains is that of farm storage of very damp cereal for fodder. The biological demand for oxygen in these grains is always very high, and storage in airtight silos under simply confined atmospheres appears to be an excellent solution from both a technical and economic point of view. The air and water tightness of the silo remain indispensable, and in whatever type of silo is used, the measured levels of residual oxygen should never exceed 0.01 to 0.02 per cent. The aerophilic microorganisms whose development is halted in anoxia will certainly develop in silos having even very slight leakage or where the plastic film is too permeable, and molds will begin progressively to develop and possibly produce mycotoxins as well. Even intense development of storage yeasts does not seem to present a problem with animal feeding, but a better knowledge of these microorganisms is absolutely necessary and further research should be rapidly undertaken in this area.

Chapter 22

The Safe Storage of Grain

Introduction

The safe storage of grain has been a common object throughout the world. However, there is still no established technique for the safe preservation of post-harvest grain for use worldwide.

To protect grain from insects and rodents, chemicals such as fumigants and insecticides have been used. However, the moisture content (MC) and the temperature of the grain have been considered to be the most important factors influencing grain quality. Thus, grain is usually dried to a MC below 15 per cent and stored either under refrigeration or aerated with naturally cold air. Both factors also affect the development of fungi and insects.

An alternative method, hermetic preservation, works on the basic principle of the restriction of the oxygen supply by placing the grain in an airtight container. This technique for preserving the grain itself is by no means a new one and many pioneering experiments in such scientific areas as Entomology and the ecology of microflora have been conducted in various countries. Furthermore, recent developments in the use of inert gases open up new and interesting possibilities.

Classically, hermetic conditions have been produced simply by making the grain container airtight. The respiration of any insect or microflora that may be present, together with the respiration of grain embryos, uses up the oxygen and produces carbon dioxide. Active studies on airtight storage have been done since the 1950s in the U.K. In the Pest Infestation Control Laboratory of the Ministry of Agriculture, Fisheries, and Food, Oxley, Hyde, Burrel and other scientists carried out many experiments on the airtight storage of both dry and very moist grains. The structures used for the airtight storage of dry grain ranged from relatively small pits to large 1000-t units made of airtight concrete or butyl rubber bags supported by metal mesh and other flexible materials. They also conducted many storage trials in tropical countries. They came to the conclusion that airtight storage is a useful method of preserving dry grain in a good condition, preventing infestation by insects and killing any insect initially present when the grain was stored without the use of toxic chemicals. The basic properties of the grain were unaffected and it could be used for milling and baking. However, in this method, even if all the oxygen initially present was used up, the carbon dioxide concentration would not exceed about 18 per cent. Moreover, a loss in grain quality occurs inevitably during the respiration step and a number of grains are damaged by the insects before they are killed.

In addition to the simple airtight method the oxygen-consuming process has, on occasion, been hastened by the burning of candles or straw in the space above the grain. In the People's Republic of China, this oxygen-consuming process has been used successfully.

Most grain-exporting nations have so far relied on chemicals to meet the strict requirements for export. In Australia, malathion, an organophophorous compound, has been used since the early 1960s. However, residual effects and resistance problems have prompted investigations on the possibility of producing lethal atmospheres by other means. Such means include the purging of silos by nitrogen or carbon dioxide gas from a cylinder or the use of oxygen-free atmospheres produced by burners or catalytic generators. In the 1970s, Bailey and the scientists at the Stored Grain Research Laboratory of the Commonwealth Scientific and Industrial Research Organization (CSIRO) in Australia, began investigations on the controlled atmosphere (CA) storage of cereal grains, although many studies had already been done on simple airtight storage. In many subsequent storage experiments, they continued to examine the most effective composition of gases for grain storage.

In Italy, Shejbal *et al.* in ASSORENI (Association of ENI Companies for Scientific Research) have been working for a number years on the use of nitrogen for grain stroage. Several experiments were carried out using minisilos, followed by 25-t capacity silos and then by 500-t metal silos for wheat and barley. They developed the hermetic storage method using nitrogen on a commercial scale.

In Japan, grain storage has been a major concern since the cultivation of rice began. Traditionally, rice has rarely been stored or handled in bulk amounts. Paddy or brown rice has been stored and transported in small bags. The packaging unit has been either 60 or 30 kg. Storage of polished rice has been avoided because it deteriorates rapidly. Although Kondo and Okamura have reported a few pioneering works on cereal preservation, few scientific studies on cereal preservation under hermetic conditions have been reported. Packaging materials, hitherto used for cereals, have been regarded only as a means of protecting the contents during transportation and a sampling spear has often been inserted into such packaging materials as straw and jute bags for the periodical inspection of the grain. Little attention has been paid to hermetic preservation. In addition, Japan is very humid during the summer and the quality of the rice deteriorates rapidly during its storage in ambient temperature warehouses. Temperature- and humidity-controlled warehouses, which are currently used to preserve the quality of freshly harvested rice, constitute less than 20 per cent of the government-designated warehouses. The storage and distribution system, controlled by the government, is well established for the annual consumption of rice. Post-harvest losses are assumed to be kept at an acceptable level. Development of a stale flavor and the loss of acceptability of surplus rice which had been stored for more than 1 year and which had been fumigated with methyl bromide caused serious social problems at the end of the 1960s. Under these social circumstances, Mitsuda and co-workers at Kyoto University made extensive studies on the hermetic storage of rice grain in a CO_2 gas atmosphere and developed a new skin-packaging technique, the so-called carbon dioxide exchange method (CEM), which is characterized by packaging grain in plastic laminated film bags with carbon dioxide gas. Hermetic storage by the CEM is now being successfully used worldwide on a commercial scale, as a form of CA storage.

Feedback Inhibition

CO_2 is the end product of the oxidation of organic materials in living organisms. Thus, the accumulation of CO_2, beyond a certain critical concentration, slows down the rate of the reactions by which it was formed, a type of control known as "feedback inhibition. For example, inhibition of succinate dehydrogenase is well known in connection with the CO_2 disease of some fruit. This enzyme is one of the members in the tricarboxylic acid cycle which takes place in the mitochondria and consists of a cyclical series of reactions in respiration. Catalase is also inhibited by a higher partial pressure of CO_2. This enzyme is often identified as a good index of viability of the grain. In addition, the activity of ribulose-1,5-diphosphate carboxylase, which is the key enzyme in the Calvin cycle, is controlled by CO_2, in a manner similar to that of hemoglobin in animal blood. The reversible nature of the CO_2 inhibition has been reported in the study on the activity and metabolism of goat sperm, stored in physiological liquid, under various partial pressures of CO_2. These biochemical results indicate the distinctive feature of CO_2 in the metabolism of stored grains.

In addition to control of the storage temperature and the MC of the grain, oxygen should be eliminated for safer storage of grain. Although the oxygen concentration in small pouches of dried foods can be reduced easily by evacuation, or by the use of oxygen scavengers, it does not seem to be practical to apply these techniques to grain storage. During the hermetic storage of grain, any oxygen initially present is exhausted within a few days, while carbon dioxide is produced by the respiration, both of the grain and of contaminating organisms. The rate at which the oxygen is exhausted depends mainly on the MC of the grain. The higher the MC of the grain, the higher the rate of exhaustion. During the period of oxygen exhaustion, however, chemical changes inevitably take place and affect the quality of the grain. Grain storage in an inert gas atmosphere seeks to avoid or reduce the changes which must be caused by the presence of oxygen. Basically, what is involved is the retardation or suppression of the respiration of the grain itself, the insects, microorganisms, etc. Normally, the gas used for this purpose is either carbon dioxide or nitrogen. Both are readily available commercial products.

Another distinctive feature of the action of CO_2 is its sterilization effect on the growth of fungi and other microorganisms. Even when there is a high level of oxygen remaining in the vessel, a higher concentration of CO_2 (more than 80 to 90 per cent) effectively prevents fungal growth, while nitrogen gas cannot do so. This property is favorable to the practical use of CO_2 in the prolonged storage of grain because the risk of leakage of oxygen into a storage vessel cannot be eliminated.

The amount of CO_2 gas adsorbed by grains depends on temperature, being larger at lower temperatures. The rate of adsorption is not affected so much by temperature.

CO_2 gas adsorption depends on the MC of rice as shown in Figure 22.1, where the maximum adsorption of CO_2 gas is measured at MCs-ranging from 0 to 122.3 per cent. Adsorption increases almost linearly in brown rice, but decreases almost linearly in paddy rice as the MC of the grain increases. Hull and defatted brown rice adsorb remarkable amounts of CO_2 gas at the lower MC. The maximum solubility level of CO_2 gas in the water contained in grains, calculated by Bunsen's absorption coefficients, is less than one half of the amount of CO_2 gas adsorbed. This suggests that the CO_2 gas adsorption phenomenon in grains is not merely caused by dissolution of gas into the water.

Many kinds of cereal grains and pulses show a similar adsorption phenomenon. The amount of CO_2 gas adsorbed by each of these grains at 20°C for 3 h. Wheat, corn, peanuts, soybeans, red beans, sesame seeds, and their flours show the same adsorption, phenomenon as rice, but the rate of adsorption and the maximum amount of CO_2 gas adsorbed vary with the kind of grain. Peanuts, soybeans, and sesame seeds are good adsorbents, and wheat, rice, and flour seem to be relatively poor ones. Compared with conventional adsorbents, such as silica gel and activated charcoal, these grains are all poor adsorbents.

Carbohydrates such as sugars, cellulose, and starch in purified form did not adsorb the CO_2 gas. There was no clear relationship

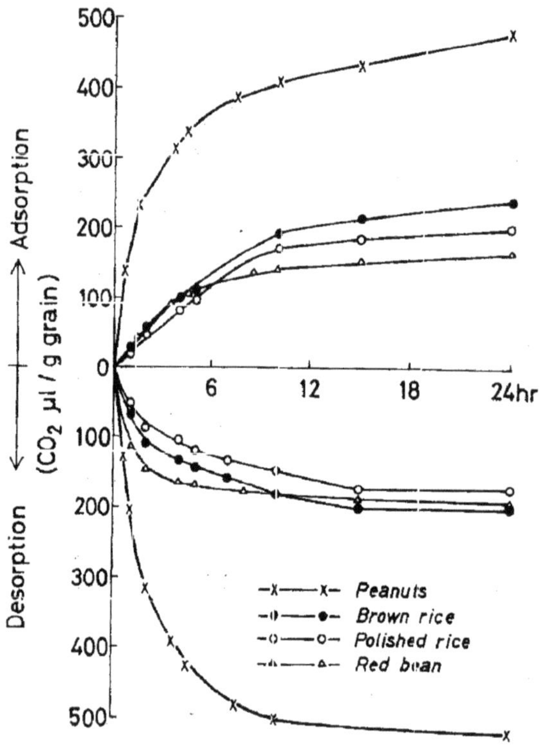

Figure 22.1: Time Course of CO_2, Gas Adsorption and Desorption by the Grains

observed between the protein content of grains and the amount of CO_2 gas adsorbed, although many kinds of pulses of high protein content are good adsorbents.

The pore volume and pore size distribution of the rice kernel in the range of 42 to 354 Å, determined with a high pressure mercury penetration porosimeter, has proved that the porosity of the grain kernel influences the CO_2 gas adsorption phenomenon. Paddy rice has a larger porosity than brown or polished rice. Defatting and/or

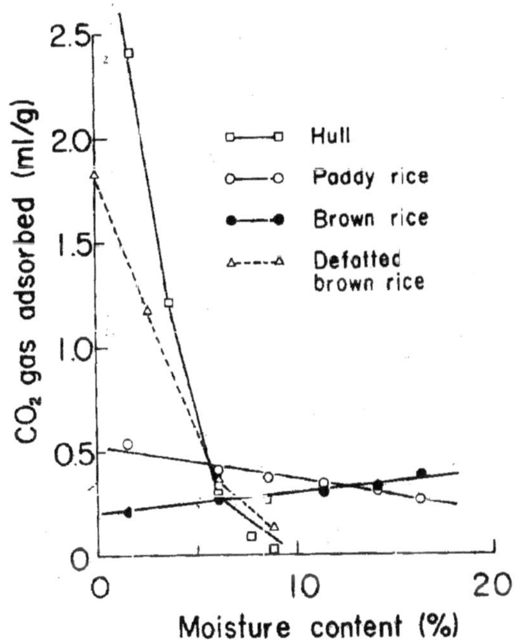

Figure 22.2: Relationship Between the MC and CO_2, Gas Adsorbed by Rice Grain

heat-drying of rice kernel, however, produced abundant pores in the intergranular spaces of their starch.

According to the classic sorption isotherm of Freundlich, the logarithm of pressure (P) and sorption volume (V) give a linear relatiow which may be written:

$$\log V = \log k + n \log P_{CO_2}$$

where k and n are constants. Experimental results obtained from the adsorption of CO_2 by brown rice at various partial pressures of CO_2 are found to agree with the isotherm.

An almost linear relation between log V and log p_{CO_2} is obtained for the higher partial pressure of the CO_2 region (p_{CO_2}), but not for the lower one. The constant n is about 1.5.

Table 22.1: The Amount of CO_2 Gas Absorbed by the Grains

Kind of Grain	*Adsorption Amount at 20°C for 3 h (ml/kg)*
Paddy rice	86
Brown rice	90
Polished rice	70
Rice flour	60
Wheat	75
Wheat flour	60
Corn	170
Peanuts	560
Soybean	440
Soybean flour	216
Red beans	64
Coffee bean	123
Sesame seed	230
Black tea	115

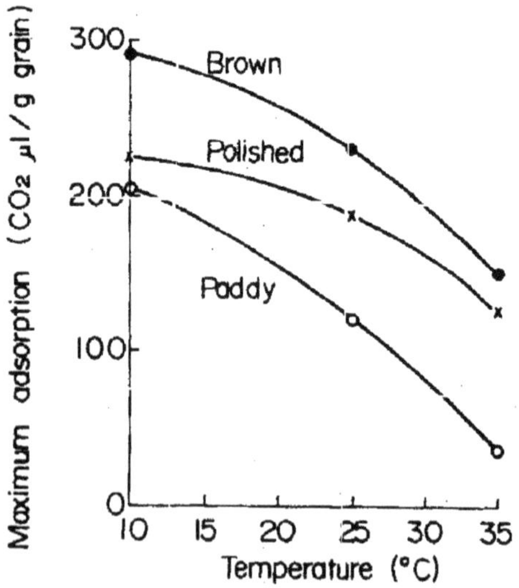

Figure 22.3: Temperature and Maximum Adsorption of CO_2 Gas

Various proteins are found to adsorb CO_2 gas when they are placed in the high partial pressure of the gas concerned. Large amounts of CO_2 gas are also adsorbed by amines and amino acids such as histamine, tyramine, L-lysine, and L-arginine. Temperature dependence, moisture dependence, and reversibility of the adsorption are, however, somewhat different from those of proteins. CO_2 binding sites in protein in the gas-solid phase system are revealed to be ε-amino, α-amino.

$$V = \frac{(C_o - C_e)v}{100 - C_e}$$

V : volume of CO_2 gas adsorbed by brown rice (ml)
C_o: original CO_2 gas composition (%)
C_e: equilibrated CO_2 gas composition (%)
v : original mixed gas volume (ml)

$$\text{Log } V = \log k + n \log P_{CO_2}$$

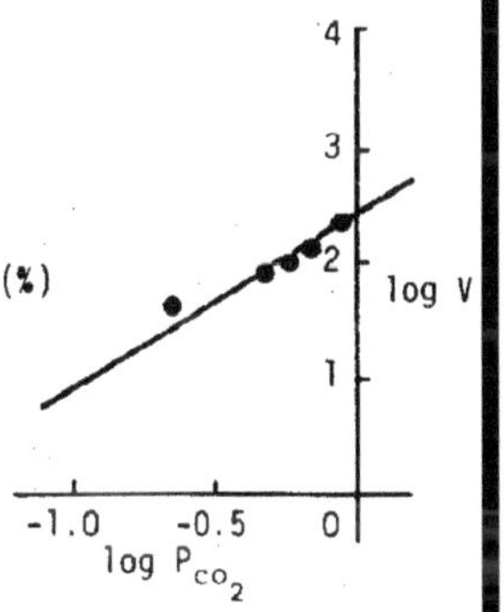

Figure 22.4: Relation Between the Amount of CO_2 Adsorbed and Partial Pressure of CO_2 (at 25°C)

Table 22.2: Comparison of Pore Volume of Rice and Hull

		Hull	Paddy	Brown	Polished
Total pore volume (12 μm~42Å)	(μl/g)	264.1	73.7	31.2	21.3
Micropore volume (354A~42Å)	(μl/g)	1322.3	38.1	15.4	11.0
Pore volume occupied by remained mercury	(μl/g)	223.2	50.7	13.0	7.6
Percentage of micropore volume	(%)	51.6	51.7	49.4	51.6
Percentage of remained mercury	(%)	84.5	68.6	41.7	35.7

Let us try to understand about CO_2 gas adsorbed by proteins in solid form. We can study some examples.

Table 22.3

Proteins	*CO_2 Gas Adsorbed (μ/g/24h)*
Casein	662
Gelatin	532
Egg albumin	267
Glutenine	231
Hemoglobin	182
Zein	150
Soybean flour	220
Whole milk	200
Chlorella algae	163
Yeast	115
Gluten	111
Rice bran	109
Raw silk	690
Fibroin, from silk	110
White rabbit hair	106

Differences between the CO_2 gas adsorption by proteins and by amino acids and amines can be best explained with the two different interactions between the amino group and CO_2. The pKa values of the amino groups in protein are lower than those in amino acids. In the solid state of protein, therefore, amino groups must be present to the unprotonized form (R-NH_2) and form carbamate salts with CO_2. This interaction is weak and reversible as is well known in the interaction of hemoglobin and CO_2 in blood. The characteristics of CO_2 gas adsorption by protein are assured to be caused by the carbamate formation. On the other hand, amino groups in free amino acids and amines seem to be present in the protonized form (R-NH_3^+), even in a solid state, as indicated by their higher pKa values. This amino group in protonized form may react with the bicarbonate ion which is formed from CO_2 and water via carbonic acid.

Table 22.4: Effect of Peptide Chain Length on the CO_2 Gas Adsorption by Amino Group

	pK Values (α-NH_2) (ε-NH_2)	*CO_2/NH_2* (*mmol/mol*)
Gly	9.78	0.024
Gly-Gly	8.25	0.111
Gly-Gly-Gly	8.09	0.465
Gly-Gly-Gly-Gly	—	1.47
Polyglycine	7.4	—
Lys	9.18 10.79	3.97
Lys-Gly	—	4.49
Lys-Glu-Ala	—	22.50
γ-Glu-Cys-Gly	8.75	1.03

Storage of Grains by CEM

Mitsuda *et al.* devised the carbon dioxide exchange method (CEM). The principle, which was based on the behavior of respiratory enzymes in cold-blooded animals in hibernation, is to lower grain respiration with both carbon dioxide and low temperatures in order to retain the original freshness of the grain even after prolonged storage. As shown in Figure 22.5, in the CEM process, cereals or beans are first placed in a bag made of plastic laminated film of low gas permeability which is being charged with carbon dioxide gas and then sealed. The laminated film often used is composed of three layers, such as oriented nylon film which shows a superior physical strength, polyvinylidene chloride film or aluminum foil which is essentially impermeable to gas and water vapor, and polyethylene film which has heat sealability. The package gradually decreases in volume as carbon dioxide gas is adsorbed by the grains and finally the skin-package. is obtained. The shape of the package can be changed, even after sealing, as necessary, and almost all the carbon dioxide gas is released whenever the package is opened.

The state of adhesion of skin-packages thus produced does not change for more than 10 years, even at an ambient temperature. The bulk density of the skin-package is more than 1.0, while that of the conventional package is below 0.9. Thus, the former sinks in water and the latter floats on the surface of water. The composition of

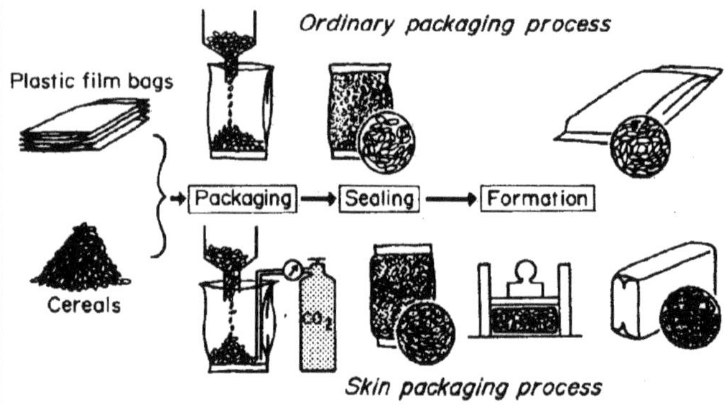

Figure 22.5: Schematic Diagram of Carbon Dioxide Exchange Method (CEM) Skin-packaging Process

gases in the CEM skin-package is about 97 per cent CO_2, 1.4 per cent N_2, 1.0 per cent H_2O, and 0.6 per cent O_2.

With the discovery of the usefulness of the CEM skin-packages for safe storage of rice, the possibility of using CEM to store large amounts of rice in steel cans as a famine reserve has been a serious concern of local governments and governmental offices in Japan. To meet these social requirements, a project team has been organized for the development of a new system of CO_2 gas storage on larger scales with the cooperation of the university, companies, and corporations and has been conducting some basic experiments on grain storage.

Polished rice has been packed in more than 40 drum cans, 200 l in volume, and stored at an ambient temperature (below 30°C) for 2 years in a CO_2 atmosphere. Chemical and sensory evaluation tests of the stored grain have been carried out and the quality of the stored polished rice has proved to be satisfactory for conventional consumer use. The results, using steel cans, have conformed to the above data using the CEM skin-packages.

The characteristics of CO_2 gas adsorption by the canned grain proved to be in satisfactory agreement with the results obtained by Warburg's manometry. These results, provide the basic knowledge for the development of automatic filling machinery and a pressure-controlling method for storage cans.

To get effective purge of intergranular air and maximum adsorption of CO_2 gas, a countercurrent-type filling machinery was newly developed by the Nippon Steel Corporation, the biggest steel company in Japan. This machinery has a filling capacity of 150 to 450 kg of grain per hour into steel cans. It operates steadily and smoothly for the production of grain cans with an adequate pressure of CO_2 gas. It has made it possible, even at present, to start making large-scale famine reserves of polished rice in steel cans and containers below the 1-t unit.

Control of rodents, birds, insects, mites, bacteria, and molds has been the primary concern in grain storage, while changes in grain quality and that of the characteristic nature during food processing have only been a secondary concern. Recently, however, changes in palatability of stored rice grain has become the primary concern in Japan.

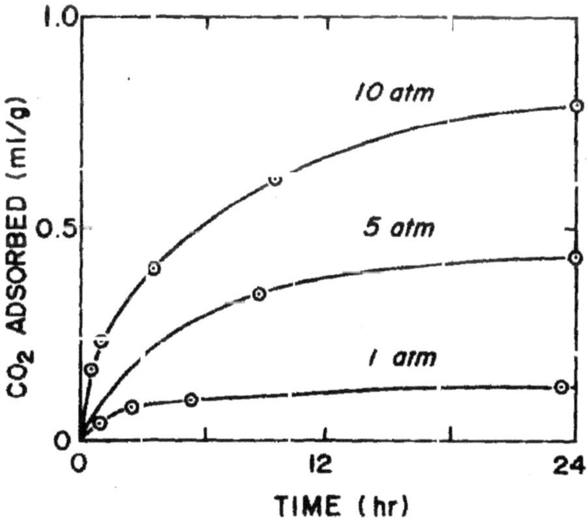

Figure 22.6: Carbon Dioxide Adsorption Rate Under Various Pressures

With the help of Figure 22.7 we can learn the changes in the inner pressures of a test vessel with or without initial high-pressure treatment.

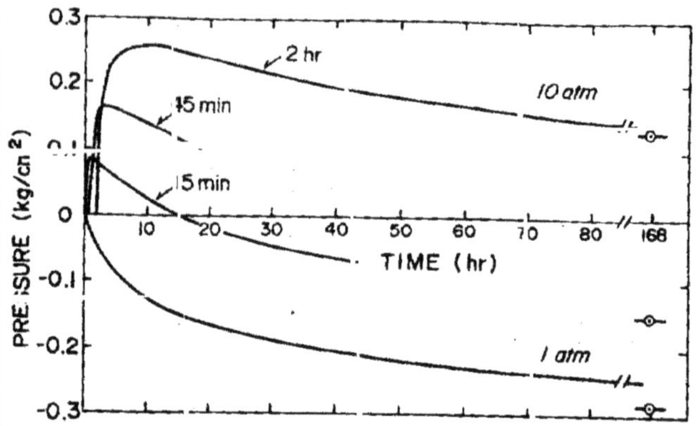

Figure 22.7: The Changes in the Inner Pressures of a Test Vessel with or without Initial High-pressure Treatment

Figure 22.8, curve diagram indicates the adsorption equilibrium in the test vessel and drum can.

Significant reduction in viability is observed in the paddy stored with a higher MC and higher storage temperatures. Compared with the data obtained with nonairtight storage, the rate of reduction in viability was two or four times slower in the high MC rice packed by the CEM. Actual germination ability in the field was in good agreement with the germination capacity in Petri dishes in the laboratory. In addition, the growth rate in the field, various plant characteristics and the frequency of abnormal growth were also examined on the Chiba Foundation Seed and Stock Farm.

The rice grain packed by the CEM had a lower water-soluble activity than that packed by conventional methods. Three strains of rice show a similar tendency. Polyethylene bags were the worst packaging material employed. The changes in free fatty acids content are less in the rice grains stored by the CEM. Changes in the free fatty acid content of rice showed the same tendency as water-soluble acidity. The CEM seemed to be effective in retarding fat degradation.

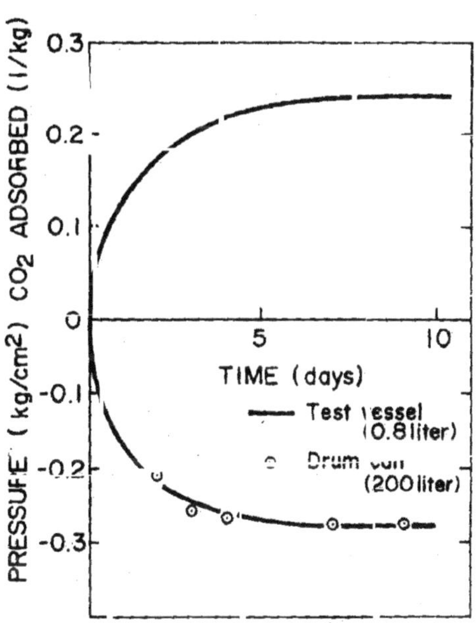

Figure 22.8: Adsorption Equilibrium in the Test Vessel

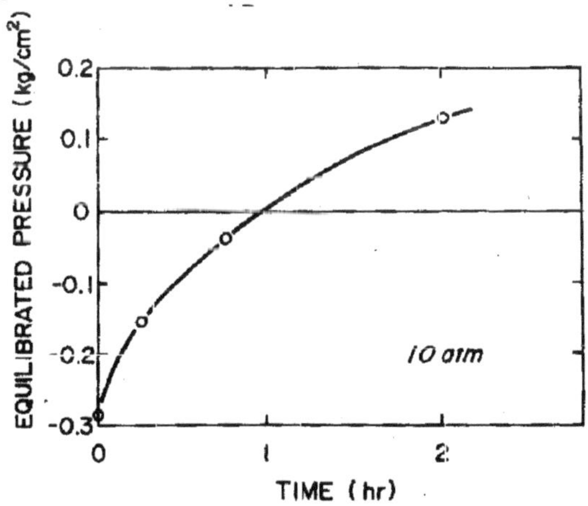

Figure 22.9: The Relationship Between Equilibrated Pressure and the Time

The rice stored under natural conditions contained significant amounts of pentanal and hexanal which are responsible for the stale flavor of cooked rice. On the other hand, the rice stored by the CEM contained smaller quantities of these compounds when comparing their peak heights.

Table 22.5: Growth of Rice Grain Stored for 5 Years in Different Storage Conditions

Moisture of Grain (%)	*Storage Temp. (°C)*	*Length of Stem (cm)*	*Length of Ear (cm)*	*Number of Ear/Stump*	*Weight of Ear (g/Stump)*
14.0	0	694 ± 3.28	122.7 ± 0.96	122.9 ± 3.04	24.5 ± 4.38
11.3	10	71.8 ± 3.01	17.4 ± 0.82	122.6 ± 2.73	25.3 ± 4.02
	0	72.6 ± 2.92	17.6 ± 1.02	122.5 ± 2.71	25.5 ± 4.23
8.7	10	71.4 ± 3.37	17.5 ± 0.96	17.7 ± 2.93	25.6 ± 4.51
	0	71.1 ± 2.86	17.2 ± 0.93	122.3 ± 2.72	24.4 ± 4.17
	Ambient	72.4 ± 3.54	17.2 ± 0.96	18.6 ± 2.95	222.8 ± 4.38
5.9	20	71.8 ± 2.95	17.0 ± 1.02	17.3 ± 2.68	25.1 ± 4.10
	10	69.6 ± 3.15	122.7 ± 0.89	122.8 ± 2.99	24.2 ± 4.82
	0	69.1 ± 3.28	17.1 ± 0.89	122.6 ± 2.85	23.5 ± 4.73
Control*		72.3 ± 3.17	17.4 ± 1.08	122.6 ± 2.52	24.6 ± 3.85

Note: M ± SD (n = 90).

* Freshly harvested grain

The Changes in Stored Wheat

When wheat grains are stored for 3 years in different storage conditions, the changes noticed as in the physical factors have been reported in Table 22.6.

Table 22.6

Storage Condition	*Storage Temp. (°C)*	*Length of Stem (cm)*	*Length of Ear (cm)*
CEM Storage	25	3	8
	15	2	4
	0	4	10
	Ambient	5	4

Contd...

Table 22.6–Contd...

Storage Condition	*Storage Temp. (°C)*	*Length of Stem (cm)*	*Length of Ear (cm)*
Conventional Storage	25	5	9
	15	7	5
	0	10	9
	Ambient	3	3
Control*		6	1

Note: M ± SD (n = 90); CEM: Carbon dioxide exchange method.

* Freshly harvested grain.

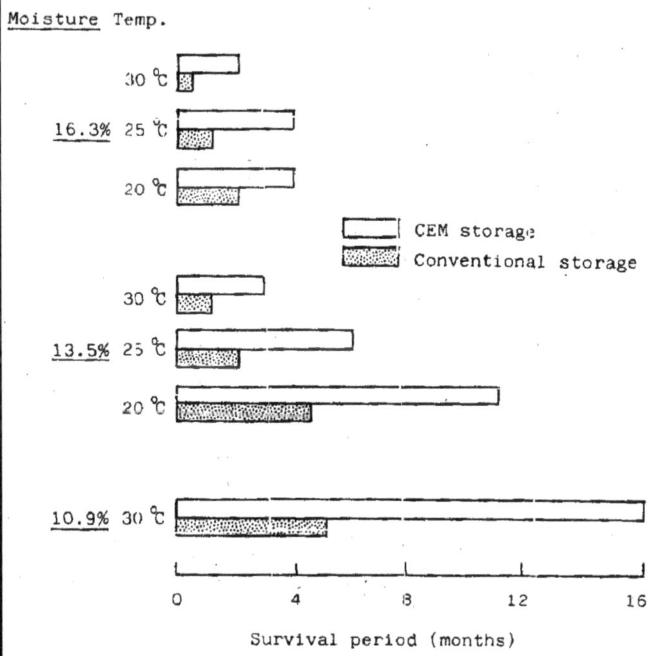

Figure 22.10: Periods for Maintaining the Germination Capacity of Paddy Rice Above 90 per cent

Table 22.7: Germination in Petri Dishes and Percent of Mature Seedling of Stored Wheat After 3 or 5 Years in Different Conditions

Storage			Germination		Mature
Condition	*Temp. (°C)*	*Period (Years)*	*Activity (%)*	*Capacity (%)*	*Seedling (%)*
CEM Storage	25	3	100	100	87
		5	97	98	21
	15	3	100	100	92
		5	96	98	80
	0	3	96	99	90
		5	90	94	85
	Ambient	3	97	98	87
		5	98	98	11
Conventional Storage	25	3	98	99	91
		5	97	97	47
	15	3	99	99	89
		5	96	97	88
	0	3	98	99	92
		5	97	98	82
	Ambient	3	98	98	90
		5	97	98	12

The Packing of Polished Rice

The CEM skin-packaging of polished rice has been accepted by various types of consumers in Japan. For example, many people who live abroad, as well as those engaged in deep sea fishing, have reported the successful retention of flavor in rice stored in the CEM skin-packages. The CEM skin-packages have also made it possible to develop the new type of polished rice which retains the germ portion, "Haigaseimai" (germ rice), which is enriched with various nutrients such as thiamine, riboflavin, pyridoxine nicotinic acid, tocopherol, Ca and Fe, and "Shingen" (vitamin-enriched rice) because the rapid deterioration of these types of polished rice is effectively prevented by this packaging and preservation method.

For the CEM storage of brown rice, which is the most common form for storage and transportation in Japan, there remain some problems. At MCs up to about 15 per cent, brown rice grain can be stored safely in the CEM skin-packages for more than 3 years at temperatures of 15°C or below, which can be obtained in naturally cold places, such as underwater and underground, or in temperature-controlled warehouses. At MCs above 15 per cent, however, anaerobic respiration has often been observed in rice grain stored at temperatures above 30°C. Yanai and Ishitani also reported in their study on the hermetic storage of brown rice that carbon dioxide gas is not considered to exhibit any particularly beneficial effect concerning the preservation of the quality of brown rice. In addition, they reported that specimens, with a high MC, stored in a pressurized carbon dioxide-filled package, displayed a strong reduction in rice quality in comparison with those stored in a pressurized nitrogen-filled or air-containing packages. In contrast, under low temperature conditions, the inert gas-filling storage method was definitely effective in preserving the rice quality as examined by the palatability test. Thus, they concluded that the use of inert gases is considered to make possible long-term storage, even of brown rice with a high MC, by effectively preventing an appreciable reduction in palatability.

Table 22.8: Composition of Volatile Carbonyl Compounds from Cooked Rice* (%)

Carbonyl Compounds	*CEM** Skin-Package*		*Kraft Paper*		*Polyethylene*		*Paper Bag (Husked)****	
	4	*8*	*4*	*8*	*4*	*8*	*4*	*8*
Acetaldehyde	47	54	44	32	38	33	40	28
Propionaldehyde	4	3	5	2	3	1	4	2
Acetone	32	28	23	28	41	44	27	31
Methylethylketone	9	1	3	2	0	2	3	2
Buthylaldehyde	2	1	0	2	0	1	3	1
Valeraldehyde	2	2	8	4	4	3	5	4
Caproaldehyde	4	11	17	30	14	16	18	32

* Prepared from grains stored for 4 and 8 months (Sasanishiki variety).

** CEM, carbon dioxide exchange method.

*** Transported in the husked slate and polished in Kyoto.

The Table 22.9 indicates the Rheological properties of polished rice determined by Amylograph.

Table 22.9: Rheological Properties of Polished Rice Determined by Amylograph

	Sasanishiki		*Toyonishiki*		*Hatsunishiki*	
	August	*April*	*August*	*April*	*August*	*April*
CEM* skin-packages						
Temperature at the start of viscosity increase (°C)	77.8	79.0	77.8	80.0	77.6	79.7
Maximum viscosity (BU)	814	101.5	779	1015	937	1054
Temperature at maximum viscosity (°C)	87.9	88.2	87.1	88.3	822.4	88.5
Minimum viscosity (BU)	397	487	399	483	429	474
Breakdown (BU)	418	528	380	532	508	580
Final viscosity at 50°C (BU)	693	838	696	857	747	792
Consistency (BU)	296	351	310	374	318	318
Kraft paper bags						
Temperature at the start of viscosity increase (°C)	722.9	81.4	78.6	81.0	77.0	81.6
Maximum viscosity (BU)	885	1195	752	1040	941	1070
Temperature at maximum viscosity (°C)	87.6	91.0	87.9	88.8	87.2	88.5
Minimum viscosity (BU)	419	509	389	450	444	456
Breakdown (BU)	467	686	363	590	497	614
Final viscosity at 50°C (BU)	713	820	683	768	779	792
Consistency (BU)	294	311	294	318	336	336
Polyethylene bags						
Temperature at the start of viscosity increase (°C)	77.9	80.2	79.1	82.3	77.9	80.6
Maximum viscosity (BU)	876	933	753	926	941	1070
Temperature at maximum viscosity (°C)	88.2	88.4	88.2	90.5	87.1	89.6
Minimum viscosity (BU)	404	438	389	465	425	486
Breakdown (BU)	472	495	364	461	499	584
Final viscosity at 50°C (BU)	690	740	684	810	745	799
Consistency	286	302	295	345	320	313

Contd...

Table 22.9–Contd...

	Sasanishiki		*Toyonishiki*		*Hatsunishiki*	
	August	*April*	*August*	*April*	*August*	*April*
Kraft paper (husked)						
Temperature at the start of viscosity increase (°C)	722.6	78.6	77.9	80.0	78.3	80.1
Maximum viscosity (BU)	949	1030	804	930	848	1020
Temperature at maximum viscosity (°C)	87.5	87.2	87.1	88.6	87.3	88.6
Minimum viscosity (BU)	423	471	388	460	408	487
Breakdown (BU)	526	559	417	470	440	533
Final viscosity at 50°C (BU)	706	780	681	788	686	820
Consistency (BU)	284	309	294	328	278	333

* CEM: Carbon dioxide exchange method.

Chapter 23

Grain Storage in Developing Countries

Introduction

Over the last 10 years extensive research has been directed, towards the controlled atmosphere grain storage.

When considering the application of controlled atmosphere grain storage within the confinement of scientists' interpretation of controlled atmosphere, *i.e.* "the control of the atmosphere within a grain storage by modification of the concentration of one or more of the naturally occurring gases," the progression into controlled atmosphere must be described as slow. Even though probably the largest single use of the system in the world is practiced at Kwinana Shipping Terminal, in Western Australia the extension of the method to the rest of the facility (total capacity of 950,000 t wheat) is now underway. The International Symposium on Controlled Atmosphere and Fumigation in Grain Storages held in Perth, Western Australia in 1983 covered practical aspects of putting this technology into commercial practice.

Vacuum Cleaning and Sealing Operation

A prerequisite of controlled atmosphere or fumigation is a suitably sealed (airtight) storage, and the major efforts of the Central Bulk Handling Authority have been directed to this area. Selection of suitable sealing materials originated by the Commonwealth Scientific and Industrial Research Organization (CSIRO) Stored Grain Research Laboratory and extended by a National Industry Coordinating Committee on Silo Sealants resulted in a number of materials being known to perform satisfactorily in Australian conditions. This work was published in a document in 1984. Further improvement in materials and their uses has been achieved by cooperation between handling authorities, sealant merchants, and application contractors.

A series of tests applied to candidate sealing systems, designed to aid in selection and acceptance of new materials, has been further refined, particularly in Western Australia. These tests, required before a new material will be considered, are not more than a guide. Results will show if a material will certainly fail, but satisfactory performance under test conditions will not show proof of field reliability. Tests are conducted on samples applied to the substrate by the testing laboratory after drying at ambient conditions for 7 d and aging for 7 d in 50°C.

Requirements of a material differ according to their use. For example, an external membrane on a corrugated iron-clad roof of a horizontal storage must have flexibility and recovery characteristics that are long lasting, yet excessive adhesion strength can create problems. An internal membrane on a vertical concrete silo requires greater adhesion strength, cohesion, and surface hardness to withstand and reduce the frictional drag of moving grain, yet demands on flexibility may be less. An external heat-reflective coating serves to protect the membrane from weathering, to reduce roof expansion and contraction, and to reduce internal bin temperatures by limiting solar and thermal radiation.

Daily internal bin temperature fluctuations have a direct influence on internal and external gas interchange and this should be kept below 5 per cent/d. A reflective topcoat is not necessary in a vertical storage sealed internally for obvious reasons, but a similar coat can be applied to provide a harder surface to the membrane to resist abrasion and reduce the drag of moving grain. Care is needed to maintain the kinematic angle of wall friction at a satisfactory level.

A white heat-reflective coating was observed to reduce aluminium cladding temperature by up to 35°C at Kwinana in Western Australia.

The first approach to sealing a horizontal, warehouse-type storage to a satisfactory level gives the impression of a formidable task. After completing one successfully, this ceases to be the case. In fact, a 25,000-t-wheat capacity storage is now completed in Western Australia in 12 man weeks.

Sealing a 2000-t-wheat capacity vertical concrete silo internally is simplified when problems of access are overcome. This has been achieved by the use of grain as a support and safety measure, together with the development of an access, preparation, and application system, demonstrated in the figures which follow.

Grain is leveled and provides access for vacuum cleaning of accumulated dust above normal grain levels. Forced fresh air circulation is provided. Safety demands strict control over the operation of the grain discharge valve at this time.

An annular platform is passed into the cell roof opening (approximately 500 × 500 mm) in pieces and bolted together inside and connected to a chain bridle feeding through a central top slab hole (75 mm) to a pneumatic hoist frame, also developed specifically for the purpose. A tarpaulin is rigged to protect the grain, to catch dropped objects, and to provide safety to operators. Grain is discharged in 2-m rests as the coating is completed. Worker access is via a specially developed cage and pneumatic hoist system. A manual winch, two-way radio communication, an observer, and fresh air recirculation all contribute to safe working procedures. During membrane application, a recirculation pipe is also fitted to the cell wall alongside the worker cage path to allow for future attention. Applicators are removed from the cell while the grain is being discharged.

Carbon dioxide in gaseous form is piped to the oversilo area and connected via flexible pipe to the internal pipe leading to the cell bottom. Concentration targets, one-shot or maintenance applications, are easily adjusted. Following carbon dioxide introduction, a trolley-mounted fan is connected for recirculation. In a multicell block, such as Kwinana, Western Australia, 144 main cells, 100 interspace cells, 12 trolley fans are sufficient to meet demands and can easily be increased. Each cell is protected against excess pressures by a negative pressure relief valve set to operate at 2000 Pa and by a separate positive pressure relief valve also set at 2000 Pa.

Figure 23.1 explains about vacuum cleaning operation.

The stages of sealing operations are indicated in Figures 23.2 and 23.3.

Control of Grain Pests

Western Australia exports about 90 per cent of grain produced.

Pre-1959 saw relatively little attention to the control of grain pests.

Large-scale use of phosphine and hydrogen cyanide in leaky storages provided short-term relief from insect infestations from 1959 to 1963.

The introduction in 1963 of chemical grain protectants applied to grains as received into the central storage system brought dramatic results:

1. All grains produced in Australia, surplus to feed and seed requirements, are delivered directly from the harvester to central storage systems.
2. The first protectant used was malathion and this lasted 17 years in Western Australia.

The "Export (Grains) Regulations" were introduced in the mid-1960s and included was a nil tolerance for live insects.

Insect resistance to chemical protectants was observed in the early 1970s and became progressively more prevalent.

Late in the 1970s, CSIRO, failed for industry assistance in the transfer of controlled atmosphere technology into commercial practice, particularly in the sealing of storages. The National Coordinating Committee for Silo Sealants was formed to evaluate materials available and to develop sealing methods.

Fenitrothion replaced malathion in Western Australia in 1980.

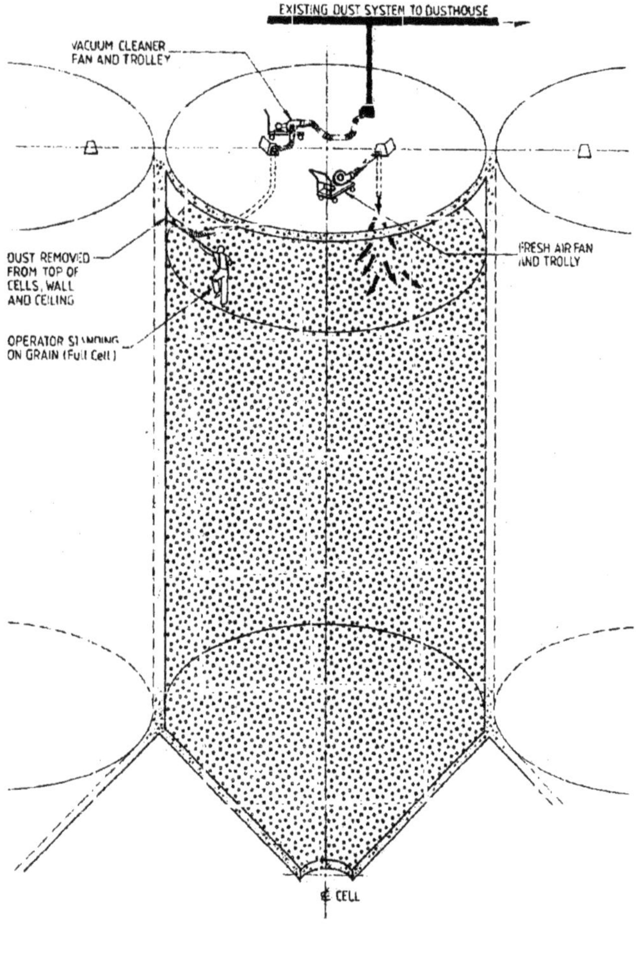

Figure 23.1: Vacuum Cleaning Operation

Grain buyers were showing increasing concern over chemical residues in grain even though these were within WHO/FAO accepted tolerances.

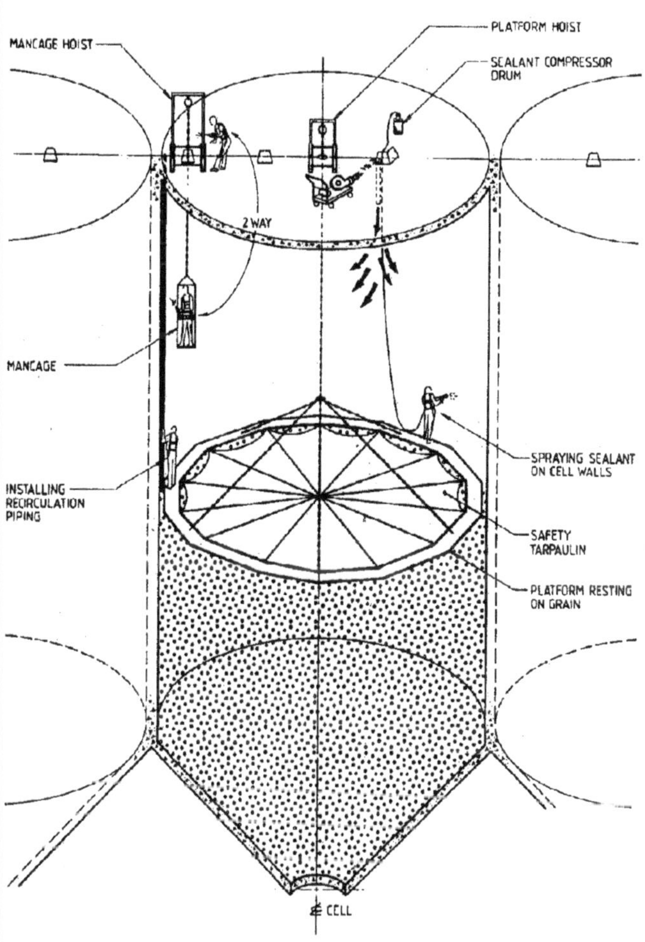

Figure 23.2: The Stages of Sealing Operations

The Board of Directors of Cooperative Bulk Handling Limited approved expenditure, for the sealing of three horizontal storages, each of 25,000-t capacity. The philosophy accepted was that:

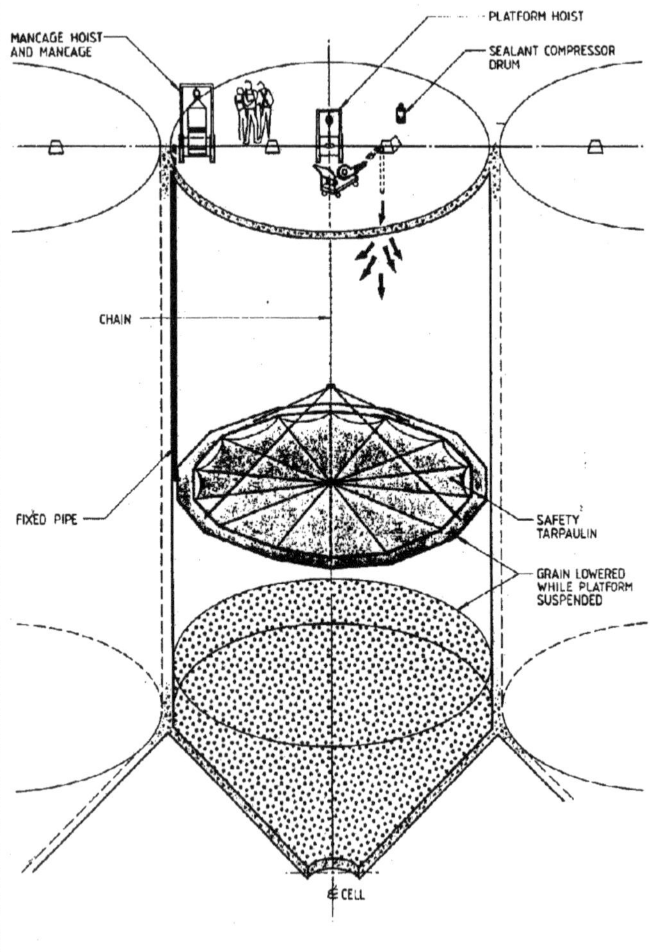

Figure 23.3: The Stages of Sealing Operations

Insect resistance and chemical residues must progressively become more serious.

The only acceptable solution in the long run was to replace residual chemicals with controlled atmosphere or fumigation. There was little differentiation locally at that time between controlled atmosphere and fumigation.

To achieve satisfactory controlled atmosphere or fumigation, storages must be sealed and, although this appeared a formidable task in a 25,000-t corrugated iron clad structure. It was realized that the problems would never be overcome if not faced.

Also, it was necessary to prove materials and methods as well as to train contractors in the art. All of this would be better achieved if not conducted in a situation of urgency, *i.e.* when there were no protectants to change to or when grain with chemical residues could not be sold.

Controlled atmosphere using carbon dioxide and fumigation with phosphine and methyl bromide were demonstrated in 1981 in the three storages sealed the previous year, during both summer and winter, and the results were excellent.

When using residual grain protectants, chemical residues and insect resistance are inevitable.

Controlled atmosphere or fumigation is effective in sealed storages. This does not exclude any other beneficial pest control method from a totally integrated pest control management system. These include hygiene, residual sprays, space sprays, aeration with or without refrigeration, thermal disinfestation,–even the more exotic biologic, radiation, microwave, and so on.

Time was needed to change from full reliance on a chemical system to a controlled environment system.

Included in the criteria considered when the decision was taken to seal all storages was the fact that *Rhyzopertha dominica* (F.) had exhibited resistance to organophosphate insecticides. The acceptable alternative was a synthetic pyrethroid, Bioresmethrin. This added the element of costs into the considerations.

It was fortunate for the growers of Western Australia that sealing of storages was started in 1982. Events have shown that:

- Grain placed in sealed storage needs no application of protectant chemicals. Chemical residues have been reduced accordingly.
- Insect resistance development in Western Australia has been contained with very little use of Bioresmethrin.

- The quantities of grain fumigated have snowballed dramatically and will continue to do so.
- Saving on grain protectants have paid for all sealing works and have shown a healthy credit.

Table 23.1: Total Capacity of Storages Sealed and Treated with Controlled Atmosphere and Fumigation

Western Australia	*Storages Sealed*		*Controlled Atmospheres Fumigations*		
	Annual	*Totals*	*Annual*	*Receivals (%)*	*Totals*
1982	621	621	72	1.4	72
1983	680	1,301	1,004	17.0	1,076
1984	820	2,121	1,050	21.25	2,126
1985	699	2,820	3,747	45.94	5,873
1986	832	3,652	3,335	64.20	9,208
1987	220	3,872	2,950	64.36	13,158
1988	22	3,894	3,237	70.15*	16,395*

* Figures are in thousand metric tonnes.

Controlled Atmosphere Project

The total facility is a controlled atmosphere project that embraces the benefits of fully sealed storages and a nearby and ample supply of carbon dioxide without the need for onsite storage vessels or mobile tankers. About 4 million t of wheat, oats, and barley will pass through this terminal annually, subject to seasonal fluctuations. The horizontal storages provide 500,000 t of long-term storage ability to balance the rail haulage system with shipping requirements or to provide carryover storage to suit marketing surges.

Carbon dioxide will be the controlled atmosphere choice with the ability to change to other gases should, say, carbonation of concrete become a problem or should there be dictates of yet unforseen circumstances. Should insects exhibit a tolerance to carbon dioxide-rich atmospheres, the concentrations used and/or the exposure period can be increased with ease. Also, the medium of controlled atmosphere or fumigation can be changed at will with the facilities provided. When insects become resistant to all available mediums, the hammer and piece of railway line method will still be effective.

Access for sealing, inspection and maintenance of the seal has been simplified.

Internal ducting avoids external manifolds, with the associated valves and controls dependent on more than one person.

Introduction of gases recirculation, and exhausting are three separate functions, removing the limitation existing in multicell blocks using a manifold system that dictates that a cell cannot be exhausted while an adjacent cell is being purged or circulated.

Controls over pressure testing, gas introduction, recirculation, and exhausintg are all in the individual cell top area.

Light-, medium-, and heavy-weight gases can be handled in the same system.

During exhausting, the total system is under negative pressure. Recirculation creates pressure slightly above and below atmospheric (about 125 Pa) which in fact is less than pressures created in a concrete unsealed structure by external breezes.

Figure 23.4 indicates the control system in developing countries

Now let us try to learn about pressure inside the sotrage.

The pressure inside the storage is raised by the gable end fans to 180 Pa, the capacity of the fans by design, limited to protect the structure and the membrane. Half-life of pressure decay must exceed 30 min (originally calculated with H.J. Banks) and this must be conducted in stable climatic conditions. The corrugated aluminium roof area of 2.8 ha (7 acres), reacting to ambient conditions, affects a very sensitive response in internal pressures when sealed.

As a result, pressure test are generally conducted between midnight and 4 a.m. The pressure test method can be refined with current technology and will be should the simplicity of the current system fail to indicate the condition of the storage seal. This is not likely because experience developed with familiarity has shown that the suitability of the pressure test can be forecast with reasonable accuracy by observation of the oil bath in the pressure relief vents after the storage is closed (sealed). This observation, particularly at sunrise, can be applied to all metal-clad structures regardless of size.

Esperance is a shipping terminal situated about 700 km southeast of Perth and carbon dioxide supply sources. Contracts

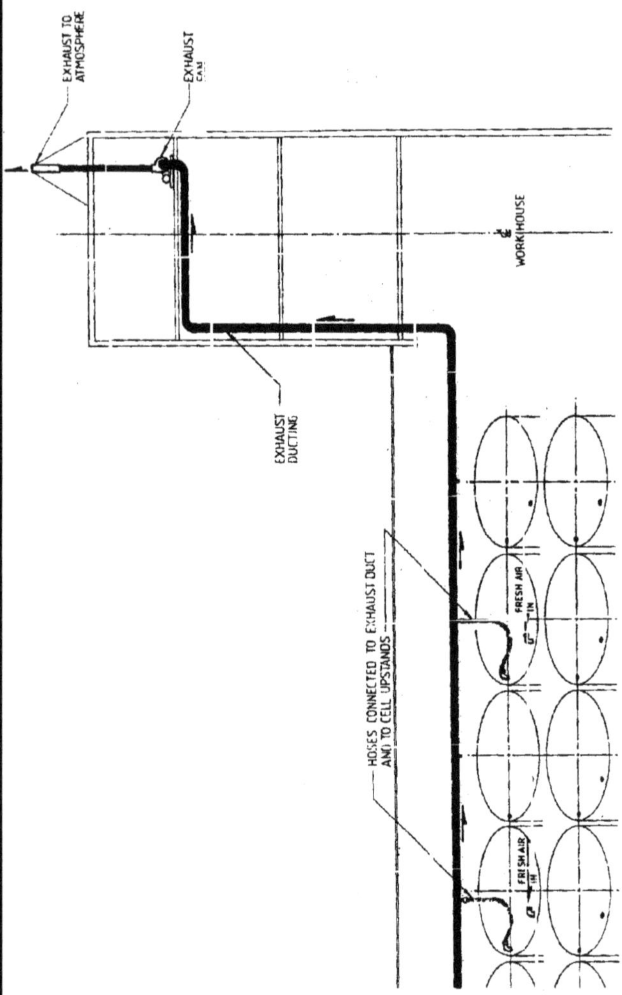

Figure 23.4: Control System in Developing Countries

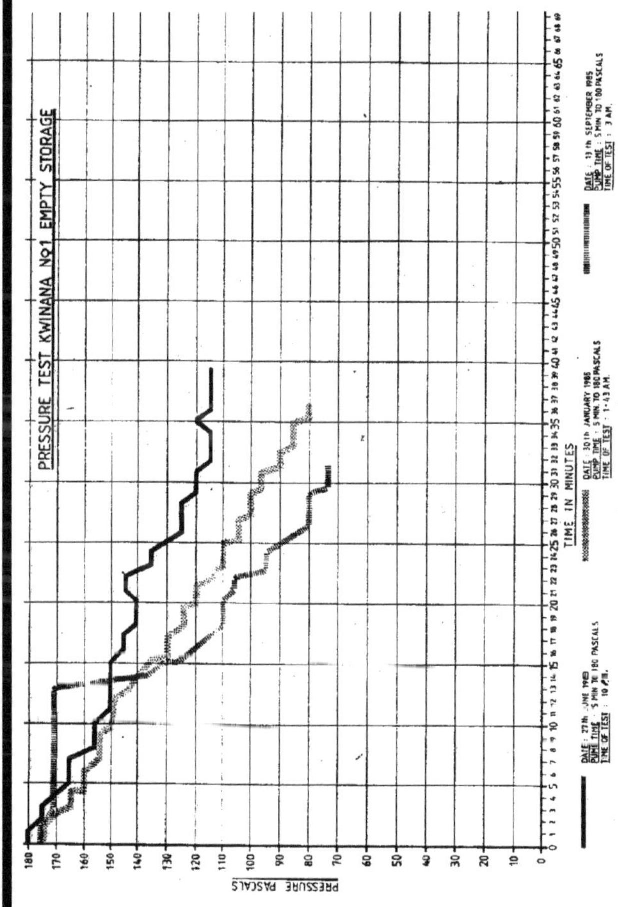

Figure 23.5: Pressure Test at Kwinana No. 1, Empty Storage

have been issued to Liquid Air (Australia) Pty Limited for the provisions of an onsite storage vessel and associated facilities. This will be a demonstration exercise of the ability to conduct a pest management program at a remote location through the medium of road tanker delivery on a continuous basis. The facility includes several horizontal storages, vertical concrete cells, and vertical welded steel cells. About 400,000 t annually will pass through the system. Although the costs of carbon dioxide usage are much higher than the currently available alternative of phosphine, the additional cost is considered as an investment into the future, so that at no time will our export ability be dependent on one only method or material.

A nitrogen generator has been purchased for the purpose of on-site generation of controlled atmosphere at remote locations to establish suitability and cost comparisons with transport of liquid gases over long distances. The generator consists of a blower, a gas feed and control system, a combustion chamber, a catalyst unit, a heat exchanger, and a spray cooler. A refrigeration unit was added by request to minimize the amount of cooling water required (now 250 l/h). The capacity of the unit is 100 m^3/h; oxygen content can be adjusted down to zero and the fuel used is liquid propane gas (LPG).

Storage units utilized were welded steel vertical cells of 5000 t with the exterior painted white. The unit was connected to one filled with barley and the exhaust of this cell was connected to the second empty cell.

The connection of the empty cell to the exhaust of the full cell and the resultant reduction of oxygen content indicated that two full cells or more could be treated together with the same unit.

The purge phase was completed in 2 d. Then a maintenance phase was applied daily, but only during the storage cooling times of the day (about 12 h/d). Fuel used was 9 kg/h and cost was 58 cents/t, estimated at 36 cents/t if two cells were treated together.

Purge time was 2 d and exposure time was 28 d. Following the exposure period, both cells were left sealed for 28 d. The oxygen content increased (loss rate) at the rate of 0.5 per cent daily in the empty cell and at 0.15 per cent daily in the full cell. This was in windy conditions and corresponds to an air ingress rate of 9 and 5 per cent, respectively. No live insects were found during outloading. The trial was directed toward control over oxygen levels throughout the storage and these were entirely satisfactory. Malting barley showed no signs of deterioration following exposure to the anaerobic conditions. Plans are to repeat the application of low-oxygen controlled atmosphere to vertical steel cells and to demonstrate the application in a 14,600-t metal-clad horizontal structure.

The change from a chemical pest control regime to a controlled atmosphere and fumigation system was motivated by the presence of chemical residues and insect resistance. It is necessary, however, to compare the costs of the two methods.

The change to Fenitrothion was a result of *Tribolium castaneum* resistance to malathion. The development of resistance to Fenitrothion of *R. dominica* dictates the use of the synthetic pyrethroid Bioresmethrin. Pyrethroids used do not control *Tribolium;* therefore, Bioresmethrin and Fenitrothion must be applied together to provide total protection.

The cost of Bioresmethrin reached $1.25/t (in Australian dollars) during 1985 when applied at 1 ppm. Fenitrothion reached 12 cents/t, making a total of $1.37/t of grain treated. Add to this the costs of carrying stock; distribution and supply of application equipment; maintenance of that equipment at remote locations; labor to operate and supervise the application; safety equipment; and control systems, including sampling, packing, recording, transporting, analyzing for residues, and processing the information. Bioresmethrin plus Fenitrothion has been in use in Australia for about 10 years, mainly in the eastern states. Resistance in *R. dominica* to organophosphates has been detected in Western Australia, but, for two reasons, the general use of Bioremethrin has been avoided: (1) strict hygiene and disinfestation measures applied at the storage involved, aided by similar activity carried out on farms by the western Australian Agricultural Protection Board and (2) the highest priority on the storage list each year, selected for the sealing program by Cooperative Bulk Handling Limited, is given to any storage where *R. dominica* organic phosphate resistance has been detected. Had the present quantity of sealed storage not been sealed, there is no doubt that Bioresmethrin would have been in general use. Further, when selection of species resistant to the combination of Bioresmethrin and Fenitrothion reaches uncontrollable proportions, another acceptable chemical must be found and so the chain of action and reaction goes on *Oryzaephilus surinamensis* has already exhibited a tolerance to this combination.

H.T. Storages

In this store the fittings are complete with recirculation systems, control fans, pressure vents, and gas input systems. Even where phosphine is used in horizontal terminal storages, a recirculation system is advisable to provide the positive control necessary over gas-clearing activities.

In Western Australia, phosphine is only used in sealed storages (this was not always the case). Tablets are used mainly; the surface-only method of application is used in all storages and the dosage rates used are the lowest known. A dosage rate of 0.2 g/t (phosphine) (0.15 g/m^3) has been used exclusively in the fumigation of more than 15 million t over the last 8 years and has been used in all types of storages with no surviving insects. Some workers condemn low dosages on the basis of potential resistance selection, but with such results experienced the validity of the argument is questionable. Sealed storage is, of course, the key to the argument with the added advantages of less gas produced (means greater safety), less pollution of the environment, and lastly as well as leastly, less costs. Costs of phosphine fumigation are 0.9 cems/t plus labor which is minimal, about 0.4 cents/t for horizontal storage and 0.8 cents/t for vertical storage, an average of 0.6 cents/t and a total cost of 1.5 cents/t.

Table 23.2: 1985 Treatment Costs at One Treatment per Year

Residual Chemicals	Carbon Dioxide			Nitrogen	Phosphine	
All Storages	CH	TH	TCC	CSC	TCC	CH/TH
150	35	10	7	239	1.5	1.5
Seal costs	25	25	2.5	—	2.5	25
Totals 150	60	35	9.5*	239**	4.0*	26.5

Figures are in cents per tonne; CH: Country horizontal storage; TH: Terminal horizontal storage; TCC: Terminal concrete cells; and CSC: Country steel cell sealed during construction.

* Ten fillings per year.

** Costs apply to trial and will reduce.

It has been calculated that the avoidance of the use of Bioresmethrin over the years has saved the shareholders of Cooperative Bulk Handling Limited $30.25 million (in Australian dollars) to January 1986 (in-house communication). It has also been estimated that to seal all sealable storages owned by Cooperative Bulk Handling Limited will cost $27.75 million at 1985 prices. Given an 8-million-t harvest next season, a further $11 million that would have been spent on pesticides if the decision to change to controlled atmosphere and fumigation had not been taken can now be directed elsewhere.

The scientists and researchers participating in the present day revival of controlled atmosphere storage of grains are deserving the highest commendation from all involved in the preservation of stored grains. People such as Bailey (Dec.): Banks, Evans, Annis, and others (Australia); Jay, Storey, and others (US); Lu, Ni, and others (China); Shejbal and others (Italy); Multon and others (France); Burrell. Calverley, Bell, and others (UK); Calderon, Navarro, and others (Israel); Bond, Muir, Sinha, and others (Canada); Mitsuda, Yamamoto, and others (Japan); Kamel and others (Egypt); those unknown to the writer have contributed their efforts toward making available sound and efficient preservation methods for the grain industry.

The writer is aware that the same applies to the many disciplines that together make the "system", but this chapter deals with controlled atmosphere and people sometimes forget their indebtedness to people.

The concept of sealed storages, modification of the atmosphere within, and control over the distribution and maintenance of that atmosphere is simple enough in itself. The difficulties in sealing of storages have been overcome and will be further improved. There are several methods of atmosphere modification available and some clarity in potential methods to be pursued. Hopefully, alternatives, will evolve over the next decade, but in the meantime fumigants will still be an accepted medium of modification. Cost differences, case of applications, and exposure times will ensure this. Environmentalists, industrial reaction, the evolution of equally attractive alternatives, or to a lesser extent residues or resistance may one day reverse the weight of choice. Distribution and maintenance of the selected atmosphere are not problems once the commitment is made. Times of exposure are similar in that adjustments in operations will overcome most of the constraints except where storage capacity is inadequate through the chain of grain movement.

Users of fumigants (or any other control method) have a responsibility to workers, the public, material suppliers, grain consumers and handlers, and to the environment. Those that are applying fumigants in unsealed storages are not living up to that responsibility and, are not achieving the potential results the materials are capable of delivering. Nonbelievers should fit sampling tubes within an unsealed storage and observe the distribution and retention characteristics during a fumigation and compare the results with the control requirements of the insect stages and species present; or conduct a pressure test on the fumigation enclosure to observe its gas-holding capabilities or place tubes into the storage, connecting them to a manometer and observing the pressure differences created by a wind blowing on the structure and imagining the airflows resulting from the pressure differentials taking fresh air in and gas out of the storage; or fill the storage with smoke and see this pour out on the lee side.

An increase in dosage rates *will not* correct fumigation failures in a leaky storage and further flaunts the responsibilities involved. The presence of numerous dead adult insects following a fumigation does not necessarily indicate an efficient fumigation. The survival of latent stages will soon renew the need for fumigation, *e.g.*, a 99 per cent kill of *T. castaneum* will allow the population to recover to former numbers in (estimated) 6 weeks at 35°C and 11 per cent MC (*R. dominica* in 10 weeks and *Sitophilus oryzae* in 12 weeks in similar conditions). In the same table, a 99.999 per cent kill of *S. oryzae* would allow the population to build back to former numbers in 30 weeks at 30°C; yet, in Western Australia during 1980 to 1988, more than 15 million t have been fumigated in sealed storages with CO_2 and phosphine and the first insect survivor is yet to be found with postfumigation periods up to 48 weeks.

Controlled atmosphere or fumigation alone will not satisfy a strict pest control requirement. A suitable integration of all available methods in varying degrees according to local conditions will provide the ideal system within the knowledge spectrum of today.

The use of chemicals as protectants, sprays, vapors, and dusts has its place. Correct use must pay heed to a fundamental principle that exposure of insects to a chemical will inevitably select individuals tolerant/resistant to that chemical.

Forced aeration with or without applied heat or refrigeration is one of the most used and valuable aids to grain preservation. It is applied in many ways (modified ambient air, flow rates, etc.) and is not likely to be replaced by any other system. Controlled atmosphere and fumigation use forced aeration in the distribution and clearing process. Hygiene strictly applied will enhance the effect of and reduce the demands on any other pest control method.

Thermal disinfestation and radiation have a value where conveyed grain requires treatment. Inspection, detection, reporting, and action are unavoidable adjuncts to any efficient system.

Countries who are net importers rather than exporters generally have infestation standards internally less rigid than most exporting countries. This should not be used as an excuse to avoid the responsibilities of a user of fumigants. Costs estimated for any potential preservation method require constant revision. Changes can result from many influences, including the value of local money, improvements in methods and technology, and social pressures on one method or the other. This is certainly the case with costs estimated by Love, in which most costs were estimated, of necessity, for non-established systems. These estimates were also complicated by "annual cost equivalents" which took into account a real discount rate on the capital over the life of the capital item, yet this was not applied to the then current use of pesticides to which the various systems were compared. This does not seem logical, particularly in the light of knowledge today that $4.00/t (in Australian dollars) sealing cost can save $1.50/t in pesticide cost each time the storage is filled (Western Australian figures) and these costs will rise. These costs were also based on an 8-year life of sealant materials and we can now look on 20 years. A review of comparative costs of available systems should be conducted and should include methods such as thermal disinfestation and refrigerated aeration. Some fumigation costs included the cost of turning the grain for the admixture of fumigants which is avoided if the storage is sealed.

Safety, to operators and consumers, in the preservation of grain under a residual chemical regime has been developed in Australia to a high degree. Working under limits of residue tolerances accepted by the WHO/FAO Codex Alimentarious Commission, Committee of Pesticide Experts, and following the acceptance of these limits by the Australian National Health and Medical Research Council

(NHMRC) and the public health department in each state, application levels and resulting residues have been controlled.

Medical management of operational staff exposed to pesticides has assisted in maintaining the level of control throughout the industry. An Operational Safety Manual, produced by a National Committee of Bulk Handling Authority representatives, on the use of all pesticides employed provides information on characteristics of the pesticide, application techniques, postapplication advice and toxicology, and first aid information. Each Bulk Handling Authority operates on the system of safety committees formed at each major work site. Safety in fumigation of grain and grain containers is aided by the Codes of Practice, endorsed by the NHMRC which are included in the Operational Safety Manual together with information on the material and procedures.

The use of controlled atmosphere in, grain stores blends nicely with established procedures, with little modification. The sealing of grain stores to the degree required however adds a new dimension.

Sealed storages generally contain gases by the restriction of the interchange of internal and external atmospheres. The effect of these gases can be noted as follows.

Grain dusts released and created by the movement of grain, particularly respirable dusts (particles below 10 μm), that may normally be carried out of the storage with atmospheric interchange this brings the dust concentration a little closer to the concentration required for the propagation of a flame front (about 20 g/m^3) which, given an ignition source and containment, could lead to a grain dust explosion. Where workers are required to be present inside the storage during filling and emptying, the added dust increases the level of worker's discomfort and the amount of dust inhaled.

The technology is available for the inclusion of controlled atmosphere storage of grain in the integrated pest management program of small or large grain-handling or processing organizations. Those that have not considered the potential of the system would be well advised to do so. Those that may have considered the possibilities available and decided against further action may benefit from a review of that decision. In the winter's experience, the system of controlled atmosphere and fumigation is a good one. It is reliable, efficient, simple, and safe and has cost advantage potential as well as being environment friendly.

Now let us try to understand how carbon dioxide is employed in fumigation practice.

Carbon dioxide is mostly supplied in liquid form, either from a storage vessel onsite or from a tanker mounted on a road trailer or rail wagon. These units would normally be under the control of the gas supplier who will be familiar with and responsible for the safe operation of this equipment during the fumigation exercise. When this is not the case, on-site storage vessels should conform to the following:

- The relevant Australian standard and statutory requirements operating within that particular state.
- The relevant specifications which are required by the gas supplier.
- Codes and specifications of the Compressed Gas Association (International) where applicable. (The equipment should be operated according to AS 1894–1976–"Safe Handling of Cryogenic Fluids" –and supplier's operating instructions.)
- Liquid carbon dioxide will be piped through a vaporizer to hasten the generation of gas and to deliver the gas into the grain store at a temperature close to that of the grain stored.
- Gaseous carbon dioxide may be supplied through a pipeline when liquid has been vaporized off the site of the grain store or has been captured, scrubbed, cooled, and dried at the point of emission of a stack discharge from the burning of hydrocarbons or from the manufacture of ammonia. Feeding this gas into the grain store will be similar to that vaporized on site.
- All valves and connections at the point of discharge from supply vessels and at the grain store entry point should be lockable.
- Piping used to convey liquid or gaseous carbon dioxide within or around a grain storage facility should be clearly identified in line with appropriate statutory requirements or standards.
- Carbon dioxide in the form of dry ice may be used at times to provide the gas required in small storages or to

supplement an initial gaseous application. Apart from the physical handling, all other aspects of grain fumigation will be similar to the use of gas or liquid. Dry ice should not be handled without special protective clothing due to its extreme temperature (–79°C). Its sublimation takes place slowly leaving no trace of its existence.

- A routine plan of treatment should be prepared after consideration of the supply method, type of storage and filling ratio, recirculation efficiency, required exposure period, whether the system to be used is a "one-shot" or a purge and maintenance system, and the post-treatment ventilation period, in relation to the peculiarities of the particular system. Once this plan has been developed, the fumigation officer responsible for routine fumigation work using the system should adhere to it strictly.
- A "one-shot" system is where a quantity of gas is applied in the first instance (purge phase) to reach concentrations within the storage that, allowing for a daily decay rate, will maintain a minimum concentration over the minimum period required for a successful disinfestation.
- A "purge and maintenance" system is when the initial purge is the application of sufficient gas to reach the minimum concentration, plus a margin for daily decay, and further applications are made periodically to replace the daily decay losses (maintenance phase) until the minimum period is reached or in the case of extended exposure periods, when the decision is made to terminate the treatment. The choice to use one or the other will be determined by supply methods available, the type and size of the storage to be treated, and the proposed exposure period.
- Application of carbon dioxide is an operation where exposure to excessive concentrations is unlikely unless accidental release of the gas or unexpectedly high leakage occurs. Detection equipment should be available at all times and routine checks for leakage should be conducted. A self-contained breathing apparatus needs only be worn when entry into areas containing concentrations above the safe level of 0.5 per cent v/v is necessary.

- The most convenient and direct reading method for the detection and determination of carbon dioxide concentrations in air involves the use of detector tubes containing a reaction chemical which exhibits a colored stain when air containing carbon dioxide is aspirated through the tube. Since commercially available tubes vary in sensitivity and accuracy, the advice of occupational health authorities should be sought before detector tubes and pumps are purchased. Whichever brand of tube is used, the manufacturer's directions should be followed implicitly.
- All persons not engaged in the application of the gas or in the monitoring of the system for leaks should be excluded from the immediate area of the storage unit where treatment is to be applied.
- All points of access, into or from the fumigation enclosure, for grain, people, or machinery should be closed, fixed, and sealed.
- The carbon dioxide should be dispensed, and when liquid is supplied, vaporized into the system, care should be taken to ensure that the correct dosage is applied.
- A point of exit for the air displaced by the applied gas should be provided to avoid pressure buildup within the storage unit.
- When the required dosage is applied, gas input and air exit points should be closed, fixed, and sealed.
- The recirculation fan(s) should be started.
- The times of commencement and completion of carbon dioxide introduction should be recorded in a log book.
- Recirculation should continue until the required period laid down in the teatment program has been completed or satisfactory distribution has been achieved.
- Times of commencement and completion of recirculation should be recorded in a log book.
- Where persons may have to work in or pass through areas immediately adjacent to a fumigation enclosure, good ventilation of the areas should be provided. In some cases,

it may be necessary to supplement with mechanical ventilation the natural ventilation induced by opening doors and windows.

- Particular attention should be paid to understorage areas such as tunnel and elevator pit areas.
- Personnel should not be permitted to descend into areas below the fumigation enclosure without positive proof by air sampling that the atmosphere is safe to breathe.

Bibliography

Abeles, F.B. and Gahagan, H.E. (1968). Abscission: The role of ethylene, ethylene analogues, carbon dioxide and oxygen, *Plant Physiol.*, **43**: 1255.

Achilea, O., Fuchs, Y., Chalutz, E. and Rot, I. (1985). The contribution of host and pathogen to ethylene biosynthesis in *Penicillium* digitatum-infected citrus fruit, *Physiol. Plant Pathol.* **27**: 55.

Adam, Z., Itzhaki, H., Borochov, A., and Mayak, S. (1985). Substrate stimulation of an enzyme converting 1-aminocyclopropane-1-carboxylic acid to ethylene, *Plant Sci.*, **42**: 89.

Adams, D.O. and Yang, S.F. (1979). Ethylene biosynthesis identification of 1-aminocyclopropane-1-carboxylic acid: An intermediate in the conversion of methionine to ethylene. *Proc. Natl. Acad. Sci., U.S.A.*, **76**: 170.

Aharoni, N. and Liberman, M. (1979). Ethylene as a regulator of senescence in tobacco leaf discs. *Plant Physiol.*, **64**: 801.

Aharoni, N., Anderson, J.D., and Liberman, M. (1979). Production and action of ethylene in senescing leaf discs effect of indoleacetic acid, kinetin, silver ion, and carbon dioxide. *Plant Physiol.*, **64**: 805.

Allenloff, N., Phillips, W.R., and Johnston, F.B., (1954). A C^{14} study of carbon dioxide fixation in the apple. I. The distribution of incorporated C^{14} in the detached McIntosh apple, *J. Sci. Food Agric.*, **5**: 231.

Allentoff, N., Phillips, W.R. and Johnston, F.B., (1954). A ^{14}C study of carbon dioxide fixation in the apple. II. Rats of carbon dioxide fixation in the detached McIntosh apple, *J. Sci. Food Agric.*, **5**: 234.

Apelbaum, A. and Barkai-Golan, R. (1977). Spore-gennination and mycelial growth of postharvest pathologens under hypobaric pressure. *Phytopathology*, **67**: 400.

Autio, W.R. and Bramlage, W.J. (1982). Effects of AVG on maturation, ripening and storage of apples. *J. Am. Soc. Hortic. Sci.*, **107**: 1074.

Bangerth, F. (1975). The effect of ethylene on the physiology of ripening of apple fruits, in *Facteurs et Regulation de la Maturation des Fruits.* Centre National de la Recherche Scientifique. Paris, pp. 183.

Banks, N.H., Elyatem, S.M., and Hammat, M.T. (1985). The oxygen affinity of ethylene production by slices of apple fruit tissue. *Acta Hortic.*, **157**: 257.

Barbosa-Canovas, G.V., Pothakamury, U.R, Palou, E. and Swanson, B.G. (1997). *Non-thermal Preservation of Foods*. Marcel Dekker Inc. New York.

Barden, L.E. and Hanan, J.J. (1972). Effect of ethylene on carnation, keeping life, *J. Am. Soc. Hortic. Sci.*, **97**: 785.

Barkai-Golan, R. and Kopeliovitch, E. (1983). Induced ethylene evolution and climacteric-like respiration in *Rhizopus* infected *rin* and *nor* tomato mutants, *Physiol. Plant Pathal.*, **22**: 357.

Bassi, P.K. and Spencer, M.S. (1982). Effect of carbon dioxide and light on ethylene production in intact sunflower plants. *Plant Physiol.*, **69**: 1222.

Beyer, E.M. (1975). C_2H_4: its incorporation and metabolism by pea seedling under aseptic conditions. *Plant Physiol.*, **56**: 273.

Blanpied, G.D., Cadun, O. and Tamura, T. (1979). Ethylene in apple and pear experimental CA charmers, *J. Am. Soc. Hortic. Sci.*, **97**: 207.

Borecka, H. and Olak, J. (1978). The effect of hypobaric storage conditions on the growth and sporulation of some pathogenic fungi, *Fruit Sci. Rep. Poland*, **5**: 39.

Bramlage, W.J., Bareford, P.H., Blanpied, G.D., Dewey, D.H., Taylor, S., Porritt, S.W., Lougheed, E.C., Smith, W.H., and McNicholas, F S. (1977). Carbon dioxide treatments for "McIntosh" apples before CA storage, *J. Am. Soc. Hortic Sci.*, **102**: 658.

Brennand, Charlotte (1994). *Home Drying of Food.* Cooperative Extension Service, Utah State University.

Brooks, C., Miller, E.V., Bratley, C.O., Cooley, J. S., Mook, P.V. and Johnson, H.B. (1932). Effect of Solid and Gaseous Carbon Dioxide upon Transit Diseases of Certain Fruits and Vegetables, *USDA Tech Bull.* U.S. Department of Agriculture, Washington, D.C., pp. 318.

Brown, W. (1922). On the germination and growth of fungi at various temperatures and in various concentrations of oxygen and of carbon dioxide. *Ann. Bot.*, **36**: 257.

Bruhn, C.M., Schutz, H.G. and Sommer, R. (1986). Attitude change toward food irradiation among conventional and alternative consumers. *Food Technology*, **40(12)**: 86-91.

Buescher, R.W. (1979). Influence of carbon dioxide on postharvest ripening and deterioration of tomatoes, *J. Am. Soc. Hortic. Sci.*, **104**: 545.

Bull, A.T. and Trinci, A.P. (9177). The physiology and metabolic control of fungal growth, *Adv. Microbiol. Physiol*, **15**: 1.

Burg, S.P. and Burg, E.A. (1966). Fruit storage at sub-atmospheric pressures. *Science*, **153**: 314.

Burg, S.P. and Burg, E.A. (1967). Molecular requirements for the biological activity of ethylene, *Plant Physiol.*, **42**: 144.

Burg, S.P. and Burg, E.A. (1969). Interaction of ethylene, oxygen and carbon dioxide in the control of fruit ripening, *Qual. Plant. Mater Veg.*, **19**: 185.

Burg, S.P. and Thimann, K.V. (1959). The physiology of ethylene formation in apples. *Proc. Natl. Acad., Sci., U.S.A.*, **45**: 335.

Burnett, J.H. (1968). *Fundamentals of Mycology.* St. Martin's Press, New York.

Burton, W.G. (1974). Some biophysical principles underlying the controlled atmosphere storage of plant material, *Ann. Appl. Biol.*, **78**: 149.

Bussel, J., Buckley, P.M., Sommer, N.F., and Kosuge, T. (1969). Ultrastructural changes in *Rhizopus stolonifer* sporangiospores in responses to anaerobiosis, *J. Bocteriol.*, **98**: 774.

Bussel, J., Sommer, N.F. and Kosuge, T. (1969). Effect of anaerobiosis upon germination and survival of *Rhizopus stolonifer* sporangiospores, *Phytopathology*, **59**: 946.

CAST (1989). Ionizing energy in food processing and pest control: II. Applications. Task Force Report No. 115. Council for Agricultural Science and Technology, Ames, IA. 98 pp.

Chaves, A.R. and Tomas, J.O. (1984). Effect of a brief CO_2 exposure on ethylene production. *Plant Physiol.*, **76**: 88.

Cochrane, V.W. (1958). *Physiology of Fungi.* John Wiley & Sons, New York.

CODEX (1991). The Codex Standard for the Labelling of Prepackaged Food (CODEX STAN 106-1985 as amended in 1991).

Couey, R.P., Jr. (1970). Effect of oxygen tension on the growth of *Phytophthora cactorum, Phytopathology.* **60**: 358.

Dhawan, K.R., Bassi, P.K., and Spencer, M.S. (1981). Effects of carbon dioxide on ethylene production and action in intact sunflower plants, *Plant Physiol.*, **68**: 831.

Dhawan, S.S. (1998). *Practical Manual on Home Scale Processing of Fruits.* CCS Haryana Agricultural University, Hisar, India.

Diehl, J.F. (1990). *Safety of Irradiated Foods.*. Marcel Dekker Inc., NY.

Diehl, J.F. and Josephson, E.S. (1994) Assessment of the wholesomeness of irradiated foods (a review). *Acta Alimentaria*, **23**: 195-214.

Dilley, D.R. (1962). Malic enzyme activity in apple fruit. *Nature.* **196**: 387.

Dukes, P.D. and Apple, J.L. (1965). Effect of oxygen and carbon dioxide tensions on growth and inoculum potential of *Phytophthora parasitica* var. *nicotianae, Phytopathology*, **55**: 666.

Eaks, I.L. (1983). Effects of chilling on respiration and ethylene production of "Hass" avocado fruit at 20°C. *Hor. Science*, **18**: 235.

Edney, K.L. (1964). The effect of the composition of the storage atmosphere on the development of rotting of Cox's Orange Pippin apples and the production of pectolytic enzymes by *Gloeosporium* spp., *Ann. Appl. Biol*, **54**: 327.

El-Goorani, M.A. and Sommer, N.F. (1979). Suppression of postharvest plant pathogenic fungi by carbon monoxide. *Phytopathology*, **69**: 834.

El-Goorani, M.A., and Sommer, N.F. (1981). Effects of modified atmospheres on postharvest pathogens of fruits and vegetables, *Hortic. Rev.*, pp. 412.

El-Kazzaz, M.K., Sommer, N.F. and Kader, A.A. (1983). Ethylene effects on *in vitro* and *in vivo* growth of certain postharvest fruit-infecting fungi, *Phytopathology*, **73**: 998.

European Parliament and Council Directive 87/107/EEC (1988) on the approximation of the laws of the Member States concerning food additives authorised for use in foodstuffs intended for human consumption. Official Journal of the European Communities, L40: 27-33.

European Parliament and Council Directive 94/35/EC (1994) on sweeteners for use in foodstuffs. Official Journal of the European Communities L237, 10.9.94, 3-12.

European Parliament and Council Directive 94/36/EC (1994) on colours for use in foodstuffs. Official Journal of the European Communities L237, 10.9.94, 13-29.

European Parliament and Council Directive 95/2/EC (1995) on food additives other than colours or sweeteners. Official Journal of the European Communities L61, 18.3.95, 1-40.

Farquhar, J.W. (1988). A Retail Overview: Steps Necessary for Commercial Application of Food Irradiation. Paper from FAO/IAEA Advisory Group Meeting on Commercial Use of Food Irradiation, Vienna, 1988. Food Marketing Institute, Washington, DC.

FDA (1986). Irradiation in the production, processing, and handling of food; final rule. 21 CFR Part 179. Federal Register, **51(75)**: 13376-13399.

FDA (1990). HHS News Release P90-27. U.S. Dept. of HHS. Rockville, MD. May 1.

FDA. Microwave Oven Radiation. U.S. Dept. of HHS. Rockville, MD. HHS Publication No. (FDA)86-8120 (1986).

Fidler, J.C. and North, C.J. (1969). Production of volatile organic compounds by apples, *J Sci. Food Agric.*, **20**: 521.

Flowerdew, D. (1999). *Food Additives: What Every Manager Needs to Know About the Law*. Chandos Publishing/The British Library.

Follstad, M.N. (1966). Mycelial growth rate and sporulation of *Alternaria tenuis, Botrytis cinerea, Cladosporium herbarum* and *Rhizopus stolonifer* in low-oxygen atmospheres, *Phytopathology*, **56**: 1098.

Forsyth, F.R., Eaves, C.A. and Lightfoot, H.J. (1969). Storage quality of McIntosh apples as affected by removal of ethylene from the storage atmosphere, *Can J. Plant Sci.*, **49**: 507.

Foster, J.W. and Davis, J.B. (1948). Anaerobic formation of fumaric acid by the mold *Rhizopus nigricans*. *J. Bacteriol*, **56**: 329.

Fuhrer, J. (1985). Production and release of ethylene from 1-aminocyclopropane-1-carboxylic acid in *Lemna minor* L. in the dark and at different carbon dioxide compensation concentrations, *J. Plant Physiol.*, **117**: 307.

Gane, R. (1934). Production of ethylene by some ripening fruit, *Nature*, **134**: 1008.

Goos, R.D. and Tschirsch, M. (1962). Effect of environmental factors on spore germination, spore survival, and growth of *Gloeosporium musarum*. *Mycologia*, **34**: 353.

Grodzinski, B., Boesel, I. and Horton, R.F. (1982). Ethylene release from leaves of *Xanthium strumarium* L and *Zea mays* L., *Exp. Bot.*, **33**: 344.

Gunner, H.B. and Alexander, M. (1964). Anaerobic growth of *Fusarium oxysoporum*. *J. Bacteriol.*, **87**: 1309.

Hansen, E. (1942). Quantitative study of ethylene production in relation to respiration of pears. *Bot Gaz*, **103**: 543.

Harris, S. (1981). Ethylene and kiwi fruit, *Orchardist N.Z.*, **54**: 105.

Harrison, D.E.F. (1976). *The Oxygen Metabolism of Microorganisms*, Meadowfield Press, Durham, England.

Hesselman, C.W. and Freebaim, H.T. Rate of ripening of initiated bananas as influenced by oxygen and ethylene, *J. Am. Soc. Hortic. Sci.,* **94**: 635.

Hesselman, C.W. and Freebairn, H.T. (1969). Rate of ripening of initiated bananas as influenced by oxygen and ethylene, *J. Am Soc Hortic Sci.,* **94**: 635.

Hislop, E.C., Hoad, G.V. and Archer, S.A. (1973). The involvement of ethylene in plant diseases, in *Fungal Pathogenicity and the Plant's Response,* Byrd, I.W. and Cutting, C.V., Eds., Academic Press, London, pp. 87.

Huang, C.L., Resurreccion, A.V.A. Florkowski, W.J. and Hung, Y. (1987). *Food Irradiation: A Revolutionary Technology at the Crossroads.* University of Georgia AES Research Report 540.

Hulme, A.C. (1956). Carbon dioxide injury and the presence of succinic acid in apples, *Nature,* **178**: 218.

Hulme, A.C. and Arthington, W. (1953). The oxidation of quinic acid. *J. Exp. Bot.,* **4**: 129.

Hulme, A.C. and Wooltorton, L.S.C. (1958). Determination and isolation of the nonvolatile acids of pome fruits and a study of acid changes in apples during storage, *Sci. Food Agric.,* **9**: 150.

IFT (1987). Perspective on food irradiation. *Food Technology,* **41(2)**: 100-101.

Ilag, L. and Curtis, R.W. (1968). Production of ethylene by fungi, *Science,* **159**: 1357.

Imolehin, E.D. and Grogan, R.G. (1980). Effects of oxygen, carbon dioxide and ethylene on growth, sclerotial production, germination, and infection by *Sclerotinia minor. Phytopathology,* **70**: 1158.

International Consultative Group on Food Irradiation (1991). Facts about Food Irradiation (set of 14 Fact Sheets covering all aspects of food irradiation issued as public information). ICGFI Fact Series 1-14. IAEA, Vienna.

Joint FAO/IAEA/WHO. (1998). Study Group on High Dose Irradiation. Weekly Epidemiological Record, Jan. 16/98.

Kadam, S.S. (1997). *Processing of Fruits and Vegetables.* Chaitanya Publications, New Delhi.

Kader, A.A., Chastagner, G.A., Morris, L.L. and Ogawa, J.M. (1978). Effects of carbon monoxide on decay, physiological responses, ripening and composition of tomato fruits, *J. Am. Soc. Hortic. Sci*, **103**: 665.

Kang, B.G., Yocum, C.S., Burg, S.P. and Ray, P.M. (1967). Ethylene and carbon dioxide: mediation of hypocotyle hook-opening response, *Science*, **156**: 958.

Kao, C.H. and Yang, S.F. (1982). Light inhibition of the conversion of 1-aminocyclopropane-1-carboxylic acid to ethylene in leaves is mediated through carbon dioxide. *Planta*. **155**: 261.

Kaul, G.L. (1986). *Horticultural Crops in India*. Anmol Publications, New Delhi.

Kendall, P. (1985). *Food Irradiation: Promises and Controversies*. Critical Issues in Food & Nutrition 3(6), Colorado State Univ. Cooperative Extension Service, Fort Collins, CO. pp.7.

Klaui, K. (1981). Some aspects of colour in man. In: *Criteria of Food Acceptance: How Man Chooses What he Eats* (Eds.) Solms, J. and Hall, R.L. Forster Verlag. AG Publishing, Zurich, pp. 82-95.

Klotz, L.J., Stolzy, L.H., and DeWolfe, T.A. (1963). Oxygen requirements of three root-rotting fungi in a liquid medium. *Phytopathology*. **53**: 302.

Knee, M. (1971). Ripening of apples during storage. II. Respiratory metabolism and ethylene synthesis in Golden Delicious apples during the climacteric and under conditions simulating commercial storage practice. *J. Sci. Food Agric.*, **22**: 368.

Knee, M. (1973). Effects of controlled atmosphere storage on respiratory metabolism of apple fruit tissue, *J. Sci. Food Agric.*, **24**: 1289.

Knee, M. (1976). The influence of ethylene on the ripening of stored apples. *J. Sci. Food Agric.*, **27**: 383.

Knee, M. (1980). Physiological responses of apple fruits to oxygen concentrations. *Ann. Appl Biol.*, **96**: 243.

Knee, M. (1980). Physiological responses of apple fruits to oxygen concentrations, *Ann. Appl. Biol*, **96**: 243.

Knee, M. (1986). Opposing effects of daminozide and ethylene on the induction of flesh softening in apple fruits, *J. Hortic. Sci.*, **61**: 15.

Knee, M. and Hatfield, S.G.S. (1981). Benefits of ethylene removal during apple storage, *Ann. Appl. Biol.*, **98**: 157.

Knee, M., Looney, N.E., Hatfield, S.G.S. and Smith, S.M. (1983). Initiation of rapid ethylene synthesis by apple and pear fruits in relation to storage temperature. *J. Exp. Bot.*, **34**: 1207.

Knee, M., Proctor, F.J. and Dover, C.J. (1985). The technology of ethylene control: use and removal in postharvest handling of horticultural commodities, *Ann Appl. Biol.*, **107**: 581.

Kollas, D.A. (1964). Preliminary investigation of the influence of controlled atmosphere storage on the organic acids of apples, *Nature*, **204**: 758.

Kushad, M.M., Richardson, D.G., and Ferro, A. J. (1983). Intermediates in the recycling of 5-methylthioribose to methionine in fruits, *Plant Physiol.*, **73**: 257.

Lau, O.L., Liu, Y., and Yang, S.F. (1984). Influence of storage atmospheres and procedures on 1-aminocyclopropane-1-carboxylic acid concentration in relating to flesh firmness in "Golden Delicious" apple. *Hort. Science*, **19**: 425.

Lecos, C. (1986). The growing use of irradiation to preserve food. *FDA Consumer*, **20(6)**: 12-15.

Lehninger, A.L. (1975). *Biochemistry*, Worth Publishers, New York.

Li, P.H. and Hansen, E. (1964). Effects of modified atmosphere storage on organic acid and protein metabolism of pears. *Proc. Am. Soc. Hortic. Sci.*, **85**: 100.

Li, Z.G., Liu, V., Dong, J.G., Xu, R.J. and Zhu, M.Z. (1983). Effect of low oxygen and high carbon dioxide on the levels of ethylene and 1-aminocyclopropane-1-carboxylic acid in ripening apple fruits, *J. Plant Growth Regulation*. **2**: 81.

Littlefield, N.A., Wankier, B.A., Salunkhe, D.K. and McGill, J.N. (1966). Fungistatic effects of controlled atmospheres, *Appl. Microbiol.*, **14**: 579.

Liu, F.W. (1977). The ethylene problem in apple storage, in *Controlled Atmosphere from the Storage and Transport of Perishable Agricultural Commodities.* Dewey, D.H., Ed., Michigan Slate University, East Lansing, pp. 87.

Liu, F.W. (1979). Interaction of daminozide, harvesting date, and ethylene in CA storage on "McIntosh" apple quality, *J. Am. Soc. Hortic. Sci.*, **104**: 599.

Liu, F.W. (1985). Low ethylene controlled-atmosphere storage of McIntosh apples, in *Ethylene and Plant Development.* Roberts, J.A. and Tucker. G.A., Eds., Butterworths, London, pp. 385.

Lockhart, C.L. (1967). Influence of controlled atmospheres on the growth of *Gloeosporium album in vitro, Can. J. Plant Sci.*, **47**: 649.

Lockhart, C.L. (1968). Influence of various carbon dioxide and oxygen concentrations on the growth of *Fusarium oxysporum in vitro, Can J. Plant Sci.*, **48**: 451.

Looney, N.E. (1975). Control of ripening in "McIntosh.' apples. II. Effect of growth regulators and CO_2 on fruit ripening, storage behavior, and shelf life. *J. Am. Soc. Hortic. Sci.*, **100**: 332.

Lurssen, K., Naumann, K., and Schroder, R. (1979). 1-Amino-cyclopropane-1-carboxylic acid: An intermediate of the ethylene biosynthesis in higher plants, *Z. Pflanzenphysiol.*, **92**: 285.

Mapson, L.W. and Robinson, J.E. (1966). Relation between oxygen tension, biosynthesis of ethylene, respiration and ripening changes in banana fruit. *J. Food Technol.*, **1**: 215.

Mapson, L.W. and Robinson, J.E. (1966). Relation between oxygen tension biosynthesis of ethylene, respiration and ripening changes in banana fruit, *J. Food Technol.*, **1**: 215.

Marcellin, P. and Chaves, A. (1983). Effects of intermittent high CO_2 treatment on storage life of avocado fruits in relation to respiration and ethylene production, *Acta Hortic.*, **138**: 155.

Marcotte, M. (1988). Consumer acceptance of irradiation to preserve food. *FDA Consumer*, **20(6)**: 12-15.

Martinoia, E., Dalling, M.J., and Matile, P. (1982). Catabolism of chlorophyll demonstration of chloroplast-localized peroxidative and oxidative activities, *Z. Pflanzenphysiol.*, **107**: 269.

McGlasson, W.B. and Wills, R.B.H. (1972). Effects of oxygen and carbon dioxide on respiration storage life and organic acids of green banana, *Aust. J. Biol. Sci.*, **25**: 35.

McRae, D.G., Coker, J.A., Legg., R.L. and Thompson, J.E. (1983). Bicarbonate/CO2-facilitated conversion of 1-aminocyclopropane-1-carboxylic acid to ethylene in model systems and intact tissue. *Plant Physiol.*, **73**: 784.

MeBean, D. McG. (1978). Drying and Processing tree fruits. CSIRO Division November 2, 1998 of Food Research Circular No. 10. CSIRO, Australia.

Meigh, D.F. (1960). Use of gas chromatography in measuring the ethylene production of stored apples, *J. Sci, Food Agric.*, **11**: 381.

Metlitskii, L.V., Sal'kova. E.G. and Marcellin, P. (1977). Biochemical and biophysical aspects of controlled atmosphere storage of fruits (survey). *Appl Biochem. Microbiol.*, **13**: 265.

Mitchell, D.J. and Zentmyer, G.A. (1971). Effects of oxygen and carbon dioxide tensions on growth of several species of *Phytophthora, Phytopathology*, **61**: 787.

Monning, A. (1983). Studies on the reaction of Krebs cycle enzymes from apple tissue (cv. Cox Orange) to increased levels of CO_2 *Acta Hortic.*, **138**: 113.

Murata, T. and Minamide, T. (1970). Studies on organic acid metabolism and ethylene production during controlled atmosphere storage of apples (*Malus pumila* Miller, cv. Rolls), *Plant Cell Physiol.*, **11**: 857.

Nichols, R. (1966). Ethylene production during senescence of flowers. *J. Hortic. Sci.*, **41**: 279.

Paruthi, J.S. (1999). *Quick Freezing Preservation of Foods*, Vol. II. Allied Publishers Ltd., New Delhi.

Pegg, C.F. (1976). The involvement of ethylene in plant pathogenesis, in *Encyclopedia of Plant Physiology*, (*New Series, Vol. 4*), Springer-Verlag, Berlin.

Pendergrass, A., Isenberg, F.M.R., Howell, C.C. and Carroll, J.E. (1975). Ethylene induced changes in appearance and hormone content of Florida grown cabbage, *Can J. Plant Sci.*, **56**: 319.

Penner, Karen. (1982). *Freezing Fruits and Vegetables.* Kansas State University Cooperative Extension Service.

Phan, C.T., Bardon, E. and Granger, R.L. (1985). Effect of CO_2 on the conservation of "Spartan" and "McIntosh" apples in temporarily sealed regular storage, *Acta Hortic.*, **157**: 121.

Potter, N.A. and Griffiths, D.G. (1947). The effects of temperature and gas mixtures on the production of volatile substances by apples during storage, *J. Pomol. Hortic. Sci*, **23**: 171.

Potter, N.N. and Hotchkiss, J.H. (1996). *Food Science.* CBS Pub. & Distributors, New Delhi.

Prince, T.A., Herner, R.C. and De Hertog, A.A. (1982). Increase in ethylene and carbon dioxide production by *Tulipa gasneriana* L. "Prominence" after completion of the cold requirement. *Sci Hortic..* **16**: 77.

Quazi, H.H. and Freebairn, H.T. (1970). The influence of ethylene, oxygen and carbon dioxide on the ripening of banana. *Bot. Gaz.*, **131**: 5.

Ranson, S.L., Walker, D.A., and Clarke, I.D. (1957). The inhibition of succinic oxidase by high CO_2 concentrations. *J. Biochem.*, **66**: 57.

Reynolds, Susan and Paulette Williams. (1993). *So Easy to Preserve*: Cooperative Extension Service. The University Of Georgia. Revised By Judy Harrison.

Robb, S.M. (1966). Reactions of fungi to exposure to atmospheres pressure of oxygen. *J. Gen. Microbiol.*, **45**: 17.

Rodale Plans. (1993). *Solar food dryer*, (Ed.) Ray Wolf. Rodale Press, 33 East Minor Street.

Rogan, A. and Glaros, G. (1988). *Food Irradiation: The Process and Implications for Dietitians.* JADA, **88(7)**: 833-838.

Saltmarsh, M. (2000). *Essential Guide to Food Additives.* Leatherhead Food RA Publishing, pp. 1-322.

Schiffman-Nadel, M. (1974). Relation between fungal attack and postharvest maturation, *Colloq. Int. C.N.R.S.* **238**: 138.

Sequeira, L. (1973). Hormone metabolism in diseased plants, *Annu. Rev. Plant Physiol.*, **24**: 353.

Singh. B., Littlefield, N.A., and Salunkhe, D.K. (1970). Effects of controlled atmosphere (CA) storage on amino acids. organic acids, sugars, and rate of respiration of "Lambert" sweet cherry fruit. *J. Am. Soc. Hortic. Sci.*, **95**: 458.

Smith, W.H. and Parker, J.C. (1966). Preventions of ethylene injury to carnations by low concentrations of carbon dioxide, *Nature*, **211**: 100.

Sommer, N.F. (1985). Role of controlled environments in suppression of postharvest diseases. *Can. J. Plant Pathol.*, 7: 331.

Sommer, N.F., Fortlage, R.J., Buchanan, J.R., and Kader, A.A. (1981). Effect of oxygen on carbon monoxide suppression of postharvest pathogens of fruits, *Plant Dis.*, **65**: 347.

Somogyi, L.P., Ramaswamy, H.S. and Hai, Y.H. (1996) *Processing Fruits: Science and Technology, Vol. 1: Biology Principles and its Applications*. Tectronic Pub. Co. Inc., Lancaster, Basel.

Srivastava, R.P. and Sanjeev Kumar (1993). Fruit and Vegetable Preservation. International Book Distribution Co., Lucknow.

Stenvers, N. and Bruinsma, J. (1975). Ripening of tomato fruits at reduced atmospheric and partial oxygen pressures, *Nature*, **253**: 532.

Stoll, K., Hauser, F. and Datwyler, D. (1974). Maturation des pommes en chambre froide a atmosphere controlee sous l'.influence d'une reduction du taux de l'ethylene dans l'atmosphere, in *Facleurs et Regulation de la Maturation des Fruits.* Centre National de la Recherche Scientifique, Paris, pp. 81.

Subrahmanyam, K.V. (1981). Economics of production and marketing of important fruits and vegetables in Karnataka, Andhra Pradesh and Tamil Nadu. Final Report, UNDP Project, IIHR, Bangalore.

Toivonen, P., Walsh, J., Lougheed, E.C. and Murr, D.P. (1982). Ethylene relationships in storage of some vegetables, in *Proc. 3rd Natl. Controlled Atmosphere Research Conference.* Richardson, D.G. and Meheriuk, M., Eds., Timber Press, Beaverton, OR, pp. 299.

Toledo, R., Steinberg, M.P. and Nelson, A.I. (1969). Heat of respiration of fresh produce as affected by controlled atmosphere. *J Food Sci.*, **34**: 261.

Ulrich, R. (1970). Organic acids, in *Biochemistry of Fruits and Their Products,* Vol. 1. Hulme. A. C. Ed. Academic Press, London.

Vir, S. Horticulture intervention: towards prevention of micronutrient deficiency in India. Paper published in the book *Strategies for Horticulture Development in India,* published by the Ministry of Agriculture, Government of India, New Delhi.

Wang, C.V. (1977). Effect of CO_2 treatment on storage and shelf life of sweet peppers, *J. Am. Soc. Hortic. Sci.,* **102**: 808.

Wang, C.Y. (1975). Effect of short-term high CO_2 treatment on storage of "d' Anjou" pear, *J Am. Soc Hortic. Sci.,* **100**: 492.

Wang, C.Y. and Adams, D.O. (1982). Chilling induced ethylene production in cucumber (*Cucumis sativus* L.) *Plant Physiol.,* **69**: 424.

Wang, S.Y., Adams, D.O. and Liberman, M. (1982). Recycling of 5'-methylthioadenosine-ribose carbon atoms into methionine in tomato tissue in relation to ethylene production, *Plant Physiol.,* **70**: 117.

Wells, J.M. (1968). Growth of *Rhizopus stolonifer* in low-oxygen atmospheres and production of pectic and cellulolytic enzymes, *Phytopathology,* **59**: 1598.

Wells, J.M. (1968). Growth of *Rhizopus stolonifer* in low-oxygen atmospheres and production of pectic and cellulolytic enzymes, *Phytopathology,* **58**: 1598.

Wells, J.M. (1974). Growth of *Erwinia carotovora, E. atroseptica,* and *Pseudomonas fluorescence* in low oxygen and high carbon dioxide atmospheres, *Phytopathology,* **64**: 1012.

Wells, J.M. and Spalding, D.H. (1975). Stimulation of *Geotrichum candidum* by low oxygen and high carbon dioxide atmospheres, *Phytopathology,* **65**: 1229.

Wells, J.M. and Vota, M. (1970). Germination and growth of five fungi in low-oxygen and high-carbon dioxide atmospheres, *Phytopathology,* **60**: 50.

Williams, M.W. and Patterson, M.E. (1964). Nonvolatile organic acids and core breakdown of Bartlett pears. *J. Agric Food Chem.,* **12**: 80.

Williamson, C.E. (1950). Ethylene, a metabolic product of diseased or injured plants, *Phytopathology,* **40**: 205.

World Health Organisation (1966). The Technical Basis for Legislation on Irradiated Food. WHO Technical Report Series No. 316.

World Health Organisation (1981). Report of the Working Party on Irradiation of Food. Joint Expert Committee on the Wholesomeness of Irradiated Food (JEFF). WHO Technical Report Series No. 659.

World Health Organisation (1987). Principles for the Safety Assessment of Food Additives and Contaminants in Food. Environmental Health Criteria 70. International Programme on Chemical Safety (IPCS) in cooperation with the Joint FAO/WHO Expert Committee on Food Additives (JECFA). World Health Organisation, Geneva.

World Health Organisation (1994). Safety and nutritional adequacy of irradiated foods. WHO, Geneva.

Wu, M.T. and Salunkhe, D.K. (1972). Fungistatic effects of subatmospheric pressures, *Experientia*, **28**: 866.

Yang, S.F. and Hoffman, N.E. (1984). Ethylene biosynthesis and it regulation in higher plant. *Annu Rev. Plant Physiol.*, **35**: 155.

Young, R.E. and Biale, J.B. (1968). Carbon dioxide effects on fruits. III. The fixation of $C_{14}O_2$, in lemon in an atmosphere enriched with carbon dioxide, *Planta*, **81**: 253.

Index

D

S

www.ingramcontent.com/pod-product-compliance
Ingram Content Group UK Ltd.
Pitfield, Milton Keynes, MK11 3LW, UK
UKHW020204290926
15470UKWH00001B/3

9 789351 240594